# DRUG REPURPOSING AND COMPUTATIONAL DRUG DISCOVERY

*Strategies and Advances*

# DRUG REPURPOSING AND COMPUTATIONAL DRUG DISCOVERY

*Strategies and Advances*

*Edited by*
**Mithun Rudrapal, PhD**

AAP | APPLE ACADEMIC PRESS

First edition published 2024

**Apple Academic Press Inc.**
1265 Goldenrod Circle, NE,
Palm Bay, FL 32905 USA

760 Laurentian Drive, Unit 19,
Burlington, ON L7N 0A4, CANADA

**CRC Press**
2385 NW Executive Center Drive,
Suite 320, Boca Raton FL 33431

4 Park Square, Milton Park,
Abingdon, Oxon, OX14 4RN UK



**Library and Archives Canada Cataloguing in Publication**

CIP data on file with Canada Library and Archives

**Library of Congress Cataloging-in-Publication Data**

CIP data on file with US Library of Congress

ISBN: 978-1-77491-277-5 (hbk)
ISBN: 978-1-77491-278-2 (pbk)
ISBN: 978-1-00334-770-5 (ebk)

# About the Editor

**Mithun Rudrapal, PhD**

Mithun Rudrapal, PhD, FIC, FICS, CChem (India), is Associate Professor at the Department of Pharmaceutical Sciences, School of Biotechnology and Pharmaceutical Sciences, Vignan's Foundation for Science, Technology & Research (Deemed to be University), Guntur, India. Dr. Rudrapal has been actively engaged in teaching and research in the field of pharmaceutical and allied sciences for more than 12 years. He has over a hundred publications in peer-reviewed international journals to his credit and has filed a number of Indian and international patents. In addition, Dr. Rudrapal is the author of a dozen published or forthcoming books. Dr. Rudrapal works in the areas of medicinal chemistry, computer-aided drug design (CADD), drug repurposing, phytochemistry, herbal drugs, and nanophytotherapeutics.

# Contents

# Contributors

**Kirti Agrawal**
Amity Institute of Molecular Medicine and Stem Cell Research (AIMMSCR), Amity University, Noida, Uttar Pradesh, India

**Vishnu Nayak Badavath**
School of Pharmacy & Technology Management, SVKM's NMIMS University, Hyderabad, Telangana, India

**Himani Chaurasia**
Bioorganic Research Laboratory, Department of Chemistry, University of Allahabad, Prayagraj, India

**Dipak Chetia**
Department of Pharmaceutical Sciences, Faculty of Science and Engineering, Dibrugarh University, Dibrugarh, Assam, India

**Karthikeyan Deivasigamani**
Department of Pharmaceutical Sciences, SNS College of Pharmacy and Health Sciences, Coimbatore, India

**Jayati Dwivedi**
Bioorganic Research Laboratory, Department of Chemistry, University of Allahabad, Prayagraj, India

**Swastika Ganguly**
Department of Pharmaceutical Science and Technology, Birla Institute of Technology, Mesra, Ranchi, India

**Johra Khan**
Department of Medical Laboratory Sciences, College of Applied Medical Sciences, Majmaah University, Al Majmaah, Saudi Arabia
Health and Basic Sciences Research Center, Majmaah University, Al Majmaah, Saudi Arabia

**Dhruv Kumar**
Amity Institute of Molecular Medicine and Stem Cell Research (AIMMSCR), Amity University, Noida, Uttar Pradesh, India

**Siddhartha Maji**
Department of Chemistry, Oklahoma State University, Stillwater, Oklahoma, USA

**Rahul Kumar Maurya**
Amity Institute of Pharmacy, Lucknow, Amity University, Noida, Uttar Pradesh, India

**Richa Mishra**
Bioorganic Research Laboratory, Department of Chemistry, University of Allahabad, Prayagraj, India

**Vishali Murthy**
Department of Pharmaceutical Sciences, SNS College of Pharmacy and Health Sciences, Coimbatore, India

**André M. Oliveira**
Department of Environment, Federal Centre of Technological Education, Contagem, MG, Brazil

**Ipsa Padhy**
Department of Pharmaceutical Analysis, School of Pharmaceutical Education and Research, Berhampur University, Berhampur, Odisha, India

**Kumar Nallasivan Palani**
Department of Pharmaceutical Sciences, SNS College of Pharmacy and Health Sciences, Coimbatore, India

**Ghanshyam Parmar**
Department of Pharmacy, Sumandeep Vidyapeeth, Vadodara, Gujarat, India

**Ashish Patel**
Ramanbhai Patel College of Pharmacy, Charusat University, Changa, Gujarat, India

**Arpita Paul**
Department of Pharmaceutical Sciences, Faculty of Science and Engineering, Dibrugarh University, Dibrugarh, Assam, India

**Sivasubramanian Piramanayagam**
Department of Pharmaceutical Sciences, SNS College of Pharmacy and Health Sciences, Coimbatore, India

**Mithun Rudrapal**
Department of Pharmaceutical Sciences, School of Biotechnology and Pharmaceutical Sciences, Vignan's Foundation for Science, Technology & Research (Deemed to be University), Guntur, India

**Chita Ranjan Sahoo**
Central Research Laboratory, Institute of Medical Sciences and SUM Hospital, Siksha 'O' Anusandhan Deemed to be University, Bhubaneswar, Odisha, India

**Malavika Saji**
Amity Institute of Molecular Medicine and Stem Cell Research (AIMMSCR), Amity University, Noida, Uttar Pradesh, India

**Ashish Shah**
Department of Pharmacy, Sumandeep Vidyapeeth, Vadodara, Gujarat, India

**Tripti Sharma**
Department of Pharmaceutical Chemistry, School of Pharmaceutical Sciences, Siksha 'O'Anusandhan Deemed to be University, Bhubaneswar, Odisha, India

**Ramendra K. Singh**
Bioorganic Research Laboratory, Department of Chemistry, University of Allahabad, Prayagraj, India

**Vishal Kumar Singh**
Bioorganic Research Laboratory, Department of Chemistry, University of Allahabad, Prayagraj, India

**Manish Kumar Tripathi**
Department of Pharmaceutical Engineering and Technology IIT (BHU), Varanasi, Uttar Pradesh, India

**Alok Shiomurti Tripathi**
Amity Institute of Pharmacy, Lucknow, Amity University, Noida, Uttar Pradesh, India

**Abd. Kakhar Umar**
Department of Pharmaceutics and Pharmaceutical Technology, Faculty of Pharmacy, Universitas Padjadjaran, Jatinangor, Indonesia

**Mohammad Yasir**
Amity Institute of Pharmacy, Lucknow, Amity University, Noida, Uttar Pradesh, India

**James H. Zothantluanga**
Department of Pharmaceutical Sciences, Faculty of Science and Engineering, Dibrugarh University, Dibrugarh, Assam, India

# Abbreviations

| | |
|---|---|
| 3D QSAR | three-dimensional quantitative structure activity relationship |
| ACD | available chemicals directory |
| ACEs | acetylcholinesterases |
| AD | atopic dermatitis |
| ADHD | attention deficit hyperactivity disorder |
| ADMET | absorption, distribution, metabolism, excretion, and toxicity |
| ADR | adverse drug reactions |
| AF | atrial fibrillation |
| AI | artificial intelligence |
| AIDS | acquired immune deficiency virus |
| AK | adenylate kinase |
| AKI | acute kidney injury |
| ALRs | AIM2 like receptors |
| ALS | amyotrophic lateral sclerosis |
| ANN | artificial neural networks |
| ANVISA | Agencia Nacional de Vigilancia Sanitaria |
| APC | activated protein C |
| APL | acute promyelocyticleukemia |
| APP | amyloid precursor protein |
| ARG1 | arginase 1 |
| AROC | area under the receiving operator curve |
| AUPRC | area under precision-recall curve |
| AUROC | area under the receiver operating characteristic |
| BBB | blood-brain barrier |
| BEAR | binding estimation after refinement |
| BiRW | bi-random walk |
| CA | carbonic anhydrase |
| CAD | coronary artery disease |
| CADD | computer-aided drug design |
| CANDO | computational analysis of novel drug opportunities |
| CASP | critical structure prediction assessment, a biennial collective project |
| CATNIP | creating a translational network for indication prediction |
| CD | Crohn's disease |

| | |
|---|---|
| CETP | cholesteryl ester transfer protein |
| CETSA | cellular thermo-stability assay |
| CHD | congenital heart disease |
| CLP | cecum ligation and puncture |
| CMC | comprehensive medicinal chemistry |
| CNS | central nervous system |
| COMT | catechol-o-methyltransferase |
| COSMIC | catalogue of somatic mutations in human cancer |
| COVID-19 | coronavirus disease 2019 |
| COX | cyclooxygenase |
| COX-2 | cyclooxygenase-2 |
| CPUs | central processing units |
| CQ | chloroquine |
| CVDHD | cardiovascular disease herbal database |
| CVDs | cardiovascular disorders |
| DAMPs | damage-associated molecular patterns |
| DAPD | diabetes-related proteins database |
| DENV | dengue virus |
| DIVA | diverse information, visualization, and analysis |
| DKD | diabetic kidney disease |
| DL | deep learning |
| DM | diabetes mellitus |
| DNA | deoxyribonucleic acid |
| DNN | deep neural networks |
| DS | discovery studio |
| DTI | drug-target interactions |
| EAD | epoxy anthraquinone derivatives |
| EBOV | Ebola virus |
| ECFPs | extended connectivity fingerprints |
| EHRs | electronic health records |
| EMEA | Europe by European Medicine Agency |
| ENL | erythema nodosumleprosum |
| EORD | European Organisation for Rare Diseases |
| EVD | Ebola virus disease |
| FALS | familial type ALS |
| FDA | Food and Drug Administration |
| FEP | free energy perturbation |
| FN | false-negative |
| FP2 | fingerprint 2D |

| | |
|---|---|
| GARD | genetic and rare diseases |
| GAU | Gaussian scoring function |
| GEO | Gene Expression Omnibus |
| GIP | glucose-dependent insulinotropic peptide |
| GLP | glucagon-like peptide |
| GLP-1 | glucagon-like peptide 1 |
| GLP1 | glucagon-linked peptide-1 |
| GM-CSF | granulocyte-macrophage-colony stimulating factor |
| GOF | gain of function |
| GP | Gaussian process regression |
| GPCR | G-protein coupled receptors |
| GPUs | graphics processing units |
| GTEx | genotype-tissue expression |
| GWAS | genome-wide association studies |
| HBA | hydrogen bond acceptor |
| HCC | hepatocellular carcinoma |
| HCQ | hydroxychloroquine |
| HCS | high content screening |
| HCV | hepatitis C virus |
| β-HEX | β-N acetylhexosaminidase hexosaminidase A |
| HGP | human genome project |
| HLA | human leucosite antigen |
| HMDB | human metabolome database |
| HMGB1 | high mobility group box 1 protein |
| HPLC | high-performance liquid chromatography |
| HTS | high-throughput screening |
| IBD | inflammatory bowel disease |
| IC | inhibitory concentration |
| ICAM-1 | intercellular adhesion molecule 1 |
| ICD | International Classification of Diseases |
| IE | information extraction |
| IgE | immunoglobulin E |
| IL | interleukin |
| IMiDs | immune modulating drugs |
| IND | investigational new drug |
| iNOS | inducible nitric oxide synthase |
| IP | intellectual property |
| IPR | intellectual property rights |
| iPSCs | induced pluripotent stem cells |

| | |
|---|---|
| IR | information retrieval |
| JEV | Japanese encephalitis |
| KD | knowledge discovery |
| KEGG | Kyoto Encyclopedia of Genes and Genomes |
| kNN | k-nearest neighbors |
| KPLS | kernel-based PLS |
| LBDD | ligand-based drug design |
| LDK | Lenaldekar |
| LINCS | library of integrated network based cellular signatures |
| LMWH | low molecular weight heparin |
| LOF | loss of function |
| LUTS | lower urinary tract symptoms |
| MAB | multi-armed bandit |
| MACCS | molecular ACCess system |
| MAO-B | monoamine oxidase-B |
| MARS | multivariate adaptive regression splines |
| MCP | monocyte chemoattractant protein |
| M-CSF | macrophage colony-stimulating factor |
| MD | molecular dynamics |
| MDR | multidrug resistance |
| MERS | Middle Eastern respiratory syndrome |
| MHLW | Ministry of Health, Labour, and Welfare |
| MIP-1α | macrophage inflammatory protein-1α |
| ML | machine learning |
| MMP | matrix metalloprotien |
| MND | motor neuron disease |
| MoA | mechanism of action |
| MOE | molecular operating environment |
| MPA | mycophenolic acid |
| MPI | message passing interface |
| mRNA | messenger RNA |
| MS | multiple sclerosis |
| MTD | maximum tolerated dose |
| mTOR | mammalian target of rapamycin |
| MTU | methylthiouracil |
| NAM | negative allosteric modulators |
| NAMD | nanoscale molecular dynamics |
| NBC | naïve Bayesian classifiers |
| NCBI | National Centre for Biotechnology Information |

| | |
|---|---|
| NCEs | new chemical entities |
| NEN | niclosamide ethanolamine |
| NER | name entity recognition |
| NF-kB | nuclear factor kappa-light-chain-enhancer of activated B cells |
| NFT | neurofibrillary tangles |
| NGS | next generation sequencing |
| NK | natural killer |
| NLRs | NOD-like receptors |
| NMR | nucleo magnetic resonance |
| NN | neural network |
| NN | neural networks |
| NORD | National Organization for Rare Disorders |
| Nrf2 | nuclear factor erythroid 2 (NFE2)-related factor 2 |
| NRTIs | nucleoside reverse transcriptase inhibitors |
| NSAIDs | nonsteroidal anti-inflammatory drugs |
| NTD | neglected tropical diseases |
| OD | orphan diseases |
| ODA | Orphan Drug Act |
| ODC | ornithine decarboxylase |
| p38 | protein kinase 38 |
| PAD | peripheral artery disease |
| PAM | positive allosteric modulators |
| PAMPs | pathogen-associated molecular patterns |
| PCB | paclitaxel coated balloon |
| PCM | paracoccidioidomycosis |
| PCM | proteochemometric |
| PCNA | proliferating cell nuclear antigen |
| PD | psychotic depression |
| PD | Parkinson's disease |
| PDB | protein database |
| PDE-5 | phosphodiesterase-5 |
| PDE5is | phosphodiesterase 5 inhibitors |
| PDIF | protein atom score contributions derived interaction fingerprint |
| PDL1 | programmed death-ligand 1 |
| PDTD | potential drug target database |
| PGD2 | prostaglandin D2 |
| PNS | peripheral nervous system |

| | |
|---|---|
| PPAR | peroxisome proliferator–activated receptor |
| PPI | protein-protein interactions |
| PPMS | primary progressive multiple sclerosis |
| PPV | positive-predictive value |
| PRISM | profiling relative inhibition simultaneously in mixtures |
| PVR | pulmonary vascular resistance |
| QED | quantitative estimate of drug-likeness |
| QSAR | quantitative structure–activity relationship |
| R&D | research and development |
| RA | rheumatoid arthritis |
| RAGE | receptor for advanced glycation end products |
| RAR-α | retinoic acid receptor alpha |
| RCSB-PDB | Research Collaboratory for Structural Bioinformatics-Protein Data Bank Website |
| RD | reverse docking |
| RDBD | Rare Disease Repurposing Database |
| RDCRN | rare diseases clinical research network |
| RDL | relative drug likelihood |
| rDNA | recombinant DNA |
| RF | random forests |
| Rg | radius of gyration |
| rGyr | radius of gyration |
| RLRs | RIG-like receptors |
| RLS | restless legs syndrome |
| RMSD | root-mean square deviation |
| RMSF | root-mean square fluctuation |
| RNA | ribonucleic acid |
| RNS | reactive nitrogen species |
| ROS | reactive oxygen species |
| RP | recursive partitioning |
| RRMS | relapsing-remitting phase |
| SALS | sporadic ALS |
| SAR | structure-activity relationship |
| SARS | severe acute respiratory syndrome |
| SARS-CoV-2 | syndrome coronavirus 2 |
| SBDD | structure-based drug design |
| SBVS | structure-based virtual screening |
| SE | standard error |
| SEA | similarity ensemble approach |

| | |
|---|---|
| SGB | stochastic gradient boosting |
| SGLT2 | sodium-glucose transporter 2 |
| SLE | systemic lupus erythematosus |
| SMA | spinal muscular atrophy |
| SMILES | simplified molecular input line entry specification |
| SNP | single-nucleotide polymorphisms |
| STEMI | ST-segment elevation myocardial infarction |
| SVM | support vector machines |
| T1D | type 1 diabetes |
| T2D | type 2 diabetes |
| TCGA | the cancer genome atlas |
| TI | thermodynamic integration |
| TLR | toll-like receptors |
| TLR2 | toll-like receptor 2 |
| TM | text mining |
| TMFS | train-match-fit-streamline |
| TMS | template modeling score |
| TNF-α | tumor necrosis factor alpha |
| TP | true- positives |
| TTD | therapeutic target database |
| TZA | thiazolidinediones |
| VCAM-1 | vascular cell adhesion protein 1 |
| VEGF | vascular endothelial growth factor |
| vHTS | virtual HTS |
| VS | virtual screening |
| WHO | World Health Organization |
| WHO | work of heart |
| WOMBAT | world of molecular bioactivity |
| ZIKV | Zika virus |
| α-Syn | α-Synuclein |

# Preface

Drug repurposing (DR) is also known as drug repositioning, drug reprofiling, drug redirection, and therapeutic switching. It can be defined as a process of identification of new pharmacological indications from old/existing/investigational/FDA-approved drugs, and the application of the newly developed drugs to the treatment of diseases other than the drug's original/intended therapeutic use.

Traditional drug discovery is a time-consuming, laborious, highly expensive, and risky process. The novel approach of drug repositioning has the potential to be employed over traditional drug discovery program by reducing the high monetary cost, longer duration of development, and increased risk of failure. In recent years, the drug repositioning strategy has gained considerable momentum with about one-third of the new drug approvals corresponding to repurposed drugs which currently generate around 25% of the annual revenue for the pharmaceutical industry.

The application of computational tools and techniques for the prediction and exploration of the pharmacological effects of developing drug candidates or lead molecules offers a significant hope in the current drug discovery programme because it is inexpensive, time-saving, less risky, and accurate. The use of computational techniques in discovery research does not only help in the discovery of new drugs from leads or existing drug molecules, but can also be useful for the repurposing of existing drugs.

This book is primarily aimed at delivering information on various computational techniques, tools and databases utilized for drug repurposing and identifying the uses of existing drug candidates on different emerging diseases. The most recent coronavirus diseases-2019 (COVID-19) pandemic is a clear indication that there is a dire need for advanced *in silico* tools and computational techniques for drug discovery.

The book titled **Drug Repurposing and Computational Drug Discovery: Strategies and Advances** provides recent advances in drug repurposing and computational approaches pertaining to drug discovery research with potential applications in various major and emerging therapeutic areas. The content of the book comprising a definite number of interesting chapters is based on some specific and relevant topics aiming at the focal theme of the book. Each chapter delineates up-to-date and in-depth information in a lucid, constructive, and unambiguous manner with adequate consistency in flow, continuity, and technical clarity.

This book primarily focuses on the state-of-the-art drug repurposing strategies and techniques currently being utilized for the discovery of therapeutic candidates from existing drug molecules by experimental (*in vitro/in vivo*) and/or computational (*in silico*) strategies. Through this book, readers/users can update their knowledge and skills with sufficient technologically advanced tools, computational techniques, and database available for the discovery of drugs by drug repurposing and computational approaches.

In this book, drug repurposing and computational approaches for the discovery and development of drugs against certain emerging and life-threatening diseases including microbial infections (bacterial, fungal, viral, COVID-19), parasitic diseases and neglected tropical diseases (NTDs), malignant diseases (cancer), inflammatory diseases, cardiovascular disorders, diabetes, and aging and neurological (CNS) disorders are covered. In addition, challenges and regulatory issues commonly encountered in drug repurposing and computational drug discovery programs are also addressed.

Some of the key highlights of the book are as follows:

- Presents/details the scopes available for the discovery of drugs by drug repurposing approaches and computational strategies with potential advantages and clinical utilities in the treatment of infectious illness, malignant diseases, rare and difficult to treat diseases.
- Summarizes latest developments on the application of drug repurposing strategies and computational approaches in drug discovery.
- Describes/depicts drug discovery approaches from existing drug candidates or lead molecules by computational/database screening and experimental screening/assays.

Having contributions from experienced academicians, scientists, and researchers from across the globe, this book is intended to be a useful resource for a wide audience, particularly working in the area of drug discovery research including drug discovery scientists, medicinal chemists, pharmacologists, toxicologists, phytochemists, biochemists, clinicians, discovery scientists (R&D), biomedical researchers, health care professionals and researchers. The most likely users of the book are postgraduate students, doctoral researchers, senior academic researchers, professors, etc. of pharmaceutical, biomedical, and allied sciences from the higher academic institutions, universities, and pharmaceutical and biotechnology companies.

## CHAPTER 1

# Drug Repurposing in Future Drug Discovery and Development

KIRTI AGRAWAL, MALAVIKA SAJI, and DHRUV KUMAR

*Amity Institute of Molecular Medicine and Stem Cell Research (AIMMSCR), Amity University, Noida, Uttar Pradesh, India*

## ABSTRACT

Rising necessity of potential drugs for treating chronic, rare, and untreated diseases is the main motive for the development of repurposed drugs. Drug repurposing or repositioning is the technique of discovering novel therapeutic uses for drugs, either already approved or under investigation for a new indication. This method has many advantages over the conventional methods of drug discovery. Reduced risk of failure is one factor that makes drug repurposing an approachable option. The number of drugs that fail in Phase III of clinical trials is disturbing, and only very few are approved for clinical use. Since repositioned drugs are already proven safe, they have less chance of failure, saving 5–7 years of safety testing. De novo drug discovery and development is a time-taken and high investment process while drug repurposing process is efficient and economical. There are generally four types of drug repurposing approaches: computational, knowledge-based, biological-experimental, and mixed approaches, which are broadly used in repurposed drug development. In this chapter, we have further elaborated traditional and high-throughput drug repurposing approaches and emphasized over their features, their validation techniques, noteworthy applications for treating a wide range of untreated diseases, and challenges of drug repositioning.

Drug Repurposing and Computational Drug Discovery: Strategies and Advances.
Mithun Rudrapal, PhD (Ed)

## 1.1 INTRODUCTION

In the face of advances in science and technology with improved knowledge of human diseases, transformation of these benefits into therapeutic developments has been slower than expected. Pharmaceutical industries are facing a lot of challenges globally to bring new drugs into the market as it is a time-consuming and high-cost process.[1] To combat these challenges, the drug repurposing strategy has been considered for screening new usages of preexisting drugs and for developing new drugs in pharmaceutical research and development (R&D). Drug repurposing/repositioning/reprofiling is an interesting approach to identify various new uses of approved or under clinical trial drugs for new medical indications. This effective approach offers numerous gains over developing a completely new drug for a specified indication.[2] The most important benefit of this strategy is the risk of failure is lesser because the reprofiled drug has previously been checked to be suitably safe in preclinical models and/or humans trials. Thus, it is less expected to fail in terms of safety in following efficacy trials. The second important factor is that the time taken for a repurposed drug development can be effectively reduced as most of the preclinical testing, safety evaluation, and formulation development (in some cases) has been already completed. Also, low investment is required that usually depends on the developing process and stage of the repurposing drug.[3] The regulatory and phase III costs may remain more or less the same for a repurposed drug as for a new drug in the same indication, but there could still be substantial savings in preclinical and phase I and II costs.[4] Together, these advantages have the potential to result in a less risky and more rapid return on investment in the development of repurposed drugs, with lower average associated costs once failures have been accounted for (indeed, the costs of bringing a repurposed drug to market have been estimated to be USD300 million on average, compared with an estimated ~USD2–3 billion for a new chemical entity). Finally, repurposed drugs may reveal new targets and pathways that can be further exploited. It takes around 12–15 years for a de novo drug approval process.[5] The importance of drug repurposing over de novo drug synthesis has been illustrated in Figure 1.1. In general, to select a candidate drug for further developmental pipeline, drug repositioning approaches require a three-step process:

- Hypothesis formulation through identifying a candidate molecule for a specified target.
- Systematic assessment of the effect of drug in preclinical models.

- Efficacy assessment in the clinical trials (phase II) after compiling adequate safety data from phase I trials.[4]

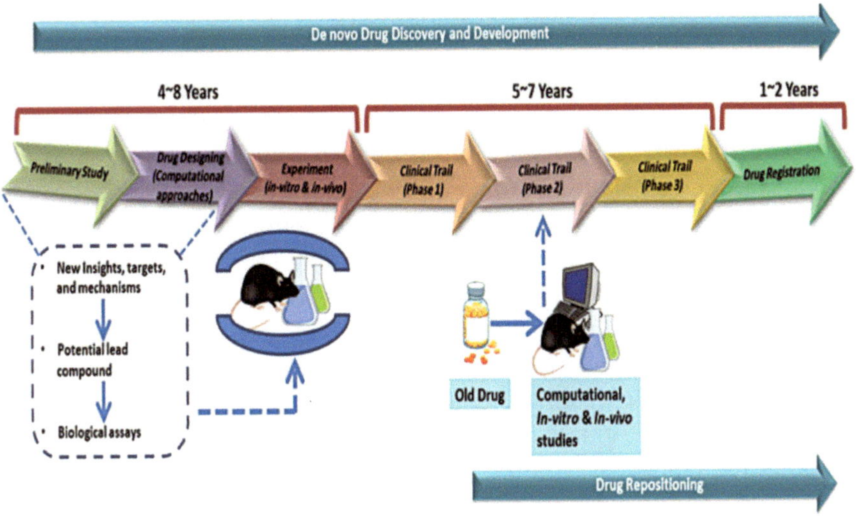


**FIGURE 1.1** Major steps and expected time in new drug discovery and development process versus repurposed drug development process.

## 1.2 ORIGIN AND SIGNIFICANCE OF DRUG REPURPOSING

However, there are some historical instances of drug repositioning that exists, but this concept has come into light in year 2004 when Ashburn and Thor, in an article, primarily defined drug repurposing as the practice of discovering new and better uses of existing drugs.[2] Sildenafil drug which was primarily used for antianginal medication was later repositioned in the middle of the 2000s to cure erectile dysfunction, and morning sickness drug thalidomide was repositioned for multiple myeloma.[6] These achievements open up new opportunities and create a vast interest in drug repositioning which stemmed in the establishment of many startup companies focusing on repurposing. Also, market research reports have confirmed that drug repositioning has an imperative share in the R&D marketing with spending 10–50%.[7] The term "polypharmacology" means one drug-multiple hits or off-target effects, and its principle has been understood since the beginning of drug discovery. Conventionally, the aim of drug discovery and development was to detect the possible therapeutic agents by means of a one drug-one target model that signifies that high selectivity would maximize the efficiency and minimalize their side effects. To find out such particular compounds, problems arise because a huge majority of compounds mediate often undesired effects.[8] Due to this observation, the theory of polypharmacology has been emerged. For better understanding the scope of drug repositioning, Nancy et al. reported a bibliometric examination by studying a few drugs. For instance, chlorpromazine drug synthesized for regulating mental disorders and as a preoperative medicine that was later tried for treating several diseases like whooping cough and for treating the symptoms of radiation therapy in cancer patients. An antimalarial compound chloroquine that was synthesized in 1934 was later used for targeting several diseases including fever, skin rash, and other parasitic diseases.[7] Numerous drug classes show polypharmacological features like antipsychotic,[9] cholinesterase inhibitors,[10] selective serotonin reuptake inhibitors,[11] and thrombolytic agents. Amantadine drug was originally developed for treating influenza but later redirected for Parkinson's disease.[12] Likewise, zidovudine was earlier used as a cancer-treating drug, but after redirection, it is in use for targeting HIV/AIDS.[13] Although every drug has the ability to hit multiple molecular targets, it is important to understand its clinical efficiency toward new targets that requires reinvestigation of prevailing drugs for therapeutic redirection.

Many strict regulations are required to develop, validate, and bring any new drug into the market and this process also needs 12–15 years with substantial investment due to diversity in physico-chemical properties of the compounds and exponential increase in the drug production. Still the whole world is trying to eradicate the COVID-19 strains; it is difficult to wait for 12–15 years for anti-COVID-19 drug development and this pandemic situation has already proved the power and efficiency of drug repurposing. This drawback promotes various pharmaceutical companies and/or research centers to quickly and proficiently employ the already-approved and existing drugs for new interventions.[14] Also, those molecules that were earlier failed to show their efficacy toward predetermined targets can typically get a good start for their reprofiling for other new indications that can be further developed as potential therapies for rare diseases that are still facing lack of diagnosis, treatment, and resources. Drug repositioning is a short-term and less expensive approach that brings effective treatments to patients in comparison with time-taken drug discovery and development practices.[15] Additionally, this method also helps to overcome the increasing overhead costs for drug development, as a result of lowering down the medication cost for patients' treatment.

## 1.3 DRUG REPURPOSING APPROACHES AND TECHNIQUES

Nowadays, the modern and progressive drug approaches have produced a huge amount of data like gene expression data, mutational analysis, drug chemical structure outlines, association between drug-disease, drug targeted proteins, etc. which could be most useful to identify the right drug for a specific target. There are various systematic approaches that are in use for developing intriguing drug repurposing strategies such as computational approaches experimental approaches, knowledge-based approaches, and mixed approaches in which computational and experimental approaches and their validation techniques have been displayed in Figure 1.2.

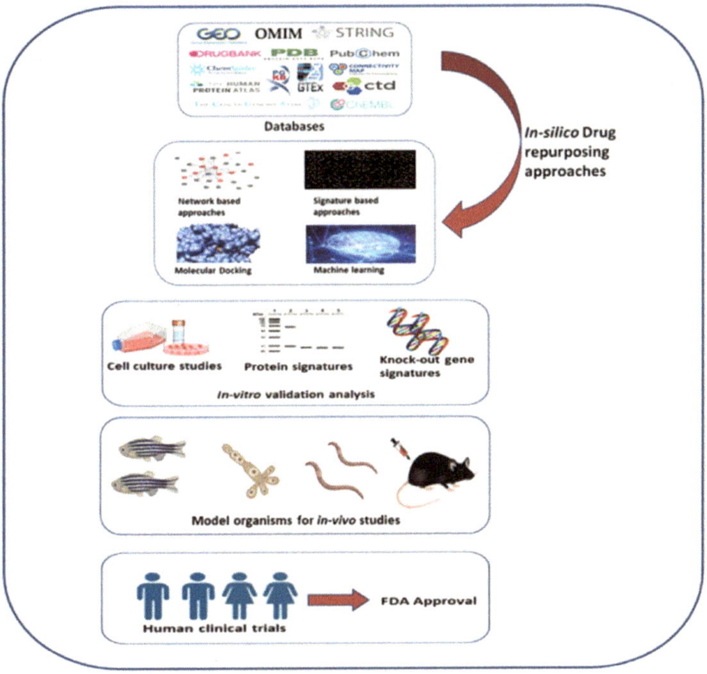


**FIGURE 1.2** Representation of drug repurposing and validation approaches.

### *1.3.1 COMPUTATIONAL APPROACHES*

Computational approaches, also called in silico approaches, are mainly based on data-driven tools that allow systematic data analysis of any type like gene expression analysis, mutational analysis, genotypic or phenotypic traits, proteomic data, chemical structure profiles, or electronic health records (EHRs), which led to generating drug repurposing hypothesis.[16] These findings would be helpful to treat incurable diseases, cancer, etc., that require essential and adequate data to carry out the proposed research. Some commonly used computational databases and resources for drug repurposing are listed in Table 1.1.

**TABLE 1.1** A Panel of Databases and Resources Used in Computational Approaches and Drug Repositioning.

| Database/ resource | Resource type | Description | URL |
|---|---|---|---|
| PubChem | Drug database | An open database holds chemical information of more than 90 million compounds with their bioactivities, gene, and protein targets | https://pubchem.ncbi.nlm.nih.gov/search/search.cgi |
| Human Protein Atlas | Disease and drug database | Mapping of all the human proteins in cells, tissues, and organs using integrative omics technology | *https://www.proteinatlas.org* |
| TCGA (The Cancer Genome Atlas Program) | Disease-related genetic/genomic data | Characterization (RNAseq, microarray, sequence information) of over 33 types of cancer | http://cancergenome.nih.gov/ |
| GEO (Gene Expression Omnibus) | Disease-related transcriptional genomics data repository | Database of next-generation sequencing (NGS), microarray, high-throughput functional, and transcriptional genomics data of diseases | http://www.ncbi.nlm.nih.gov/geo/ |
| ChEMBL | Compound information resource | Database for more than 2.1 million bioactive compounds with drug-like properties | https://www.ebi.ac.uk/chembl/ |
| ClinicalTrials | Drug–disease associations | Web-based resource of public- and private-supported clinical data of diseases and conditions | https://clinicaltrials.gov |
| Allen Brain Atlas | Disease database | Gene expression database for human and mouse brain diseases and conditions | http://www.brain-map.org/ |

**TABLE 1.1** *(Continued)*

| Database/ resource | Resource type | Description | URL |
|---|---|---|---|
| COSMIC (Catalogue of somatic mutations in human cancer) | Cancer mutation database | An online database of somatic mutations in human cancer. | http://cancer.sanger.ac.uk/cosmic |
| GTEx (Genotype-Tissue Expression) | Gene expression database | Catalog of relationship between genotype and gene expression in multiple human tissues | https://www.gtexportal.org/home/ |
| STRING | Protein association database | Web resource of identified and predicted protein–protein interactions (PPI), network analysis | https://string-db.org/cgi/input.pl |
| Drug Bank | Drug database | Covers 11,000 drugs that contain more than 200 data fields of chemical data and drug targets | https://www.drugbank.ca/ |
| e-Drug3D | Drug database | It allows exploring FDA-approved drugs and active metabolites | http://chemoinfo.ipmc.cnrs.fr/MOLDB/index.html |
| ChemSpider | Drug database | Database of 64 million drug chemical structures | http://www.chemspider.com/ |
| SIDER | Drug database | Data of marketed medicines and their documented adverse reactions (side effect frequency, drug–target relation) | http://sideeffects.embl.de/ |
| Drug Central | Drug database | Online drug information resource contains info on chemical entities, active ingredients, drug mode of action, pharmaceutical products, indications, etc. | http://drugcentral.org/ |

### *1.3.1.1 TEXT MINING-BASED APPROACHES*

Text mining techniques are useful to find out the data associated with drugs, genes, diseases, and then to organize the appropriate information from the retrieval data based on the co-occurrence among the applicable entities or by employing natural language processing techniques. Along with a lot of information that is available about drug repurposing, enormous relationship conceptual data of novel biological entity are accessible from publications. Text mining allows us to discover fresh data by mining from numerous published resources with computational method. Hearst proposed a common definition for text mining (TM) as "It is the discovery of novel, formerly unknown information through automatic information extraction from various written sources using the computer machine.[17]" Usually, TM in the field of biology comprises four steps as follows:

- Data/information retrieval (IR), in which biological study-related documents are mined from the literature. These related documents require filtration due to the presence of some useless conceptions in the documents.
- Biological name entity recognition (NER), in which useful biological conceptions are recognized using precise vocabularies.
- Biological information extraction (IE); and
- Knowledge discovery (KD)-relevant information is mined for knowledge discovery about biological theories to eventually build a knowledge graph with the extracted information.[18]

Text mining also helps to determine the relationship between drug–target and drug–disease. The concept of text mining techniques in the medical field was originated from the Swanson (ABC) model. For example, if the drug A is related to the gene B, and the gene B is associated with the disease C, then the drug A may have a new association with the disease C.[19,20] Based on this concept, Li et al. established a method to build disease-definite drug–protein connection maps through linking network and text mining. Primarily, they mined the relationship between disease–protein using network mining, followed by searching of drug terms that are indirectly linked with particular diseases like Alzheimer's disease in PubMed using text mining.[21] Lastly, the drugs and proteins could be connected through disease–protein or drug–disease relationships.

#### 1.3.1.2 SEMANTICS-BASED APPROACHES

Semantics-based computational approaches are broadly used to retrieve information, image, and other fields including drug repurposing. The pipeline of semantic technology mainly comprises three steps:

- Biological entity relationships are mined from previous data from large medical databases to create the semantics network.
- Construction of semantics networks using ontology networks through the addition of the previously acquired information.
- Lastly, for the prediction of new relationships in the semantics networks, mining algorithms are planned to calculate new relationships in the semantics networks.[22] Based on a hypothesis in which similar drugs are correlated with similar targets and similar targets are connected to similar drugs, Guillermo et al., in their research, illustrated an algorithm to calculate drug–target relationships using semantic link prediction techniques and edge partition approaches and built a semantic network containing (drug–drug, target–target) relationships which made precise predictions of drug–target association.[23] Likewise, Chen et al. constructed a statistical-based model of semantic linked subnetwork between drug and target for the prediction of drug–target association. The suggested model successfully recognized some known and random drug–target pairs with high accuracy and also identified drugs for repurposing. For instance, a drug named barbiturate earlier used for curing migraines was predicted for treating insomnia with biological literature support.[24] Similarly, a study demonstrated that tamoxifen drug that was used in treating breast cancer can also cure ovarian cancer that was also supported by the literature.[25]

#### 1.3.1.3 MACHINE LEARNING

Drug repurposing through computational approaches has progressed over the past two decades starting from drug similarity attempts that usually used single source of medical or biological data into a pioneering application area for machine learning techniques. Similar to machine learning models, computational drug reprofiling necessitates wide-ranging data to disclose the fundamental relationships between the biomedical and the biological entity. The workflow of machine learning normally includes four steps: (1)

preprocessing of data, (2) feature extraction, (3) model fitting, and (4) evaluation.[26] Machine learning techniques for drug repurposing include logistic regression, neural network, random forest, support vector machine, and deep learning for multiclass classification, binary classification, and values prediction.[27] For logistic regression, similarity-based machine learning framework like "PREDICT" uses integrated drug–drug similarity and disease–disease similarity values as features by applying logistic regression by assuming that similar drugs are generally linked with similar diseases and vice versa.[28] Similarly, "SPACE" also used logistic regression for therapeutic drug class prediction by incorporating multiple data sources.[29] Likewise, Luo et al. proposed a new computational approach named "MBiRW" that uses some broad similarity measures and algorithm of bi-random walk (BiRW) to predict novel indications for specific drug by incorporating features information of drug or disease with identified drug–disease associations. The similarity measures were developed to predict similarity of drugs and diseases followed by their similarity network construction that was then integrated into heterogeneous network with known drug–disease association. Based on this network, the MBiRW algorithm predicts novel drug–disease interaction.[30] For support vector machine (SVM) techniques, Napolitano et al., in their work, predicted drug therapeutic class based on similarity in gene expression, drug chemical structure, and molecular target.[31] Aliper and Plis used deep learning techniques with gene expression data to predict therapeutic drug categories and showed that deep neural networks (DNN) exceeded SVM after 10-fold cross-validation that signified a proof for employing deep learning for drug discovery.[32]

### *1.3.2 EXPERIMENTAL APPROACHES*

Experimental approaches comprise target screening, cell-based assays, animal model, and clinical aspects.

#### *1.3.2.1 BINDING ASSAY*

Protein-based techniques such as mass spectrometry, affinity chromatography have been employed to detect the binding associates for drugs. Cellular thermo-stability assay (CETSA) is used for target mapping engaged in cells using bio-physical principles that detect thermal stability of target proteins through drug-like ligand which have suitable cellular affinity.[4] Chemical genetics can similarly offer an improved understanding of the binding and efficacy relationship in the cellular perspective. Many industries focus on high-throughput direct binding/catalytic assays to analyze small-molecule-kinase binding using a range of in vitro assays and organism-based assays to create heat maps of biological interactions.[4] Karaman et al. performed an in-vitro assay to analyze 38 kinase inhibitors contrary to 317 different human protein kinases and as a result, they identified "3175" binding interactions.[33]

#### *1.3.2.2 IN VITRO CELL-BASED ASSAY*

In phenotypic screening, sequence of cell-based in vitro assays in 96 or 384 well-plates is used.[34] For instance, Corsello et al. utilized "PRISM" (profiling relative inhibition simultaneously in mixtures) method that tracks the cell proliferation during drug treatment by unique molecular barcodes. Cell line proliferation will decrease with higher drug efficacy that ultimately results in the reduction of a definite molecular barcode. They performed a viability assessment of 578 samples of human cancer cell-lines from different tumor types when given a treatment of more than 4500 drugs comprising hundreds of nononcology drugs.[35]

#### *1.3.2.3 ANIMAL MODEL SCREENING ASSAY*

Animal model screening assays can also be exploited in drug repositioning. This approach not only recognized the drugs against diseases but also produced organ-toxicity and pharmaco-kinetic results as compared with a cell-based screening assay. In a study, Ridges et al. utilized genetically engineered T-cells comprising zebra fish as an animal model for the assessment of effectiveness of about 26,000 small molecules against leukemia and reported that Lenaldekar (LDK), a T-cell proliferation inhibitor, exhibited a noteworthy activity against several hematologic malignancies.[36]

## 1.4 VALIDATION OF COMPUTATIONAL DRUG REPURPOSING MODELS

Validation of drug repositioning outcomes can be done computationally and/or experimentally. Computational validation can be done in a straightforward manner to evaluate AUROC (area under the receiver operating characteristic) values, positive-predictive value (PPV)[37] sensitivity, and specificity. Moreover, drug validity can be evaluated by comparing predicted targets in Clinical Trials, PubMed, or electronic health records (EHRs). Experimental validation includes in vivo and in vitro cell-based targeting assays in a controlled situation and in an animal model such as clinical trials.[27] For instance, albendazole was identified as a repurposed candidate drug for cancer treatment by validating through in vitro and in vivo experiments to treat liver and ovarian cancer.[38]

## 1.5 CURRENT AND FUTURE APPLICATIONS OF POTENTIAL REPURPOSED DRUGS IN DRUG DISCOVERY

Repurposed drugs are safe and cost less, making them market-friendly, take less time for development, and have increased chances of approval and success rates. Many drugs are currently being repurposed and are under various stages of their clinical trials listed in Table 1.2 to prove their efficacy against various diseases. Some of the approved and currently investigated drugs for various diseases are discussed below.

### *1.5.1 DRUG REPURPOSING AGAINST CANCER*

According to WHO statistics, 19.3 million new cases were reported in 2020, and if the current trends continue, an estimated 30.2 million people will be affected by 2040. Extensive research is being conducted in search of anticancer drugs, and more than 10,000 clinical trials are being conducted in a year, of which only 5% of the drugs entering Phase I are approved.[39] Nelfinavir, an inhibitor of HIV protease used to treat AIDS, has shown anti-carcinogenic properties, which are now being extensively studied.[40] Studies have shown nelfinavir regulating cell cycle and inhibiting proliferation of tumors in ovarian cancer cells by decreasing the levels of PCNA (proliferating cell nuclear antigen) and proteins involved in the cell cycle.[41] Aspirin, commonly used as an antipyretic and analgesic, is an NSAID (nonsteroid anti-inflammatory drug)[42] with well-established cardiovascular protective properties. Aspirin is also known to effectively prevent thrombosis and platelet aggregation via COX-1 inhibition and is currently being used to treat thromboembolism in patients prone to cardiovascular diseases.[43] From recent studies, aspirin is known to inhibit the carcinogenesis-promoting enzyme cyclooxygenase (COX),[44] and numerous other studies support the antineoplastic property of aspirin. Several clinical studies have shown an increase in survival chances in colorectal cancer patients using aspirin.[45] Disulfiram, approved for the treatment of chronic alcoholism, has proven an antineoplastic effect. It manifests its effect by interfering with processes involving copper and zinc, and copper is vital for tumor angiogenesis.[46] Disulfiram treatment has also been effective against breast cancer cells and glioblastoma cells and showed increased efficacy in cis-platin sensitive cancer cells in clinical trials.[47] Metformin, used for the treatment of type II diabetes, has been linked with reducing cancer incidence through many studies. Treatment with metformin has decreased cancer incidence and mortality rate compared to patients in other forms of diabetic treatment.[48] Clinical studies on prostate cancer cells showed an increased apoptosis rate,[49] while decreased cell proliferation was observed in breast cancer cells.[50] Thalidomide, first marketed as a morning sickness remedy during pregnancy, was later approved for the treatment of ENL (erythema nodosumleprosum) in 1998 by FDA.[51] Later, thalidomide was shown effective against multiple myeloma.[52] Celecoxib, an NSAID approved by FDA in 1999 for arthritis treatment, has shown promising antitumor activity in various types of cancer, especially breast cancer.[53] Another drug, prazosin, approved for hypertension,[54] now being used for the treatment of multiple conditions, including congestive heart failure,[55] has also shown effectiveness against pheochromocytoma[56] and in the inhibition of glioblastoma growth by inhibiting the AKT pathway.[57] Artemisinins, due to their anti-inflammatory property, are being investigated for their possible action against respiratory disorders in lung cancer models.[58]

### *1.5.2 DRUG REPURPOSING AGAINST CNS DISORDERS*

Drug development for diseases affecting Central Nervous System (CNS) is one of the most extensively studied areas. Still, the failure rate of developed drugs is comparatively higher than in other areas, with most available ones focusing on short-term management of symptoms instead of tackling the cause. In degenerative diseases like Alzheimer's disease and Parkinson's disease (PD), the available medications focus on reliving the observable symptoms, while the degeneration of neurons and central cells continues, worsening the condition.[59] Also, among the drugs synthesized, the majority fail due to their ineffective penetration across the BBB (blood–brain barrier). They are the most significant limiting factor in successful CNS drug discovery as it excludes almost 98% of the small-molecule drugs and 100% of large-molecule drugs from reaching the brain.[60] Repurposing drugs already intended for other CNS disorders have the additional benefit of having a greater chance of crossing the BBB and more chance of success.[61] Mifepristone, approved by the FDA in 2000 as a pregnancy-terminating agent due to its progesterone-inhibiting properties, is also known to be an effective selective inhibitor of glucocorticoids and later discovered to possess anticancer activity. Evidence suggests mifepristone is effective against glioma, one of the most commonly occurring tumors in the CNS, with minimal cytotoxicity on healthy cells.[62] Studies showed increased apoptotic activity and decreased proliferation of cancer cells using mifepristone in combination with temozolomide in glioblastoma patients.[63] Mifepristone is also found to be effective against PD (psychotic depression), which is characterized by elevated levels of cortisol, differentiating it from other subtypes of depression. Currently, there are no FDA-approved drugs for its treatment, and with promising Phase 2/3 results, mifepristone could serve as a potential therapeutic agent for the treatment of PD in the future.[64] Mifepristone was repurposed to treat Cushing's syndrome, caused by elevated glucocorticoid levels and was approved by FDA in 2012 after many clinical trials.[65] Amantadine, the antiviral approved for influenza, was later repurposed for Parkinson's based on various case studies. Patients showed decreased tremor and akinesia and increased rigidity with controllable side effects upon its administration.[66] Atomoxetine, initially intended for treating Parkinson's, was repurposed for treating ADHD (attention-deficit hyperactivity disorder) due to its selective norepinephrine reuptake inhibition in the pre-synapses.[67] Ropinirole, an agonist of dopamine D2, initially intended for hypertension,[68] was later repurposed to treat advanced and early Parkinson's disease.[69] Ropinirole has also shown effectiveness against RLS (restless legs syndrome) in a 52-week study where it was found safe for long-term use and was successful in improving the symptoms.[70] Mecamylamine was first approved for hypertension due to its ability to block impulse transmission in the ganglia.[71] Infliximab was approved for the market in 1998 to treat CD (Crohn's disease).[72] Infliximab is a well-established antagonist of TNF-α and was used to treat CD patients unresponsive to conventional therapy. Increased production of TNF-α is associated with various conditions like Parkinson's and Alzheimer's diseases. It was showed that infliximab, with its TNF blocking property, could reduce the risk of Alzheimer's disease[73] and needs to be explored further to be of clinical use.

### 1.5.3 DRUG REPURPOSING AGAINST CARDIOVASCULAR DISEASES

Cardiovascular diseases (CVDs) are the leading cause of death worldwide, according to WHO statistics. The number has steadily increased from 2 million to 8.9 million from 2000 to 2019. Repurposing of diabetic and anti-inflammatory drugs has been especially successful, as both are a significant factor in many CVDs and are a promising area for repurposing.[74] Paclitaxel, approved by the FDA in 1992 for the treatment of ovarian cancer, is recently found helpful in controlling restenosis in patients who underwent coronary intervention. Restenosis occurs due to narrowing of the lumen due to the increased proliferation in the artery.[75] Results from various randomized trials show paclitaxel to be effective and safe, with reduced occurrence of restenosis.[76] PCB (paclitaxel coated balloon) angioplasty was also found effective long term and had the potential to be developed into a first-line therapy for restenosis.[77]

Colchicine, approved for managing gout, has established anti-inflammatory properties through the disruption of microtubules.[78] It has been shown effective against pericarditis and was approved in 2015 for its first-line treatment. Recent meta-analysis data shows reduced pericardial effusion and recurrence risk of pericarditis.[79] Currently, many trials are going on to evaluate the effect of colchicine on other cardiovascular disorders like STEMI (ST-segment elevation myocardial infarction), AF (atrial fibrillation), CAD (coronary artery disease), and percutaneous coronary intervention. Drospirenone is an oral contraceptive with anti-mineralocorticoid activity. Meta-analysis data shows its effectiveness against hypertension in postmenopausal women. Estrogen deficiency in postmenopausal women leads to an increased risk of cardiovascular diseases as a result of hypertension. Lowering of systolic and diastolic pressure was observed in hypertensive women with very few side effects.[80] Donepezil, approved to treat Alzheimer's disease due to its cholinesterase-inhibiting property, was shown to have a cardiovascular protective effect. Many PDE5is (phosphodiesterase 5 inhibitors) have cardiomyocyte heterotrophy inhibition properties and cardioprotective effects. Sildenafil, tadalafil, is PDE5is initially approved for treating erectile dysfunction but is now being investigated for their cardio protectiveproperties.[81] Currently, clinical studies are examining the effects of sildenafil on PVR (pulmonary vascular resistance) and tadalafil on cardiac stress and change of ventricular torsion.[82]

### 1.5.4 DRUG REPURPOSING FOR OTHER DISEASES

#### 1.5.4.1 INFLAMMATORY DISEASES

Inflammatory diseases constitute a significant area of concern. They often lead to other conditions like cancer, neurodegenerative diseases (Alzheimer's disease, multiple sclerosis, Parkinson's disease, etc.), and metabolic disorders (type II diabetes, obesity, etc.).[83] Developing therapeutic methods that can regulate chronic inflammation can help manage these diseases to an extent.

Tofisopam, an active CNS compound, was marketed for anxiety treatment and had antipsychotic properties.[84] Dextofisopam (enantiomer of tofisopam) was found to be effective against IBD (inflammatory bowel disease) in clinical trials.[85] Cefadroxil, an antibiotic, targets many proteins related to IBD like endothelin-1 receptor, alanine aminopeptidase, and peptide transporter 1, making it a promising candidate for IBD drug studies.[86] Budesonide has anti-inflammatory properties and is used in the treatment of asthma.[87] Recent studies also showed them effective against ulcerative colitis, and a significant reduction in inflammation was observed.[88] Penicillamine was first approved for the treatment of Wilson's disease and was repurposed and approved for the treatment of RA (rheumatoid arthritis) a decade later.[89] Sirolimus, another immunosuppressive approved by the FDA in 1999, is currently being studied for its anti-inflammatory action in patients with SLE (Systemic Lupus erythematosus) and has promising results after Phase I/II studies.[90] Rituximab, initially approved for lymphoma, ustekinumab approved for psoriasis, and certolizumab approved for Crohn's disease are some of the drugs repurposed and approved for RA in the last 10 years.[91]

#### 1.5.4.2 INFECTIOUS DISEASES

Infectious diseases have been an ongoing problem for centuries all around the globe. Their ability to grow, spread, and mutate fast, along with their increasing resistance to existing treatments and drugs, make them challenging to control. Drug repurposing became particularly important during the outbreak of COVID-19, for which no specific treatment is available and repurposing existing drugs was the fastest and safest solution. WHO recommended a combination therapy of the antimalarial drugs, chloroquine and hydroxychloroquine, the HIV drugs ritonavir and lopinavir, and the antiviral remdesivir as a method of treatment.[92]

The anticancer drug cisplatin was found to be effective against *Pseudomonas aeruginosa.* The drug was found to inhibit the endotoxin secretion and DNA replication mechanism of the pathogen through various studies.[93] The antibiotic daptomycin, the immunosuppressant mycophenolic acid, and the hypertensive drug manidipine are some drugs that have shown promise against Zika virus infection.[94] The FDA-approved drug dapsone was intended for treating leprosy but is now being studied for its antimalarial potential.[95] Studies reveal that the antibiotics Biapenem and Tabipenem, the anti-inflammatory drug ebselen, the anticancer drug bortezomib and elesclomol, and the cardiovascular drug verapamil are effective against *M. tuberculosis* infection. Bortezomib was also found effective against JEV (Japanese encephalitis), for which no effective treatment is available. Studies on mice models showed reduced mortality and brain damage, opening up a new possibility of JEV treatment.[96]

**TABLE 1.2** List of Some Repurposed Drugs with Their Old and New Indication[92,68] (https://clinicaltrials.gov/).

| Name | Old indication | New indication |
|---|---|---|
| Amantadine | Influenza | Parkinson's |
| Antomexetine | Parkinson's | ADHD |
| Arbidol | Antiviral | COVID-19 |
| Arsenic | Syphilis | Leukemia |
| Artemisinin | Malaria | Cancer, respiratory disorders |
| Aspirin | Antipyretic, analgesic | CVD, cancer |
| Auranofin | RA | Cancer |
| Azathioprine | Immunosuppressant | Rheumatoid arthritis |
| Biapenem | Antibiotic | *M. tuberculosis* infection |
| Bortezomib | Cancer | *M. tuberculosis* infection |
| Budesonide | Asthma | Ulcerative colitis |
| Cefadroxil | Antibiotic | Inflammatory bowel disease |
| Celecoxib | Arthritis | Cancer |
| Certolizumab | Crohn's disease | Rheumatoid arthritis |
| Chloroquine | Malaria | Cancer, COVID-19 |
| Ciclosporin | Immunosuppressant | Rheumatoid arthritis |
| Cisplatin | Cancer | Antibacterial |
| Colchicine | Gout | Pericarditis, STEMI, CAD |
| Dapsone | Leprosy | Malaria |

**TABLE 1.2** *(Continued)*

| Name | Old indication | New indication |
|---|---|---|
| Daptomycin | Antibiotic | Zika virus |
| Disulfiram | Chronic alcoholism | Cancer |
| Donepezil | Alzheimer's disease | Cardiovascular diseases |
| Doxycycline | Antibacterial | COVID-19 |
| Drospirenone | Oral contraceptive | Postmenopausal hypertension |
| Ebselen | Inflammation | *M. tuberculosis* infection |
| Elesclomol | Cancer | *M. tuberculosis* infection |
| Favipiravir | Antiviral | COVID-19 |
| Hydroxychloroquine | Malaria | Cancer, COVID-19 |
| Indomethacin | Anti-inflammatory | Cancer |
| Infliximab | Crohn's disease | Alzheimer's disease |
| Ivermectin | Antiparasitic drug | COVID-19 |
| Lopinavir | HIV | Zika virus |
| Manidipine | Hypertension | Zika virus |
| Mecamylamine | Hypertension | ADHD |
| Metformin | T2DM | Cancer |
| Mifepristone | Pregnancy termination | Cancer, PD, Cushing's syndrome |
| Mycophenolic acid | Immunosuppressant | Zika virus |
| Nafamostat | Pancreatitis | COVID-19 |
| Nelfinavir | AIDS | Cancer |
| Paclitaxel | Ovarian cancer | Restenosis |
| Penicillamine | Wilson's disease | Rheumatoid arthritis |
| Prazosin | Hypertension | Cancer |
| Rapamycin | Immunosuppressant | Cancer |
| Remdesvir | Ebola | Zika virus, COVID-19 |
| Ribavirin | Hepatitis C | COVID-19 |
| Ritonavir | HIV | Zika virus |
| Rituximab | Lymphoma | Rheumatoid arthritis |
| Ropinirole | Hypertension | Parkinson's disease, restless leg syndrome |
| Sildenafil | Erectile dysfunction | Pulmonary vascular resistance |

**TABLE 1.2** *(Continued)*

| Name | Old indication | New indication |
|---|---|---|
| Sirolimus | Immunosuppressant | Systemic lupus erythematosus |
| Sofosbuvir | Hepatitis C | COVID-19 |
| Tabipenem | Antibiotic | *M. tuberculosis* infection |
| Tadalfil | Erectile dysfunction | Cardiac stress |
| Thalidomide | Pregnancy-related morning sickness | ENL, cancer |
| Tocilizumab | Inflammation | COVID-19 |
| Tofisopam | Anxiety | Inflammatory bowel disease |
| Ustekinumab | Psoriasis | Rheumatoid arthritis |
| Verapamil | Cardiovascular disease | *M. tuberculosis* infection |

## 1.6 CURRENT CHALLENGES IN DRUG REPURPOSING

The core aim of drug repurposing is to use old drugs for new indications that are already in extensive use in cancer therapy. In the face of advantages like validated anticancer pharmaco-kinetic properties, safety, and acceptability in humans, yet there is still a probability of failure in late phases of clinical trials due to the competition from efficacious new drug development. Other obstacles in repurposed drug development include legal issues like intellectual property (IP) issue and unfair prescription charges. The IP concern stops certain repurposed drugs from entering into the market. Filing secondary patents offers a chance to find new targets for existing drugs. Hopefully, such obstacles will prove resolvable.[34]

## 1.7 CONCLUSION

Drug repurposing holds excellent promise as a reliable mode of drug discovery. By identifying a new indication to an old drug, we can speed up the process and provide a safe, more efficient, and economical solution to novel challenges. Also, with new investigations unraveling hitherto unknown facts about the pathophysiology of diseases, it is necessary to revise the mode of action of existing drugs and screen them for their potential new indications.

## KEYWORDS


- **animal model screening**
- **cell-based assay**
- **drug repurposing approaches**
- **machine learning**
- **network analysis**
- **text mining**


## REFERENCES


1. Scannell, J. W.; Blanckley, A.; Boldon, H.; Warrington, B. Diagnosing the Decline in Pharmaceutical R&D Efficiency. *Nat. Rev. Drug Discov.* **2012,** *11* (3), 191–200.
2. Ashburn, T. T.; Thor, K. B. Drug Repositioning: Identifying and Developing New Uses for Existing Drugs. *Nat. Rev. Drug Discov*. **2004,** *3* (8),673–683.
3. Breckenridge, A.; Jacob, R. Overcoming the Legal and Regulatory Barriers to Drug Repurposing. *Nat. Rev. Drug Discov*. **2018,** *18* (1), 1–2.
4. Pushpakom, S.; Iorio, F.; Eyers, P. A.; Escott, K. J.; Hopper, S.; Wells, A.; Doig, A.; Guilliams, T.; Latimer, J.; McNamee, C.; Norris, A.; Sanseau, P.; Cavalla, D.; Pirmohamed, M. Drug Repurposing: Progress, Challenges and Recommendations. *Nat. Rev. Drug Discov*. **2018,** *18* (1), 41–58.
5. Wouters, O. J.; McKee, M.; Luyten, J. Estimated Research and Development Investment Needed to Bring a New Medicine to Market, 2009-2018. *JAMA* **2020,** *323* (9), 844–853.
6. Novac, N. Challenges and Opportunities of Drug Repositioning. *Trends Pharmacol. Sci.* **2013,** *34* (5), 267–272.
7. Kumar, R.; Harilal, S.; Gupta, S. V.; Jose, J.; Thomas, D. G.; Uddin, M. S.; Shah, M. A.; Mathew, B. Exploring the New Horizons of Drug Repurposing: A Vital Tool for Turning Hard Work into Smart Work. *Eur. J. Med. Chem*. **2019,** *182* (15), 111602.
8. Lee, H. M.; Kim, Y. Drug Repurposing Is a New Opportunity for Developing Drugs Against Neuropsychiatric Disorders. *Schizophr. Res. Treatment* **2016,** *2016*, 6378137.
9. Bianchi, M. T. Promiscuous Modulation of Ion Channels by Anti-Psychotic and Anti-Dementia Medications. *Med. Hypotheses* **2010,** *74* (2), 297–300.
10. Zhang, H. Y.; Tang, X. C. Neuroprotective Effects of Huperzine A: New Therapeutic Targets for Neurodegenerative Disease. *Trends Pharmacol. Sci.* **2006,** *27* (12), 619–625.
11. Bianchi, M. T. Non-Serotonin Anti-Depressant Actions: Direct Ion Channel Modulation by SSRIs and the Concept of Single Agent Poly-Pharmacy. *Med. Hypotheses* **2008,** *70* (5), 951–956.
12. Schwab, R. S.; England, A. C.; Poskanzer, D. C.; Young, R. R. Amantadine in the Treatment of Parkinson's Disease. *JAMA* **1969,** *208* (7), 1168–1170.
13. Mitsuya, H.; Weinhold, K. J.; Furman, P. A.; St Clair, M. H.; Lehrman, S. N.; Gallo, R. C.; Bolognesi, D.; Barry, D. W.; Broder, S. 3'-Azido-3'-Deoxythymidine (BW A509U): An Antiviral Agent That Inhibits the Infectivity and Cytopathic Effect of Human

T-Lymphotropic Virus Type III/Lymphadenopathy-Associated Virus in Vitro. *Proc. Natl. Acad. Sci. U. S. A.* **1985,** *82* (20), 7096–7100.
14. Parvathaneni, V.; Kulkarni, N. S.; Muth, A.; Gupta, V. Drug Repurposing: A Promising Tool to Accelerate the Drug Discovery Process. *Drug Discov. Today* **2019,** *24* (10), 2076–2085.
15. Jourdan, J. P.; Bureau, R.; Rochais, C.; Dallemagne, P. Drug Repositioning: A Brief Overview. *J. Pharm. Pharmacol.* **2020,** *72* (9), 1145–1151.
16. Hurle, M. R.; Yang, L.; Xie, Q.; Rajpal, D. K.; Sanseau, P.; Agarwal, P. Computational Drug Repositioning: From Data to Therapeutics. *Clin. Pharmacol. Ther.* **2013,** *93* (4), 335–341.
17. Hearst, M. A. Untangling Text Data Mining, 1999; pp 3–10.
18. Zhu, F.; Patumcharoenpol, P.; Zhang, C.; Yang, Y.; Chan, J.; Meechai, A.; Vongsangnak, W.; Shen, B. Biomedical Text Mining and Its Applications in Cancer Research. *J. Biomed. Inf.* **2013,** *46* (2), 200–211.
19. Weeber, M.; Klein, H.; De Jong-Van Den Berg, L. T. W.; Vos, R. Using Concepts in Literature-Based Discovery: Simulating Swanson's Raynaud-Fish Oil and Migraine-Magnesium Discoveries. *J. Am. Soc. Inf. Sci. Technol.* **2001,** *52* (7), 548–557.
20. Jarada, T. N.; Rokne, J. G.; Alhajj, R. A Review of Computational Drug Repositioning: Strategies, Approaches, Opportunities, Challenges, and Directions. *J. Cheminform.* **2020,** *12* (1), 46.
21. Li, J.; Zhu, X.; Chen, J. Y. Building Disease-Specific Drug-Protein Connectivity Maps from Molecular Interaction Networks and PubMed Abstracts. *PLoS Comput. Biol.* **2009,** *5* (7), e1000450.
22. Xue, H.; Li, J.; Xie, H.; Wang, Y. Review of Drug Repositioning Approaches and Resources. *Int. J. Biol. Sci.* **2018,** *14* (10), 1232–1244.
23. Wang, H.; Wu, T.; Qi, G.; Ruan, T. On Publishing Chinese Linked Open Schema. In *Lecture Notes in Computer Science (Including Subseries Lecture Notes in Artificial Intelligence and Lecture Notes in Bioinformatics)*, 2014; vol 8796, pp 293–308.
24. Chen, B.; Ding, Y.; Wild, D. J. Assessing Drug Target Association Using Semantic Linked Data. *PLoS Comput. Biol.* **2012,** *8* (7), e1002574.
25. Lee, J. Y.; Shin, J. Y.; Kim, H. S.; Heo, J. I.; Kho, Y. J.; Kang, H. J.; Park, S. H.; Lee, J. Y. Effect of Combined Treatment With Progesterone and Tamoxifen on the Growth and Apoptosis of Human Ovarian Cancer Cells. *Oncol. Rep.* **2012,** *27* (1), 87–93.
26. Yella, J. K.; Yaddanapudi, S.; Wang, Y.; Jegga, A. G. Changing Trends in Computational Drug Repositioning. *Pharmaceuticals* **2018,** *11* (2), 57.
27. Park, K. A Review of Computational Drug Repurposing. *Transl. Clin. Pharmacol.* **2019,** *27* (2), 59–63.
28. Gottlieb, A.; Stein, G. Y.; Ruppin, E.; Sharan, R. PREDICT: A Method for Inferring Novel Drug Indications With Application to Personalized Medicine. *Mol. Syst. Biol.* **2011,** *7*, 496.
29. Liu, Z.; Guo, F.; Gu, J.; Wang, Y.; Li, Y.; Wang, D.; Lu, L.; Li, D.; He, F. Similarity-Based Prediction for Anatomical Therapeutic Chemical Classification of Drugs by Integrating Multiple Data Sources. *Bioinformatics* **2015,** *31*, 1788–1795.
30. Luo, H.; Wang, J.; Li, M.; Luo, J.; Peng, X.; Wu, F. X.; Pan, Y. Drug Repositioning Based on Comprehensive Similarity Measures and Bi-Random Walk Algorithm. *Bioinformatics* **2016,** *32*, 2664–2671.

31. Napolitano, F.; Zhao, Y.; Moreira, V. M.; Tagliaferri, R.; Kere, J.; D'Amato, M.; Greco, D. Drug Repositioning: A Machine-Learning Approach Through Data Integration. *J. Cheminform.* **2013,** *5* (1), 30.
32. Aliper, A.; Plis, S.; Artemov, A.; Ulloa, A.; Mamoshina, P.; Zhavoronkov, A. Deep Learning Applications for Predicting Pharmacological Properties of Drugs and Drug Repurposing Using Transcriptomic Data. *Mol. Pharm.* **2016,** *13* (7), 2524–2530.
33. Karaman, M. W.; Herrgard, S.; Treiber, D. K.; Gallant, P.; Atteridge, C. E.; Campbell, B. T.; Chan, K. W.; Ciceri, P.; Davis, M. I.; Edeen, P. T.; Faraoni, R.; Floyd, M.; Hunt, J. P.; Lockhart, D. J.; Milanov, Z. V.; Morrison, M. J.; Pallares, G.; Patel, H. K.; Pritchard, S.; Wodicka, L. M.; Zarrinkar, P. P. A Quantitative Analysis of Kinase Inhibitor Selectivity. *Nat. Biotechnol.* **2008,** *26* (1), 127–132.
34. Zhang, Z.; Zhou, L.; Xie, N.; Nice, E. C.; Zhang, T.; Cui, Y.; Huang, C. Overcoming Cancer Therapeutic Bottleneck by Drug Repurposing. *Signal Transduct. Target. Ther.* **2020,** *5* (1), 113.
35. Corsello, S. M.; Nagari, R. T.; Spangler, R. D.; Rossen, J.; Kocak, M.; Bryan, J. G.; Humeidi, R.; Peck, D.; Wu, X.; Tang, A. A.; Wang, V. M.; Bender, S. A.; Lemire, E.; Narayan, R.; Montgomery, P.; Ben-David, U.; Garvie, C. W.; Chen, Y.; Rees, M. G.; Lyons, N. J.; McFarland, J. M.; Wong, B. T.; Wang, L.; Dumont, N.; O'Hearn, P. J.; Stefan, E.; Doench, J. G.; Harrington, C. N.; Greulich, H.; Meyerson, M.; Vazquez, F.; Subramanian, A.; Roth, J. A.; Bittker, J. A.; Boehm, J. S.; Mader, C. C.; Tsherniak, A.; Golub, T. R. Discovering the Anticancer Potential of Non-Oncology Drugs by Systematic Viability Profiling. *Nat. Cancer* **2020,** *1* (2), 235–248.
36. Ridges, S.; Heaton, W. L.; Joshi, D.; Choi, H.; Eiring, A.; Batchelor, L.; Choudhry, P.; Manos, E. J.; Sofla, H.; Sanati, A.; Welborn, S.; Agarwal, A.; Spangrude, G. J.; Miles, R. R.; Cox, J. E.; Frazer, J. K.; Deininger, M.; Balan, K.; Sigman, M.; Müschen, M.; Perova, T.; Johnson, R.; Montpellier, B.; Guidos, C. J.; Jones, D. A.; Trede, N. S. Zebrafish Screen Identifies Novel Compound With Selective Toxicity Against Leukemia. *Blood* **2012,** *119* (24), 5621–5631.
37. Guney, E.; Menche, J.; Vidal, M.; Barábasi, A. L. Network-Based in Silico Drug Efficacy Screening. *Nat. Commun.* **2016,** *7,* 10331.
38. Lim, H.; Poleksic, A.; Yao, Y.; Tong, H.; He, D.; Zhuang, L.; Meng, P.; Xie, L. Large-Scale Off-Target Identification Using Fast and Accurate Dual Regularized One-Class Collaborative Filtering and Its Application to Drug Repurposing. *PLoS Comput. Biol.* **2016,** *12* (10), e1005135.
39. Fortmeyer, R. The Zero Effect. *Archit. Rec.* **2007,** *195* (3), 153.
40. Subeha, M. R.; Telleria, C. M. The Anti-Cancer Properties of the HIV Protease Inhibitor Nelfinavir. *Cancers* **2020,** *12* (11), 3437.
41. Brüning, A.; Burger, P.; Vogel, M.; Rahmeh, M.; Gingelmaier, A.; Friese, K.; Lenhard, M.; Burges, A. Nelfinavir Induces the Unfolded Protein Response in Ovarian Cancer Cells, Resulting in ER Vacuolization, Cell Cycle Retardation and Apoptosis. *Cancer Biol. Ther.* **2009,** *8* (3), 226–232.
42. Aronoff, D. M.; Neilson, E. G. Antipyretics: Mechanisms of Action and Clinical Use in Fever Suppression. *Am. J. Med.* **2001,** *111* (4), 304–315.
43. Miner, J.; Hoffhines, A. The Discovery of Aspirin's Antithrombotic Effects. *Texas Hear. Inst. J.* **2007,** *34* (2), 179–186.
44. Sostres, C.; Gargallo, C. J.; Lanas, A. Aspirin, Cyclooxygenase Inhibition And Colorectal Cancer. *World. J. Gastrointest. Pharmacol. Ther.* **2014,** *5*(1), 40.

45. Reimers, M. S.; Bastiaannet, E.; Van Herk-Sukel, M. P. P.; Lemmens, V. E. P.; Van Den Broek, C. B. M.; Van De Velde, C. J. H.; De Craen, A. J. M.; Liefers, G. J. Aspirin Use After Diagnosis Improves Survival in Older Adults With Colon Cancer: A Retrospective Cohort Study. *J. Am. Geriatr. Soc.* **2012,** *60* (12), 2232–2236.
46. Chen, D.; Cui, Q. C.; Yang, H.; Dou, Q. P. Disulfiram, a Clinically Used Anti-Alcoholism Drug and Copper-Binding Agent, Induces Apoptotic Cell Death in Breast Cancer Cultures and Xenografts via Inhibition of the Proteasome Activity. *Cancer Res.* **2006,** *66* (21), 10425–10433.
47. Wiggins, H. L.; Wymant, J. M.; Solfa, F.; Hiscox, S. E.; Taylor, K. M.; Westwell, A. D.; Jones, A. T. Disulfiram-Induced Cytotoxicity and Endo-Lysosomal Sequestration of Zinc in Breast Cancer Cells. *Biochem. Pharmacol.* **2015,** *93* (3), 332–342.
48. Noto, H.; Goto, A.; Tsujimoto, T.; Noda, M. Cancer Risk in Diabetic Patients Treated With Metformin: A Systematic Review and Meta-Analysis. *PLoS One* **2012,** *7* (3), 1–9.
49. Yang, J.; Wei, J.; Wu, Y.; Wang, Z.; Guo, Y.; Lee, P.; Li, X. Metformin Induces ER Stress-Dependent Apoptosis Through MiR-708-5p/NNAT Pathway in Prostate Cancer. *Oncogenesis* **2015,** *4* (6), 1–8.
50. Hadad, S.; Iwamoto, T.; Jordan, L.; Purdie, C.; Bray, S.; Baker, L.; Jellema, G.; Deharo, S.; Hardie, D. G.; Pusztai, L.; Moulder-Thompson, S.; Dewar, J. A.; Thompson, A. M. Evidence for Biological Effects of Metformin in Operable Breast Cancer: A Pre-Operative, Window-of-Opportunity, Randomized Trial. *Breast Cancer Res. Treat.* **2011,** *128* (3), 783–794.
51. Matthews, S. J.; McCoy, C. Peginterferon Alfa-2a: A Review of Approved and Investigational Uses. *Clin. Ther.* **2004,** *26* (7), 991–1025.
52. Rajkumar, S. V. Thalidomide in the Treatment of Multiple Myeloma. *Expert Rev. Anticancer Ther.* **2001,** *1* (1), 20–28.
53. Li, J.; Hao, Q.; Cao, W.; Vadgama, J. V.; Wu, Y. Celecoxib in Breast Cancer Prevention and Therapy. *Cancer Manag. Res.* **2018,** *10*, 4653–4667.
54. Kirkendall, W. M.; Hammond, J. J.; Thomas, J. C.; Overturf, M. L.; Zama, A. Prazosin and Clonidine for Moderately Severe Hypertension. *JAMA* **1978,** *240* (23), 2553–2556.
55. Lang, C. C.; Choy, A. M. J.; Rahman, A. R.; Struthers, A. D. Renal Effects of Low Dose Prazosin in Patients With Congestive Heart Failure. *Eur. Heart J.* **1993,** *14* (9), 1245–1252.
56. Nicholson, J. P.; Vaughn, E. D.; Pickering, T. G.; Resnick, L. M.; Artusio, J.; Kleinert, H. D.; Lopez-Overjero, J. A.; Laragh, J. H. Pheochromocytoma and Prazosin. *Ann. Intern. Med.* **1983,** *99* (4), 477–479.
57. Assad Kahn, S.; Costa, S. L.; Gholamin, S.; Nitta, R. T.; Dubois, L. G.; Fève, M.; Zeniou, M.; Coelho, P. L. C.; El-Habr, E.; Cadusseau, J.; Varlet, P.; Mitra, S. S.; Devaux, B.; Kilhoffer, M.; Cheshier, S. H.; Moura-Neto, V.; Haiech, J.; Junier, M.; Chneiweiss, H. The Anti-hypertensive Drug Prazosin Inhibits Glioblastoma Growth via the PKC Δ-dependent Inhibition of the AKT Pathway . *EMBO Mol. Med.* **2016,** *8* (5), 511–526.
58. Cheong, D. H. J.; Tan, D. W. S.; Wong, F. W. S.; Tran, T. Anti-Malarial Drug, Artemisinin and Its Derivatives for the Treatment of Respiratory Diseases. *Pharmacol. Res.* **2020,** *158*, 104901.
59. Gribkoff, V. K.; Kaczmarek, L. K. The Need for New Approaches in CNS Drug Discovery: Why Drugs Have Failed, and What Can Be Done to Improve Outcomes. *Neuropharmacology* **2017,** *120*, 11–19.

60. Pardridge, W. M. The Blood-Brain Barrier: Bottleneck in Brain Drug Development. **2005,** *2* (1), 3–14.
61. Morofuji, Y.; Nakagawa, S. Drug Development for Central Nervous System Diseases Using In Vitro Blood-Brain Barrier Models and Drug Repositioning. *Curr. Pharm. Des.* **2020,** *26* (13), 1466–1485.
62. Llaguno-Munive, M.; Vazquez-Lopez, M. I.; Jurado, R.; Garcia-Lopez, P. Mifepristone Repurposing in Treatment of High-Grade Gliomas. *Front. Oncol.* **2021**, *11* , 606907.
63. Llaguno-Munive, M.; Romero-Piña, M.; Serrano-Bello, J.; Medina, L. A.; Uribe-Uribe, N.; Salazar, A. M.; Rodríguez-Dorantes, M.; Garcia-Lopez, P. Mifepristone Overcomes Tumor Resistance to Temozolomide Associated With DNA Damage Repair and Apoptosis in an Orthotopic Model of Glioblastoma. *Cancers (Basel).***2019**, *11* (1).
64. Block, T. S.; Kushner, H.; Kalin, N.; Nelson, C.; Belanoff, J.; Schatzberg, A. Combined Analysis of Mifepristone for Psychotic Depression: Plasma Levels Associated With Clinical Response. *Biol. Psychiatry* **2018,** *84* (1), 46–54.
65. Morgan, F. H.; Laufgraben, M. J. Mifepristone for Management of Cushing's Syndrome. *Pharmacotherapy* **2013,** *33* (3), 319–329.
66. Schwab, R. S.; England, A. C.; Poskanzer, D. C.; Young, R. R. Amantadine in the Treatment of Parkinson ' s Disease Amantadme, 2012; pp 8–10.
67. Ghanizadeh, A. Atomoxetine for Treating ADHD Symptoms in Autism: A Systematic Review. *J. Atten. Disord.* **2013**, *17* (8), 635–640.
68. Jiménez-ruiz, C. A.; López-padilla, D.; Alonso-arroyo, A.; Aleixandre-benavent, R. Since January 2020 Elsevier Has Created a COVID-19 Resource Centre with Free Information in English and Mandarin on the Novel Coronavirus COVID- 19 . The COVID-19 Resource Centre Is Hosted on Elsevier Connect , the Company ' s Public News and Information . *www.archbronconeumol.org Orig.* **2020,** *14* (4), 337–339.
69. Pahwa, R.; Lyons, K. E.; Hauser, R. A. Ropinirole Therapy for Parkinson ' s Disease. Expert Rev. Neurother. **2004,** *4* (4), 581–588.
70. Garcia-Borreguero, D.; Grunstein, R.; Sridhar, G.; Dreykluft, T.; Montagna, P.; Dom, R.; Lainey, E.; Moorat, A.; Roberts, J. A 52-Week Open-Label Study of the Long-Term Safety of Ropinirole in Patients With Restless Legs Syndrome. *Sleep Med.* **2007,** *8* (7–8), 742–752.
71. Shytle, R. D.; Penny, E.; Silver, A. A.; Goldman, J.; Sanberg, P. R. Mecamylamine (Inversine®): An Old Antihypertensive With New Research Directions. *J. Hum. Hypertens.* **2002,** *16* (7), 453–457.
72. Lichtenstein, G. R.; Feagan, B. G.; Cohen, R. D.; Salzberg, B. A.; Safdi, M.; Popp, J. W.; Langholff, W.; Sandborn, W. J. Infliximab for Crohn's Disease: More Than 13 Years of Real-World Experience. *Inflamm. Bowel Dis.* **2018**, *24* (3), 490–501.
73. Torres-Acosta, N.; O'Keefe, J. H.; O'Keefe, E. L.; Isaacson, R.; Small, G. The Rapeutic Potential of TNF-α Inhibition for Alzheimer's Disease Prevention. *J. Alzheimer's Dis.* **2020,** *78* (2), 619–626.
74. Schubert, M.; Hansen, S.; Leefmann, J.; Guan, K. Repurposing Antidiabetic Drugs for Cardiovascular Disease. *Front. Physiol.* **2020,** *11*, 568632.
75. Zhang, D.; Yang, R.; Wang, S.; Dong, Z. Paclitaxel: New Uses for an Old Drug. *Drug Des. Devel. Ther.* **2014,** *8*, 279–284.
76. Park, S.-J.; Shim, W. H.; Ho, D. S.; Raizner, A. E.; Park, S.-W.; Hong, M.-K.; Lee, C. W.; Choi, D.; Jang, Y.; Lam, R.; Weissman, N. J.; Mintz, G. S. A Paclitaxel-Eluting Stent for the Prevention of Coronary Restenosis. *N. Engl. J. Med.* **2003,** *348* (16), 1537–1545.

77. Lansky, A.; Grubman, D.; Scheller, B. Paclitaxel-Coated Balloons: A Safe Alternative to Drug-Eluting Stents for Coronary in-Stent Restenosis. *Eur. Heart J.* **2020,** *41* (38), 3729–3731.
78. Bhattacharyya, B.; Panda, D.; Gupta, S.; Banerjee, M. Anti-Mitotic Activity of Colchicine and the Structural Basis for Its Interaction With Tubulin. *Med. Res. Rev.* **2008,** *28* (1), 155–183.
79. Lutschinger, L. L.; Rigopoulos, A. G.; Schlattmann, P.; Matiakis, M.; Sedding, D.; Schulze, P. C.; Noutsias, M. Correction to: Meta-Analysis for the Value of Colchicine for the Therapy of Pericarditis and of Postpericardiotomy Syndrome (BMC Cardiovascular Disorders (2019) 19 (207) DOI: 10.1186/S12872-019-1190-4). *BMC Cardiovasc. Disord.* **2019,** *19* (1), 1–11.
80. Zhao, X.; Zhang, X. F.; Zhao, Y.; Lin, X.; Li, N. Y.; Paudel, G.; Wang, Q. Y.; Zhang, X. W.; Li, X. L.; Yu, J. Effect of Combined Drospirenone With Estradiol for Hypertensive Postmenopausal Women: A Systemic Review and Meta-Analysis. *Gynecol. Endocrinol.* **2016,** *32* (9), 685–689.
81. Korkmaz-Icöz, S.; Radovits, T.; Szabó, G. Targeting Phosphodiesterase 5 as a Therapeutic Option against Myocardial Ischaemia/Reperfusion Injury and for Treating Heart Failure. *Br. J. Pharmacol.* **2018,** *175* (2), 223–231.
82. Gelosa, P.; Castiglioni, L.; Camera, M.; Sironi, L. Drug Repurposing in Cardiovascular Diseases: Opportunity or Hopeless Dream? *Biochem. Pharmacol.* **2020,** *177*, 113894.
83. Guo, H.; Callaway, J. B.; Ting, J. P. Y. Inflammasomes: Mechanism of Action, Role in Disease, and Therapeutics. *Nat. Med.* **2015,** *21* (7), 677–687.
84. Rundfeldt, C.; Socała, K.; Wlaź, P. The Atypical Anxiolytic Drug, Tofisopam, Selectively Blocks Phosphodiesterase Isoenzymes and Is Active in the Mouse Model of Negative Symptoms of Psychosis. *J. Neural Transm.* **2010,** *117* (11), 1319–1325.
85. Leventer, S. M.; Raudibaugh, K.; Frissora, C. L.; Kassem, N.; Keogh, J. C.; Phillips, J.; Mangel, A. W. Clinical Trial: Dextofisopam in the Treatment of Patients With Diarrhoea-Predominant or Alternating Irritable Bowel Syndrome. *Aliment. Pharmacol. Ther.* **2008,** *27* (2), 197–206.
86. Grenier, L.; Hu, P. Computational Drug Repurposing for Inflammatory Bowel Disease Using Genetic Information. *Comput. Struct. Biotechnol. J.* **2019,** *17*, 127–135.
87. Jenkins, C. R.; Bateman, E. D.; Sears, M. R.; O'Byrne, P. M. What Have We Learnt about Asthma Control From Trials of Budesonide/Formoterol as Maintenance and Reliever? **2020,** *25* (8), 804–815.
88. Lázaro, C. M.; de Oliveira, C. C.; Gambero, A.; Rocha, T.; Cereda, C. M. S.; de Araújo, D. R.; Tofoli, G. R. Evaluation of Budesonide–Hydroxypropyl-β-Cyclodextrin Inclusion Complex in Thermoreversible Gels for Ulcerative Colitis. *Dig. Dis. Sci.* **2020,** *65* (11), 3297–3304.
89. Jaffe, I. A. Penicillamine : An Anti-Rheumatoid Drug. Am. J. Med. **1983**, *75* (6), 63–68.
90. Lai, Z. W.; Kelly, R.; Winans, T.; Marchena, I.; Shadakshari, A.; Yu, J.; Dawood, M.; Garcia, R.; Tily, H.; Francis, L.; Faraone, S. V.; Phillips, P. E.; Perl, A. Sirolimus in Patients with Clinically Active Systemic Lupus Erythematosus Resistant to, or Intolerant of, Conventional Medications: A Single-Arm, Open-Label, Phase 1/2 Trial. *Lancet* **2018,** *391* (10126), 1186–1196.
91. Kingsmore, K. M.; Grammer, A. C.; Lipsky, P. E. Drug Repurposing to Improve Treatment of Rheumatic Autoimmune Inflammatory Diseases. *Nat. Rev. Rheumatol.* **2020,** *16* (1), 32–52.

92. Singh, T. U.; Parida, S.; Lingaraju, M. C.; Kesavan, M.; Kumar, D.; Singh, R. K. Drug Repurposing Approach to Fight COVID-19. *Pharmacol. Rep.* **2020,** *72* (6), 1479–1508.
93. Yuan, M.; Chua, S. L.; Liu, Y.; Drautz-Moses, D. I.; Hoong Yam, J. K.; Aung, T. T.; Beuerman, R. W.; Santillan Salido, M. M.; Schuster, S. C.; Tan, C. H.; Givskov, M.; Yang, L.; Nielsen, T. E. Repurposing the Anticancer Drug Cisplatin With the Aim of Developing Novel Pseudomonas Aeruginosa Infection Control Agents. *Beilstein J. Org. Chem.* **2018,** *14*, 3059–3069.
94. Mercorelli, B.; Palù, G.; Loregian, A. Drug Repurposing for Viral Infectious Diseases: How Far Are We? *Trends Microbiol.* **2018,** *26* (10), 865–876.
95. Standing, J. F.; Wong, I. C. K.; Winstanley, P. Chlorproguanil-dapsone for Malaria. *Lancet* **2004**, 1753–1754.
96. Lv, B. M.; Tong, X. Y.; Quan, Y.; Liu, M. Y.; Zhang, Q. Y.; Song, Y. F.; Zhang, H. Y. Drug Repurposing for Japanese Encephalitis Virus Infection by Systems Biology Methods. *Molecules* **2018,** *23* (12), 3346.

# CHAPTER 2

# Approaches, Strategies, and Advances in Computational Drug Discovery and Drug Repurposing

TRIPTI SHARMA[1], IPSA PADHY[2], and CHITA RANJAN SAHOO[3]

[1]*Department of Pharmaceutical Chemistry, School of Pharmaceutical Sciences, Siksha 'O'Anusandhan Deemed to be University, Bhubaneswar, Odisha, India*

[2]*Department of Pharmaceutical Analysis, School of Pharmaceutical Education and Research, Berhampur University, Berhampur, Odisha, India*

[3]*Central Research Laboratory, Institute of Medical Sciences and SUM Hospital, Siksha 'O' Anusandhan Deemed to be University, Bhubaneswar, Odisha, India*

## ABSTRACT

Drug discovery is a challenging, expensive, and time-consuming procedure that has an extremely low success rate. When it comes to the early phases of drug discovery, computational techniques are very beneficial since it substantially reduce attrition rates in the drug development process. The use of artificial intelligence, particularly machine learning and deep learning methodologies, has become in grained in the drug development process. Computational drug discovery and development is experiencing tremendous advancement in recent times. These approaches effectively exploits known targets, drugs, pathways or disease biomarkers by utilizing

Drug Repurposing and Computational Drug Discovery: Strategies and Advances.
Mithun Rudrapal, PhD (Ed)

various bioinformatics, chemo-informatics, system biology and network biology tools. Ligand-based and structure-based approaches are widely used computational methods in the drug discovery process. Three-dimensional quantitative structure activity relationship (3D QSAR) and pharmacophore modeling are most commonly used techniques in ligand-based approaches for giving predictive models for lead generation and optimization. Structure-based approaches use structural data obtained experimentally or through computational homology modeling. Molecular docking, structure-based virtual screening (SBVS) and molecular dynamics (MD) are frequently used SBDD strategies for analysis of molecular recognition events of binding energetic, molecular interactions and induced conformational changes. Drug repurposing is the program of drug discovery, which involves finding new indications for pre-existing marketed drugs, failed drugs or withdrawn drugs. Drug repurposing has lately acquired recognized as an effective alternative capable of delivering medication. The chapter highlights diverse drug repurposing tactics and overviews commonly used resources, open source databases/tools, workflow systems, pipelines in the form of codes, software tools. Computer based methodologies that are comprehensively used in drug repurposing studies have been summarized. Various challenges and limitations met in computational drug repurposing studies are also addressed along with further research directions.

## 2.1 INTRODUCTION

Computational approaches in new drug discovery and development are experiencing tremendous progression, globally. This rapid growth in computational techniques has been plausible due to development of powerful hardware, advances in software, and availability of biological data. Indeed, software and tools provide high quality in prediction, simulations, reliability, and versatility for different operating systems making them convenient to use. Increase in availability of biological data, crystal structure of biological targets, and several databases in the past few decades further add to the ease of the process.[1] Furthermore, the development of latest central processing units (CPUs) and graphics processing units (GPUs) has scaled up the calculation speed and therefore the performance. High-speed performance, increased flexibility, and capability of GPUs along with high-level programming languages such as OpenCL, CUDA make the approach very convenient.[2,3] Computational strategies in drug development are helpful for the researchers to generate and evaluate several molecules against different disease pathways simultaneously.[4]

Drug repurposing or repositioning is the program of drug discovery, which involves finding new indications for pre-existing marketed drugs, failed drugs, or withdrawn drugs.[5] It is the safer and faster alternative for drug development, when the potential treatment is not available or recommended. Since the preclinical and clinical studies of the repurposed drug are well established, it decreases the cost and time required for the molecule to reach the market, and the risk of failure is limited. Furthermore, the advantage of this approach is enhanced patent life of the drug molecule. All these advantages account for the interest of pharmaceutical companies for drug repurposing.[6] Recently, about 30% of the new drugs and vaccines approved by FDA are repurposed of old drugs and almost 170 repositioned drugs entered the drug development pipeline during the year 2010–2017.[7] Experimental and computational-based approaches are the two ways for drug repurposing. Computational-based approaches in drug discovery effectively exploit known targets, drugs, pathways, or disease biomarkers by utilizing various bioinformatics, chemo-informatics, system biology, and network biology tools.[8] The advances during the last few years in the fields of computational approaches have aided in drug discovery process.[9,10] But the current scenario demands an integrated application of various computational tools that will be beneficial at every point in the drug discovery pipeline.[11]

## 2.2 COMPUTATIONAL APPROACHES IN DRUG DISCOVERY

The computational approaches simulate the interactions between the desired biomolecular targets such as enzymes, receptors, or transporters and selected scaffold that further helps in designing complementary compound databases for the selected target. The compound databases are then screened to identify and optimize lead molecules, thereby propelling the drug discovery process one step ahead.[12] Ligand-based and structure-based approaches are widely used computational methods in the drug discovery process (Fig. 2.1). Ligand-based approach helps to find a molecule with a specific pharmacological activity by extensive database searching and matching the fingerprint sequences of the repositioned molecule(s). More specifically, the selected compound is converted into a numeric string that is then matched with the databases of compounds with similar biological activity. Ligand-based software and databases either stand alone or online tools are utilized for this purpose.[13] In the structure-based approach the investigational compound is tested toward a set of biological targets to determine its binding affinity. Several software tools and databases are available to the researchers for the purpose, namely, AUTODOCK, GOLD, GLIDE, and a few more.[14]

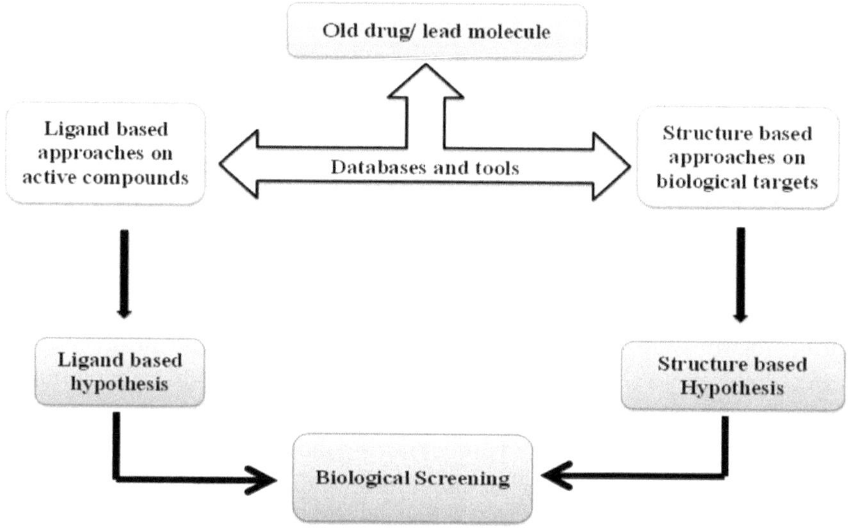


**FIGURE 2.1** Computational approaches in drug discovery.

### 2.2.1 *LIGAND-BASED APPROACHES*

Ligand-based drug design (LBDD) plays an approachable role in the drug discovery process when no information is available regarding the three-dimensional structure of potential drug targets.[15] Ligand-based approaches evaluate ligands/molecular scaffolds that are known to interact with the targets of interest. Basically, strategies employed by this method are assortment of chemical species having chemical resemblance to known ligands with a couple of similarity measures and constructing QSAR models predicting biological activity from the chemical makeup. Of note, this method is primarily used for in silico design and screening of novel ligands as potential drug candidates. The approach is a valuable tool for optimizing the ADMET properties of drug likely candidates. Three-dimensional quantitative structure activity relationship (3D QSAR) and pharmacophore modeling are the most commonly used techniques in ligand-based approaches for giving predictive models for lead generation and optimization.[16] Target fishing and reverse docking methods are also being used in the ligand-based drug discovery process.[17]

#### 2.2.1.1 QSAR MODELING

QSAR analysis is the most efficient tool in establishing a statistically significant correlation between chemical structures and their pharmacological activities.[18] The structural information is defined in terms of a series of parameters known as molecular descriptors.[19] These molecular descriptors are classified as topological, geometrical, thermodynamic, electronic, and constitutional types.[20] Table 2.1 lists out the type of software programs available for computation of molecular descriptors. Linear regression, multiple regressions, partial least squares, and principal component analysis or regression are the statistical approaches covered under linear methods of QSAR analysis. Artificial neural networks (ANN), *k*-nearest neighbors (*k*NN), and Bayesian neural networks are the nonlinear methods of QSAR modeling.[21] QSAR model construction involves a sequential preprocessing and transformation of data sets division, feature/descriptor selection, model development and validation, and finally followed by interpretation and domain analysis.[19] To generate a perfect QSAR model, proper selection of compounds (minimum 20 with reliable and comparable bioactivity data), molecular descriptors for ligands with no autocorrelation to avoid over fitting, and use of internal and external model validation methods for determining applicability and productivity is too crucial.[22]

**TABLE 2.1** Software for Computation of Molecular Descriptors.

| Software | Description | Website | Availability status |
|---|---|---|---|
| ADAPT | Geometrical, topological, physicochemical, electronic | http://research.chem.psu.edu/pcjgroup/adapt.html | Free |
| ADMET Predictor | Constitutional, functional group counts, topological, E-state, 3D descriptors, molecular patterns, acid–base ionization, empirical estimates of quantum | www.simulations-plus.com | Commercial |
| ALOGPS2.1 | log P, log S | www.vcclab.org | Free |

**TABLE 2.1** *(Continued)*

| Software | Description | Website | Availability status |
|---|---|---|---|
| ADRIANA. Code | Topological, constitutional, functional group counts, E-state, meylanflags, moriguchi molecular patterns, 3D descriptors, etc. | www.molecularnetworks.com | Commercial |
| ACD/logP | log S, log P, log D, pKa | www.acdlabs.com | Commercial |
| CDK | Topological, geometrical, electronic, constitutional | http://cdk.github.io | Free |
| Dragon | Topological, constitutional, 2dautocorrelations, geometrical, GETAWAY, WHIM, RDF, functional groups, etc. | www.talete.mi.it | Commercial |
| JOElib | Topological, counting, geometrical properties, etc. | www.ra.cs.uni-tuebingen.de | Free |
| MOE | Topological, structural keys, physical properties, etc. | www.chemcomp.com | Commercial |
| MOLD2 | 1D, 2D | www.fda.gov | Free |
| MOLGEN-QSPR | Constitutional, topological, geometrical, etc. | www.molgen.molgenqspr.html | Commercial |
| Open BABEL | 166-bit, MOLPRINT2D MACCS, structural key fingerprints, daylight fingerprint (FP2) | www.openbabel.org | Free |
| PADEL | Molecular fingerprints, 1D, 2D, 3D descriptors | www.padel.nus.edu.sg | Free |
| PowerMV | Constitutional, atom pairs, fingerprints, BCUT | www.niss.org/PowerMV | Free |

### *2.2.1.2 COMPUTATIONAL CHEMISTRY*

Data quarrying and analysis methodologies aid in speeding up the process of target assessment.[23,24] Big data analysis and artificial intelligence methods (deep neural networks) are embroidering to the computational approaches, which has resulted in advanced and fruitful drug development.[25] Free sources such as Open-Targets,[26] UniProt, and ChEMBL[27] offer useful starting point to cover areas of disease association, protein annotation, and potential ligands, respectively. In recent times, machine learning algorithms are applied for target identification and drug discovery.[28]

### 2.2.1.3 CHEMICAL SPACE

Polypharmacological studies indicate that most drug molecules are not target specific and hit multiple targets.[29] This possibility has led to the concept of the "chemical space" to entitle and consider all ensembled organic molecules in the quest for finding new drugs.[30] Information regarding all the organic molecules can be extracted from various databases such as Pubchem,[31] ZINC,[32] Chemspider,[33] ChemDB,[34] BindingDB,[35] ChEMBL,[27] NCI open,[36] CTD,[37] HMDB,[38] SMPDB,[39] Drugbank.[40]

### 2.2.1.4 CHEMIOINFORMATICS

Chemioinformatics deals with collection, storage, analysis, and manipulation of large quantities of chemical data.[41] Computer-aided drug design, structural representation, and chemmetrics are the primary elements of the cheminformatics.[42] The chemical structure representations can be linear, 2D or in 3D format, or as SMILES (simplified molecular input line entry specification).[43] CAS Draw, DIVA (diverse information, visualization, and analysis), Structure Checker Accord, MarvinSketch, PowerMV, ArgusLab, Babel, Chimera, CLIFF, Dragon, Grace, JOELib, ORTEP, Packmol, Polar, Biosoft, Q-chem, KOWWIN, etc., are some commonly used chemioinformatic software tool packages.[44]

### 2.2.1.5 PHARMACOPHORE MODELING

The atomic and electronic groups in a molecule are transformed into pharmacophoric features and designated as hydrogen bond donors or acceptors, cationic, anionic, aromatic, or hydrophobic attributes called as pharmacophore fingerprints.[45] The pharmacophore model is generated first by translating protein–ligand interactions. Ligand-based 3D pharmacophore models can be developed when no 3D structure of the macromolecule target is available.[46] 3D pharmacophore models are used for hit and lead discovery and optimization.[47] Accelrys' (CA, USA), Discovery Studio Schrödinger's (NY, USA) PHASE, Chemical Computing Group's (QC, Canada), Molecular

Operating Environment (MOE), and Inte:Ligand's (Maria Enzersdorf, Austria) LigandScout are some frequently utilized pharmacophore modeling software. Advanced approaches employing 3D pharmacophore concept such as AutoDock Bias,[48] PyRod,[49] SILCS-Pharm,[50] PharmIF,[51] and Pharm-mapper[52] are popular among researchers.

### 2.2.1.6 *IN SILICO SCREENING*

The concept of "drug-likeness" riddles out molecules with unsolicited properties from screening libraries and reduces attrition risk at the later stages of drug discovery. A variety of machine learning techniques are available to the modeling of drug-likeness like artificial neural networks (ANN), support vector machine (SVM), and recursive partitioning.[53] Quantitative estimate of drug-likeness (QED),[54] relative drug likelihood (RDL),[55] and Gaussian scoring function (GAU)[56] are few quantitative approaches used for defining druggability. Recently, artificial intelligence (AI) approaches are applied for model construction for specific ADMET endpoint evaluation,[53] like Naïve Bayesian Classifiers (NBC) approach[57] and recursive partitioning (RP) approaches.[58] ADMET studies are indispensable steps in any drug discovery program as it provides pharmacokinetic and pharmacodynamic, that is, toxicity (T) properties of lead molecules and reduce the risk of experimental cost, time, and drug failure. Progresses in combinatorial chemistry and high-throughput screening have comprehensively amplified the number of small molecules for which early data on ADMET are available as reference.[59] Table 2.2 depicts some software application tools for ADMET screening. Figure 2.2 describes sequential steps in toxicity prediction modeling.

**TABLE 2.2** Software Application Tools for ADMET Screening.

| Software tools | Feature application | Web address |
|---|---|---|
| ADMET lab | Systematic ADMET screening | http://admet.scbdd.com/ |
| ADMET predictor | ADMET property screening | https://www.simulations-plus.com/software/admetpredictor/ |
| ADVERpred | Prediction of adverse effects of drugs | http://www.way2drug.com/adverpred/ |
| eMOLTOX | Prediction of molecular toxicity | http://xundrug.cn/moltox |
| LIVERTOX | Screening of hepatotoxicity | https://livertox.nih.gov/ |
| Molinspiration | Screening of molecular properties | http://www.molinspiration.com/ |

**TABLE 2.2** *(Continued)*

| Software tools | Feature application | Web address |
|---|---|---|
| Mousetox | Small molecule cytotoxicity prediction | http://enalos.insilicotox.com/ MouseTox/ |
| SOM prediction | Knowledge based prediction of ADMET properties | http://www.scfbio-iitd.res.in/ software/ drugdesign/som.jsp |
| QikProp | Schrodinger tool for ADMET prediction | https://www.schrodinger.com/ qikprop |
| SwissADME | Assessment of ADMET parameters | http://www.swissadme.ch/ |

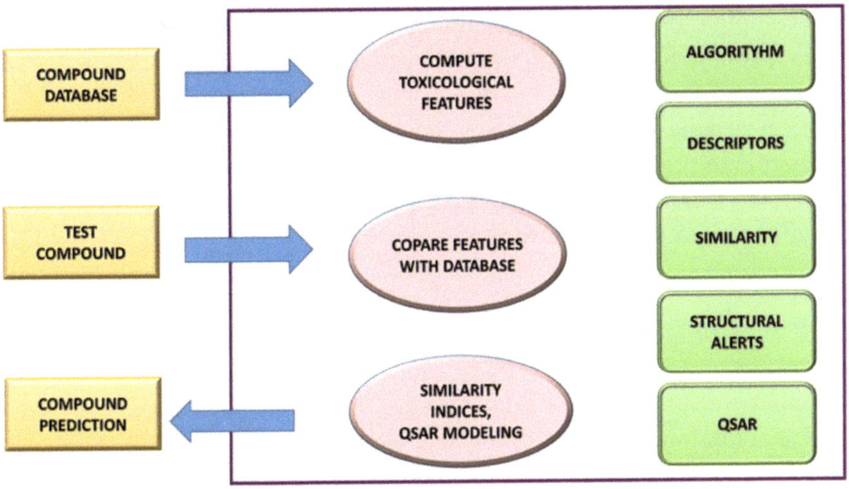


**FIGURE 2.2** Sequential toxicity modeling.

## 2.2.2 STRUCTURE-BASED APPROACHES

Structure-based drug design (SBDD) methods are a prominent component of modern medicinal chemistry.[60] Molecular docking, structure-based virtual screening (SBVS), and molecular dynamics (MD) are frequently used SBDD strategies for analysis of molecular recognition events of binding energetic, molecular interactions, and induced conformational changes.[61] Structure-based approaches use structural data obtained experimentally or through computational homology modeling. The accessibility of 3D macromolecular targets allows a meticulous inspection of the binding site topology along with electrostatic properties, such as charge distribution as well. Current methods allow for the design of ligands containing the necessary features for efficient modulation of the target receptor.[62]

### 2.2.2.1 MOLECULAR MODELING AND PROTEIN STRUCTURE PREDICTIONS

The advances in computational biology have made molecular modeling a reliable method in the drug discovery pipeline.[63] Comparative modeling or homology modeling uses target sequence information for the prediction of three-dimensional structures.[64] The approach of molecular modeling represents thoughtful applications of algorithms of protein structure prediction.[65] The computational methods for protein structure prediction include comparative modeling, fold recognition, first-principles methods with database information, and first-principles methods without database information.[66] CASP (critical structure prediction assessment, a biennial collective project) plays a key role in protein structure prediction.[67] RasMol, PyMOL, Chimera, etc., are some visualization tools for visualizing and analyzing predicted models defining the protein structure.[68]

### 2.2.2.2 MOLECULAR DOCKING

Docking precisely fits ligand into the specific receptor-binding site and evaluates their binding strength.[69] Molecular docking can be flexible or rigid. Flexible docking is a well-thought-out good approach with better prediction than conventional docking.[70] Molecular docking is divided into three sections; ligand–receptor preparation based on force field estimations followed by applying flexible or rigid docking methods and setting search strategies for ligand ratification.[71] Mapping of binding site is done following ligand and receptor preparation by GRID calculations.[72] Various algorithms are applied in the docking process such as fragment-based, Monte Carlo and molecular dynamic based. empirical-based, and knowledge-based scoring functions are usually applied for docking studies.[73,74] Some frequently used software tools for docking studies are depicted in Table 2.3. Analyzing docking outputs is essential as it explains the number and nature of bond present in the ligand protein complex.[75] LigPLOT (2D format),[76,77] PyMOL,[78] and Chimera (3D format)[79] are highly used software tools for analyzing docking results.

**TABLE 2.3** Some Frequently Used Software Tools for Docking Studies.

| Software | Algorithms | License |
|---|---|---|
| AUTODOCK | Genetic Lamarckian | Open, GNU, GPL |
| SWISSDOCK | Evolution based optimization | Open, academic |
| AUTODOCKVINA | Local optimization | Apache |
| GOLD | Genetic algorithms | Commercial |
| GLIDE | Hybrid optimization | Commercial |
| CDOCKER | Hybrid optimization and shape mapping | Commercial |
| iGEMDOCK | Genetic | Commercial |
| Arguslab | Hybrid optimizations | Open |
| MOE-DOCK | Hybrid optimizations | Commercial |

#### *2.2.2.3 MOLECULAR DYNAMICS*

Molecular dynamics simulations are useful tools for the drug discovery process. It is advantageous over molecular docking studies as it facilitates the evaluation of the binding energetic and kinetics of the receptor–ligand interactions at atomic levels.[80] Numerous tools are available to explore the atomic level changes in the biomolecules exhausting the molecular dynamic theories.[81] GROMACS,[82] AMBER,[83] CHARMM-GUI,[84] NAMD,[85] and LAMMPS,[86] etc. are commonly used software packages for molecular dynamic simulations.

### ***2.2.3 SYSTEMS-BASED APPROACHES***

Merging of genomics, proteomics, and metabolomics databases with system pharmacology models aims at generating a disease-specific network that would build confidence in particular target identification that is the crucial step in drug discovery and identification process.[87] Systems-based approaches connect protein targets to their physiological arena, seeing a much wider systemic viewpoint of their environment in conjunction of their molecular details.[88] Drug targeting for diseases with complex pathophysiology involving complicated signaling transduction cascades mostly benefits from a system-based approach.[87]

#### *2.2.3.1 NETWORK PHARMACOLOGY*

Quantitative systems pharmacology or network pharmacology is an emerging discipline that aims to integrate methods of systems biology with pharmacodynamic concepts to foster advanced small-molecule and macromolecule drug discovery and development.[89] 40% of current drug discoveries are based on the concept of network pharmacology.[90] Establishing drug target disease network by utilizing high-throughput omics technologies is the major goal of network pharmacology.[91] Through unbiased investigations on potential target spaces network pharmacology endeavors to discover newer leads and targets and also repurpose prevailing drug molecules for diverse therapeutic conditions.[92] The area of network pharmacology has been broadly divided into the experimental and computational approaches. Graph theory, statistical methods, data mining, modeling, and information visualization methods are the computational approaches. The experimental approaches include integrated clinical trials and high-throughput omics techniques. Data mining, big data analytics, network construction, interactions prediction, and network analysis are key methodologies in the network pharmacology approach.[93] Data mining in network pharmacology can be done through public databases and further experimental analysis.[91]

### *2.2.3.2 PROTEOCHEMOMETRIC MODELING*

In contrast to QSAR modeling proteochemometric (PCM) modeling is based on the similarity of a group of ligands and a group of targets.[94] Commonly PCM has mainly been applied to data sets of G protein-coupled receptors (GPCRs), in particular the rhodopsin-like class A receptors, dopamine, histamine, adrenergic, and melanocortin receptors.[95] This versatile approach is deployed for hit identification for orphan targets, as well as modeling of orthosteric and allosteric ligands.[96] Table 2.4 summarizes a few more PCM-applied studies. Repurposing of mebendazole and celecoxib was done using a proteochemometric modeling known as TMFS (train-match-fit-streamline) method applying a variety of descriptors.[97]

**TABLE 2.4** Summary of Few PCM Applied Studies.

| Target protein | Method applied | No. of proteins studied | No. of ligand molecules screened | Reference |
|---|---|---|---|---|
| Aromatase | Bayesian linear regression, random forest, support vector machine | 1 | 10 | [98] |
| Carbonic anhydrases | k-nearest neighbor | 9 | 549 | [99] |

**TABLE 2.4** *(Continued)*

| Target protein | Method applied | No. of proteins studied | No. of ligand molecules screened | Reference |
|---|---|---|---|---|
| Cyclooxygenases | Gradient boosting machine, elastic net, random forest | 11 | 3228 | [100] |
| Dengue virus NS2B-NS3 proteases | Partial least squares | 4 | 45 | [101] |
| HIV-reverse transcriptases and proteases | Support vector machine | 2 | 48,221 | [95] |
| Kinases | Deep neural network, random forest, gradient boosting machine | 284 | 19.9 million | [102] |
| Nuclear receptors | Decision tree, logistic regression, ridge classifier, random forest | 11 | 7267 | [103] |
| Serine proteases | Partial least squares, random forest | 24 | 5863 | [104] |

#### *2.2.3.3 PATHWAY ANALYSIS*

Pathway analysis is also known as enrichment analysis which examines data acquired by high-throughput technologies for correlating physiological mechanism to macromolecular targets and comprises input phase, analysis phase, and output phase.[105] Null hypothesis generation is the important step in the input phase.[106] Analysis phase comprises mathematical computations defined by a specific set of algorithms that are used by widely available open-source software programs such as BioConductor[107] and GitHub[108] projects. Finally, the output phase comprises visualization and analysis of results. The relevant pathways are ranked arranged in a hierarchal manner by applying statistical attributes.

### *2.2.4 CASE STUDY*

Rath et al. (2021) used the computational approach and designed the novel mesalamine–coumarin conjugate. The macromolecular targets chosen for docking (using AUTODOCKVINA) were COX-2(cyclooxygenase-2),

TNF-α (tumor necrosis factor), MMP-9 (matrix metalloprotien), and MPO human myelo peroxidase. Druglikliness (Molinspiration and Molsoft) and ADMET (pkCSM) were evaluated by online computational tools. The active site on target proteins was analyzed (CASTp 3.0). Molecular dynamic simulations (GROMACS 2018.1 software and amber algorithms) revealed a stable ligand–protein complex for mesalamine and coumarin derivatives in the aqueous system along with highest binding affinity toward all the proteins investigated in comparison with mesalamine alone. Further, these computational results were confirmed by in vivo studies on the acetic acid induced ulcerative colitis rat model which revealed that the synthesized mesalamine-coumarin diazo derivatives were potent in reducing ulcerative colitis.[109]

## 2.3 COMPUTATIONAL APPROACHES AND DRUG REPURPOSING

### *2.3.1 TARGET-BASED APPROACH*

There are several computational processes, including visualization tools, which assist the drug design/discovery process decision systems using target-based drug design methodologies.[110–112] The prospective lead or therapeutic compounds are shaped mostly by biological targets. Target-based drug discovery has had some success. It is not only for small molecules but also has an active role in identifying antibody medicines, protein biologics, gene therapies, and nucleic acid-based treatments.[113,114] Moreover, these approaches are often quicker, simpler, and less expensive to build and operate. It would target specific active site of the candidate by the principle of SARs.[115] Figure 2.3 depicts the schematic representation of a target-based drug discovery approach.

### *2.3.2 KNOWLEDGE-BASED APPROACH*

The information-driven recent trends in the drug development method are becoming more prevalent. When combined with structural knowledge about their target proteins, structural, physicochemical, and ADMET property with standard or potent ligands have shown to be greatly significant for early-stage drug development etiquette.[116,117] The integument of advance chemo-informatics tools, which are used to evaluate the structures and characteristics of effective compounds, has also grown dramatically in the last several years. Recently, authentic databases are playing a key role in the knowledge-based approach system.[118] These are categorized with various forms in Table 2.5. Preliminarily, emphasis on the physicochemical characteristics of drug is most important for clinical approval. These parameters are molecular weight, H-bond acceptor, donor, cLogP, TPSA; number of rotatable bonds and rings. Among all, H-bond accepter parameter is a major factor in determining efficacies of drug.[119]

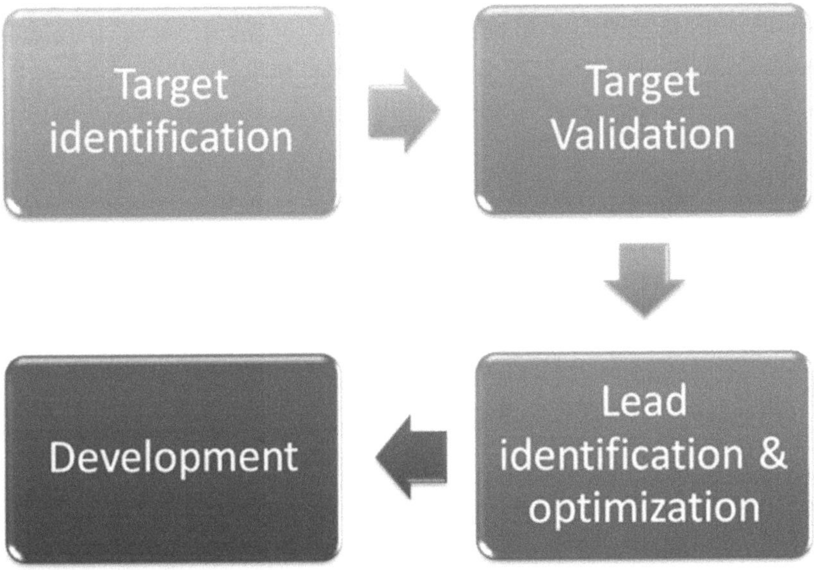


**FIGURE 2.3** Schematic representation of target-based drug discovery approach.

**TABLE 2.5** Knowledge-Based Drug Discovery Repository.

| Sl No. | Database | Link |
|---|---|---|
| | World drug index | https://www.daylight.com/about/index.html |
| | MDDR database | http://www.akosgmbh.de/accelrys/databases/mddr.htm |
| | SuperDrug | http://bioinf.charite.de/superdrug |
| | Comprehensive medicinal chemistry (CMC) | http://www.akosgmbh.de/accelrys/databases/cmc-3d.htm |
| | GPCR-PEnDB | https://gpcr.utep.edu/database |

### 2.3.3 SIGNATURE-BASED APPROACH

In the beginning, the drug discovery and development depends upon the disease of interest. The screening of a leading compound is now the most important stage by the identification of their drug-likeness properties and toxicity using high-throughput technology.[113,120] Moreover, one of the most important requirements in compound profiling and drug discovery is to evaluate the effects of these compounds on cellular activities by utilizing a large number of key proteins. Firstly, the unique molecular state of a phenotypic that differs from the wild-type or healthy state is referred to as the disease signature. Generally, heterogeneous biological data, such as gene expression in organs and protein composition in the microbiome, may be used to describe these signatures. Second, drug signatures are similar to disease signatures (Fig. 2.4). Drug fingerprints are perturbed by therapy exposure rather than disease condition.[113,120,121]

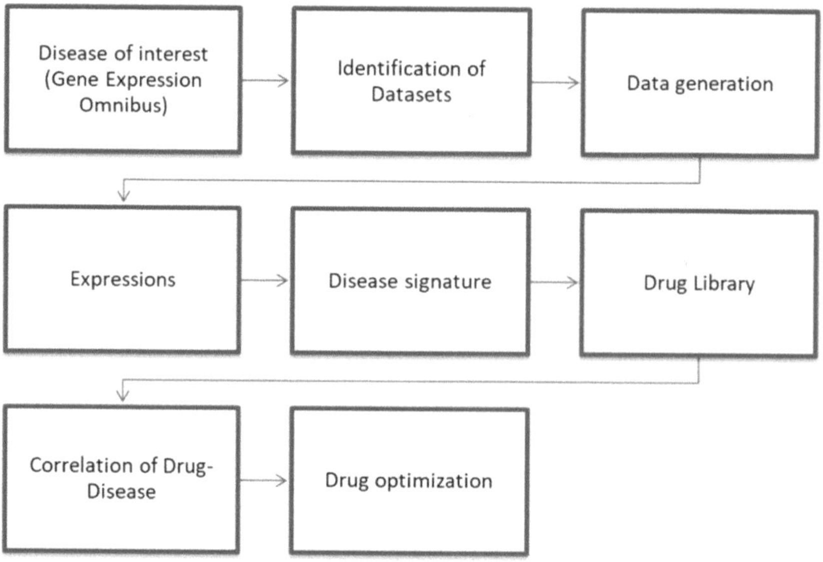


**FIGURE 2.4** Schematic representation of the disease of interest.

### 2.3.4 NETWORK-BASED APPROACH

For the development of effective inhibitors, a new technique that incorporates networks-based methods is being explored. For the purpose of determining the quantitative structure–activity relationship (QSAR) among the known inhibitors by using the first-principles quantum mechanical approach, physical characteristics of neural networks, such as electronegativity and molar volume, would be plausible.[122] This would be effective for early-stage drug development. Figure 2.5 shows schematic representation of network-based drug design.

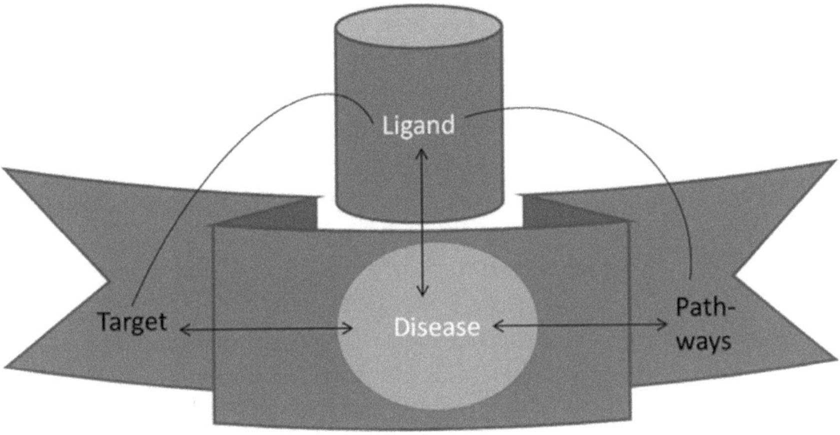


**FIGURE 2.5** Schematic representation of network-based drug design.

Moreover, the integration of omics experiments with route- and network-based methods for early drug development is intended to bridge the gap between fundamental research on pathway models and the actual requirements of the early drug development pipeline, evidently. Topology-based route analysis techniques may aid in the reduction of false positives during the target analysis process, the prioritization of target validation, the selection of optimum hits, and the progression from hit to lead during the hit to lead step. There are very few limits and difficulties to overcome, despite the fact that including pathway- and network-based drug development may increase the rate of success for finding new drugs. For example, network topology databases provide a large number of contradictory findings, and the accuracy of these reports is questionable in many cases. Network-based computational techniques will benefit from the advances made possible by the development and distribution of disease- and drug-specific omics data.[123] This will provide a new route toward a more evolved drug development pipeline with reduced attrition rates.[124]

### 2.3.5 TARGET MECHANISM-BASED APPROACH

Considering that it is a multifactorial disease, the paradigm shifts from "one drug-one target" called target-based drug development. Moreover, the main goal of either the multitarget strategy is to provide single-molecular entities with a broader pharmacological spectrum through the use of multiple drugs. When it comes to the development of new therapeutic armaments against disease, target-based drug design offers a strong strategy.[125,126]

However, the druggability of hybrid compounds that have been created by rationally integrating different pharmacophoric characteristics may be severely compromised, particularly if their molecular weight is raised and their water solubility is reduced. In this regard, recent advancements in early-stage ADMET will help to ease some of the difficulties associated with target-based drug design and development. On the other hand, it must still confirm the efficacy of multitarget treatment in fact before it can be implemented. This makes this task very appealing, but it also explains why "Pharmaceutical industries" are disengaged from target-based drug creation, particularly for different indications. However, the integration of a chemobioinformatic method may aid in bridging the gap that currently exists between the demand for disease-oriented drug design and the desire for target-based drug design.[127,128] When developing a multitarget drug, it is important to consider the fundamental molecular characteristics that are required for successful interaction with each intended biological target. To accomplish this objective, several drug design methods, most of which are inspired by ligand-based and target-based approaches, have been proposed. Significant people have attempted in past few years to integrate the vast quantity of diverse information in an attempt to develop predictive multi-target designs by integrating the information.

### 2.3.6 EXAMPLES OF SUCCESSFUL DRUG REPOSITIONING: CASE STUDIES

Research into drug repositioning often involves integrating existing knowledge about diseases, pathways, targets, and ligands into new studies. This endeavor seems to be very difficult, given the wide range of terms and circumstances under which experimental and clinical data may be collected in various formats. Because of this, substantial and coordinated community efforts are required for effective use of existing biological and clinical information as well as extraction of knowledge from this information, which may aid in the better repositioning of already available medicines in the marketplace. A large number of research are presently ongoing that are aimed at the annotation, curation, and integration of information regarding chemical–biological interactions and their mechanisms.[129, 130]

In computational drug repositioning is gaining popularity across the world these days, owing to the availability of a huge quantity of information on protein structures, pharmacophores, illness data, clinical studies, and gene expression profiles of pharmaceutical compounds. Furthermore, the proliferation of public social networking technologies, as well as the availability of computational access to genetic information, has significantly helped the computer methods in their attempts to anticipate novel indicators. As a result, current bioinformatics or computational tools are being used by the majority of pharmaceutical firms to reposition drugs from a variety of chemical regions. Any pharmaceutical business wishes to profit from the improved speed and lower costs offered by strong in silico technology. This is the ultimate goal of every pharmaceutical company. New computational techniques for focused profiling of small compounds have been created in response to the rise in drug-related data. These methods have higher levels of recall and accuracy than previous methods.[8]

These techniques enhance the repositioning process by using chemoinformatics, bioinformatics, network biology, systems biology, or genomic information to uncover previously undiscovered targets or processes of authorized medicines in a shorter amount of time than traditional methods. As a result, computational drug repositioning techniques may be generally classified into the following categories: target-based, knowledge-based, signature-based, network-based, and targeted-mechanism-based. Based on the computational findings, a selection of compounds may be selected for additional experimental testing to ensure that the computational insights are correct. A hybrid method that incorporates both computational and experimental tests is thus required to repurpose medicines for new illnesses, and the majority of pharmaceutical companies have already embraced this strategy to assess the therapeutic effectiveness of new indications.[131,132] Table 2.6 lists some successful drug repositioning.

**TABLE 2.6** List of Some Successful Drug Repositioning.

| Drug | Initial | Suggested | References |
|---|---|---|---|
| 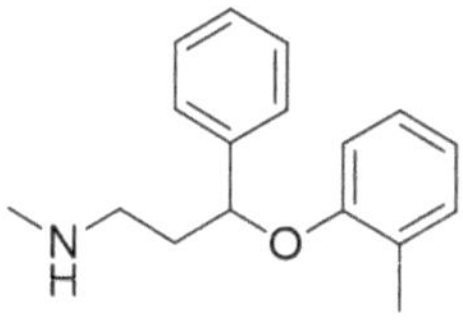 Atomoxetine | Parkinson's disease | Attention deficit hyperactivity disorder | [133] |
| 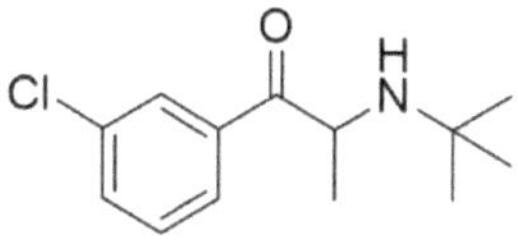 Bupropion | Depression | Obesity | [134,144] |
| 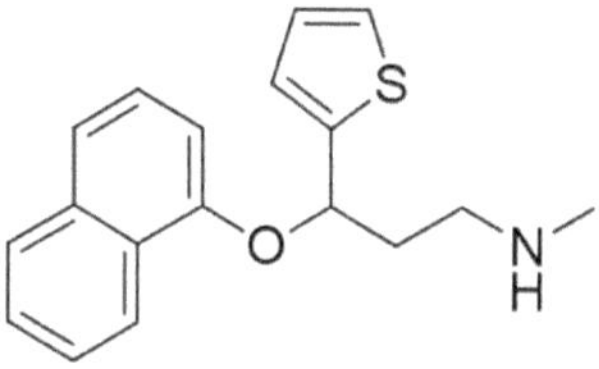 Duloxetine | Depression | Diabetic-neuropathy | [135] |
| 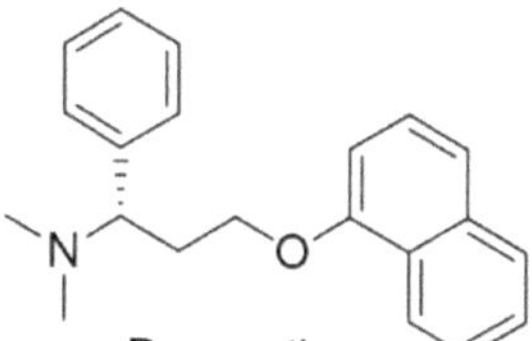 Dapoxetine | Analgesia and depression | Premature ejaculation | [136] |
| 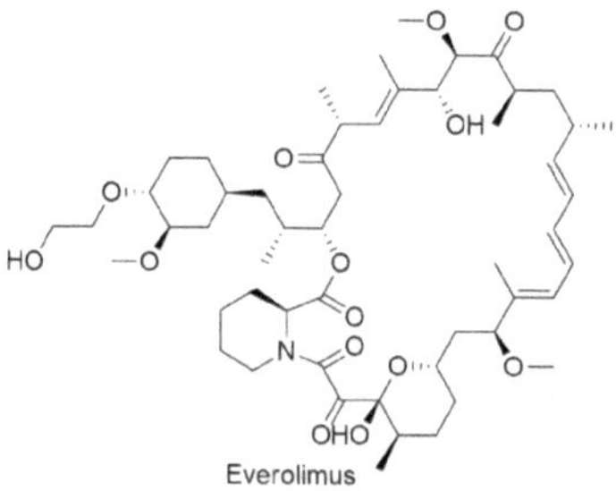 Everolimus | Immunosuppressant | Pancreatic neuroendocrine tumors | [137] |

**TABLE 2.6** *(Continued)*

| Drug | Initial | Suggested | References |
|---|---|---|---|
| 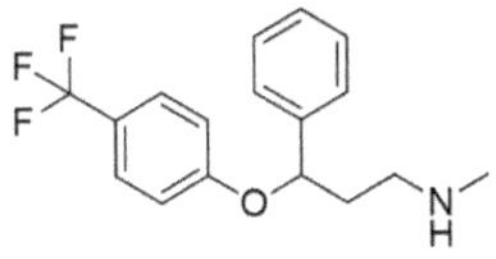 Fluoxetine | Depression | Gastrointestinal stromal tumor | [138] |
| 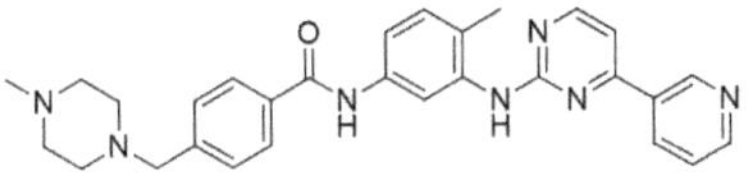 Imatinib | Chronic myeloid leukemia | | [139] |
| 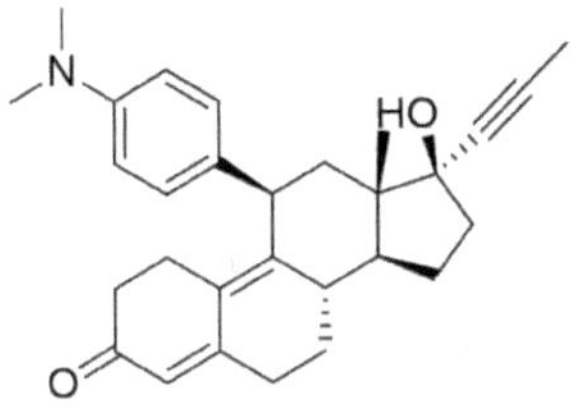 Mifepristone | Pregnancy termination | Cushing's syndrome | [140] |
| 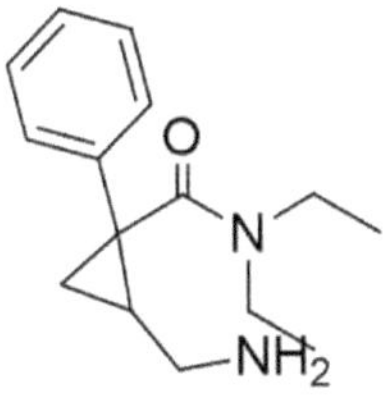 Milnacipran | Depression | Fibromyalgia syndrome | [141] |
| 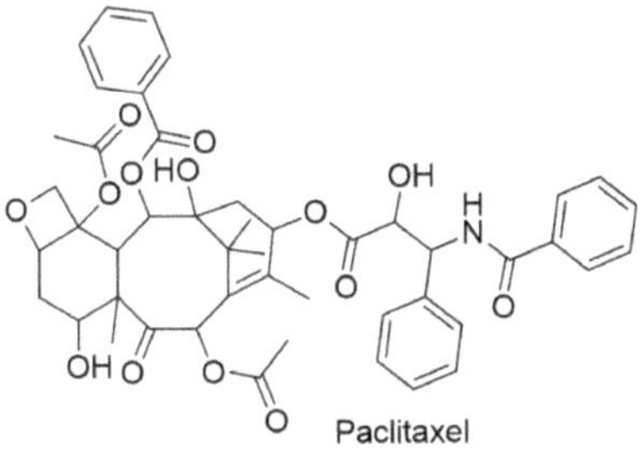 Paclitaxel | Cancer | Restenosis | [142] |

**TABLE 2.6** *(Continued)*

| Drug | Initial / Suggested | References |
|---|---|---|
| 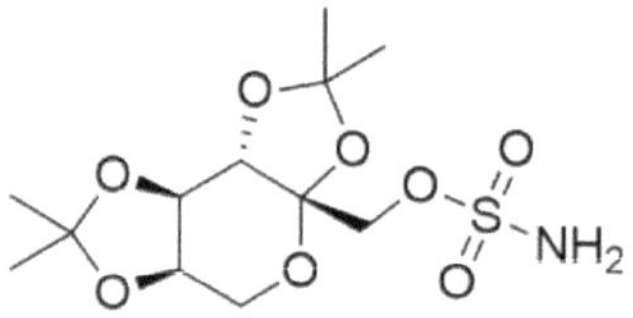 Topiramate | Epilepsy<br>Migraine | [143] |
| 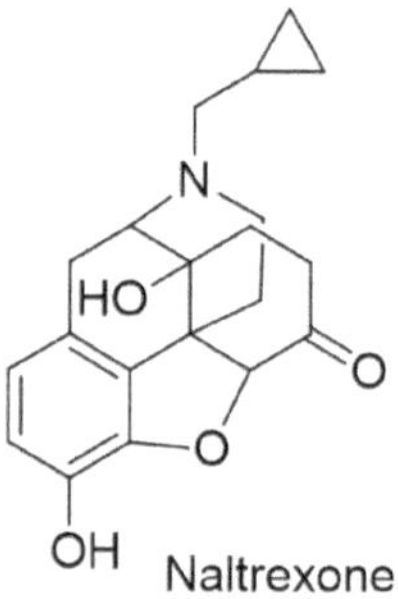 Naltrexone | Opioid addiction<br>Obesity | [144] |

## 2.4 CHALLENGES AND LIMITATIONS OF COMPUTATIONAL DRUG DISCOVERY

Despite availability of a wide range of tools and resources in computational drug discovery, developing a robust computational model is a complex process and associated with many challenges. Besides, complexity of mapping these theoretical approaches to predict behavior of living beings, inaccurate, missing, or biased data are some of the major challenges in putting these approaches into action. For example, it may be quite difficult to define a reliable gene expression signature profile because of variations in experimental conditions such as patient's age, environment factors across different experiments; this may lead to discrepancies in gene expression data. When such inaccurate or biased data are utilized to build models it may lead to unreliable results. Further, unavailability of high-resolution structural data of drug targets makes it difficult to identify interactions between drug and target. Lack of gold-standard datasets to evaluate models performance is yet another challenge in computational drug discovery. Therefore, researchers have to either split their dataset into test, training, and validation sets and the use k-fold cross validation and evaluation metrics or build their own dataset and use prevalent metrics such as specificity, sensitivity, recall, F1 score, accuracy, etc., to evaluate the model and to avoid over fitted model.[145]

Despite all these challenges, integration of multisource data related to disease, drugs, and how these drug and diseases affect the body is important aspect in computational drug discovery models. Availability and integration of these datasets could improve the performance of the models. However, there are several diseases that still lack treatments and inspire the researchers all over the world to develop novel and leading candidates for treatment of unknown and rare diseases.[146]

## 2.5 CONCLUSION AND FUTURE PERSPECTIVES

In summary, computational drug discovery and repurposing can be of immense benefit for speeding up the drug development process as well as increasing the plausibility for the failed or withdrawn drug to get a second chance to reach the market. While extra supports toward computational drugs repurposing are the need of the hour for the researchers and scientists to make further efforts to come up with new finding in this area. With this in mind, availability of open-source databases/tools, workflow systems, pipelines in the form of codes, software tools that are more robust, reproducible, and easy access to validate the analysis could be helpful for the computational drug discovery process.

## KEYWORDS

- **bioinformatics**
- **chemoinformatics**
- **drug discovery**
- **drug repurposing**
- **network biology**

## REFERENCES

1. Dudley, J. T.; Desphande, T.; Butte, A. J. Exploiting Drug–Disease Relationships for Computational Drug Repositioning. *Brief Bioinform.* **2011,** *12* (4), 303–311.

2. Stone, J. E.; Phillips, J. C.; Freddolino, P. L.; Hardy D. J.; Trabuco, L. G.; Schulten, K. Accelerating Molecular Modeling Applications With Graphics Processors. *J. Comput. Chem.* **2007,** *28* (16), 2618–2640.
3. Stone, J. E.; Hardy, D. J.; Ufimtsev, I. S.; Schulten, K. GPU Accelerated Molecular Modeling Coming of Age. *J. Mol. Graph. Model* **2010,** *29* (2), 116–125.
4. Waldman, S. A.; Terzic, A. Systems-Based Discovery Advances Drug Development. *Clin. Pharmacol. Ther.* **2013,** *93* (4), 285–287.
5. Barratt, M. J., Frail, D. E. Drug Repositioning: Bringing New Life to Shelved Assets and Existing Drugs; John Wiley & Sons, 2012; pp 389–431.
6. Pushpakom, S.; Iorio, F.; Eyers, P. A.; Escott, K. J.; Hopper, S.; Wells, A.; Doig, A.; Guilliams, T.; Latimer, J.; McNamee, C.; Norris, A. Drug Repurposing: Progress, Challenges and Recommendations. *Nat. Rev. Drug Discov.* **2019,** *18* (1), 41–58.
7. Polamreddy, P.; Gattu, N. The Drug Repurposing Landscape From 2012 to 2017: Evolution, Challenges, and Possible Solutions. *Drug Discov. Today* **2019,** *24* (3), 789–795.
8. Park, K. A Review of Computational Drug Repurposing. *Transl. Clin. Pharmacol.* **2019,** *27* (2), 59–63.
9. Azencott, C. A.; Ksikes, A.; Swamidass, S. J.; Chen, J. H.; Ralaivola, L.; Baldi, P. One-to Four-Dimensional Kernels for Virtual Screening and the Prediction of Physical, Chemical and Biological Properties. *J. Chem. Inf. Model* **2007,** *47* (3), 965–974.
10. Cheng, F., Sutariya, V. Applications of Artificial Neural Network Modeling in Drug Discovery. *Clin. Exp. Pharmacol.* **2012,** *2* (3), 1–2.
11. Ou-Yang, S. S.; Lu, J. Y.; Kong, X. Q.; Liang, Z. J.; Luo, C.; Jiang, H. Computational Drug Discovery. *Acta Pharmacol. Sin.* **2012,** *33* (9), 1131–1140.
12. Hung, C. L.; Chen, C. C. Computational Approaches for Drug Discovery. *Drug Dev. Res.* **2014,** *75* (6), 412–418.
13. Teo, C. Y.; Rahman, M. B. A.; Chor, A. L. T.; Salleh, A. B.; Ballester, P. J.; Tejo, B. A. Ligand-Based Virtual Screening for the Discovery of Inhibitors for Protein Arginine Deiminase Type 4 (PAD4). *Metabolomics* **2013,** *3* (1),118—122.
14. Batool, M.; Ahmad, B.; Choi, S. A Structure-Based Drug Discovery Paradigm. *Int. J. Mol. Sci.* **2019,** *20* (11), 2783.
15. Acharya, C.; Coop, A. E.; Polli, J. MacKerell, D. A. Recent Advances in Ligand-Based Drug Design: Relevance and Utility of the Conformationally Sampled Pharmacophore Approach. *Curr. Comput.-Aided Drug Des.* **2011,** *7* (1), 10–22.
16. Kaushik, A. C.; Kumar, A.; Bharadwaj, S.; Chaudhary, R.; Sahi, S. Ligand-Based Approach for In-silico Drug Designing. In *Bioinformatics Techniques for Drug Discovery*; Springer: Cham, 2018; pp 11–-19.
17. Gajipara, J.; Georrge, J. J. In *Tools for Ligand Based Drug Discovery*, Proceedings of the National Science Symposium on Recent Trends in Science and Technology-2018, Christ College, Rajkot & Gujarat Council on Science and Technology (GUJCOST), Govt. of Gujarat, February 11, 2018.
18. Neves, B. J.; Braga, R. C.; Melo-Filho, C. C.; Moreira-Filho, J. T.; Muratov, E. N.; Andrade, C. H. QSAR-Based Virtual Screening: Advances and Applications in Drug Discovery. *Front. Pharmacol.* **2018,** *9*, 1275.
19. Peter, S. C.; Dhanjal, J. K.;Malik, V.; Radhakrishnan, N.; Jayakanthan, M.; Sundar, D. Jayakanthan, M. *Encyclopedia of Bioinformatics and Computational*

*Biology*; Ranganathan, S., Grib-skov, M., Nakai, K., Schönbach, C., Eds.; 2018; pp 661–676.

20. Danishuddin; Khan, A. U. Descriptors and Their Selection Methods in QSAR Analysis: Paradigm for Drug Design. *Drug Discov. Today* **2016,** *21* (8), 1291–1302.
21. Verma, J.; Khedkar, V. M. Coutinho, E. C. 3D-QSAR in Drug Design-A Review. *Curr. Top. Med. Chem.* **2010,** *10* (1), 95–115.
22. Melo-Filho, C. C.; Braga, R. C.; Andrade, C. H. 3D-QSAR Approaches in Drug Design: Perspectives to Generate Reliable CoMFA Models. *Curr. Comput. Aided Drug Des.* **2014,** *10* (2), 148–159.
23. Katsila, T.; Spyroulias, G. A.; Patrinos, G. P.; Matsoukas, M. T. Computational Approaches in Target Identification and Drug Discovery. *Comput. Struct. Biotechnol. J.* **2016,** *14*, 177–184.
24. Cumming, J. G.; Davis, A. M.; Muresan, S.; Haeberlein, M.; Chen, H. Chemical Predictive Modelling to Improve Compound Quality. *Nat. Rev. Drug Discov.* **2013,** *12* (12), 948–962.
25. Jing, Y.; Bian, Y.; Hu, Z.; Wang, L.; Xie, X. Q. S. Deep Learning for Drug Design: An Artificial Intelligence Paradigm for Drug Discovery in the Big Data Era. *AAPS J.* **2018,** *20* (3), 1–10.
26. Koscielny, G.; An, P.; Carvalho-Silva, D.; Cham, J. A.; Fumis, L.; Gasparyan, R.; Dunham, I. Open Targets: A Platform for Therapeutic Target Identification and Validation. *Nucleic Acids Res.* **2017,** *45* (D1), D985–D994.
27. Gaulton, A.; Bellis, L. J.; Bento, A. P.; Chambers, J.; Davies, M.; Hersey, A.; Overington, J. P. ChEMBL: A Large-Scale Bioactivity Database for Drug Discovery. *Nucleic Acids Res.* **2012,** *40* (D1), D1100–D1107.
28. Vamathevan, J.; Clark, D.; Czodrowski, P.; Dunham, I.; Ferran, E.; Lee, G.; Li, B.; Madabhushi, A.; Shah, P.; Spitzer, M.; Zhao, S. Applications of Machine Learning in Drug Discovery and Development. *Nat. Rev. Drug Discov.* **2019,** *18*, 463–477.
29. Brown, J. B.; Okuno, Y. Systems Biology and Systems Chemistry: New Directions for Drug Discovery. *Chem. Biol.* **2012,** *19* (1), 23–28.
30. Reymond, J. L.; Ruddigkeit, L.; Blum, L.; van Deursen, R. The Enumeration of Chemical Space. *Wiley Interdiscip. Rev. Comput. Mol. Sci.* **2012,** *2* (5), 717–733.
31. Wang, Y.; Xiao, J.; Suzek, T. O.; Zhang, J.; Wang, J.; Bryant, S. H. PubChem: A Public Information System for Analyzing Bioactivities of Small Molecules. *Nucleic Acids Res.* **2009,** *37* (suppl_2), W623–W633.
32. Irwin, J. J.; Shoichet, B. K. ZINC - A Free Database of Commercially Available Compounds for Virtual Screening. *J. Chem. Inf. Model* **2005,** *45*, 177–182.
33. Williams, A. J. Public Chemical Compound Databases. *Curr. Opin. Drug Discov. Dev.* **2008,** *11* (3), 393.
34. Chen, J.; Swamidass, S. J.; Dou, Y., Bruand, J.; Baldi, P. ChemDB: A Public Database of Small Molecules and Related Chemoinformatics Resources. *Bioinformatics* **2005,** *21* (22), 4133–4139.
35. Liu, T.; Lin, Y.; Wen, X.; Jorissen, R. N.; Gilson, M. K. BindingDB: A Web-Accessible Database of Experimentally Determined Protein–Ligand Binding Affinities. *Nucleic Acids Res.* **2007,** *35* (suppl_1), D198–D201.
36. Voigt, J. H.; Bienfait, B.; Wang, S.; Nicklaus, M. C. Comparison of the NCI Open Database With Seven Large Chemical Structural Databases. *J. Chem. Inf. Comput. Sci.* **2001,** *41* (3), 702–712.

37. Davis, A. P.; King, B. L.; Mockus, S.; Murphy, C. G.; Saraceni-Richards, C.; Rosenstein, M.; Mattingly, C. J. The Comparative Toxicogenomics Database: Update 2011. *Nucleic Acids Res.* **2010,** *39* (suppl_1), D1067–D1072.
38. Wishart, D. S.; Knox, C.; Guo, A. C.; Eisner, R.; Young, N.; Gautam, B.; Hau, D. D.; Psychogios, N.; Dong, E.; Bouatra, S.; Mandal, R.; Sinelnikov, I.; Xia, J.; Jia, L.; Cruz, J. A.; Lim, E.; Sobsey, C. A.; Shrivastava, S.; Huang, P.; Liu, P.; Fang, L.; Peng, J.; Fradette, R.; Cheng, D.; Tzur, D.; Clements, M.; Lewis, A.; De Souza, A.; Zuniga, A.; Dawe, M.; Xiong, Y.; Clive, D.; Greiner, R.; Nazyrova, A.; Shaykhutdinov, R.; Li, L.; Vogel, H. J.; Forsythe, I. HMDB: A Knowledgebase for the Human Metabolome. *Nucleic Acids Res.* **2009,** *37*, D603–D610.
39. Frolkis, A.; Knox, C.; Lim, E.; Jewison, T.; Law, V.; Hau, D. D.; Liu, P.; Gautam, B.; Ly, S.; Guo, A. C.; Xia, J.; Liang, Y.; Shrivastava, S.; Wishart, D. S. SMPDB: The Small Molecule Pathway Database. *Nucleic Acids Res.* **2010,** *38*, D480–D487.
40. Knox, C.; Law, V.; Jewison, T.; Liu, P.; Ly, S.; Frolkis, A.; Pon, A.; Banco, K.; Mak, C.; Neveu, V.; Djoumbou, Y.; Eisner, R.; Guo, A. C.; Wishart, D. S. DrugBank 3.0: A Comprehensive Resource for 'Omics' Research on Drugs. *Nucleic Acids Res.* **2011,** *39*, D1035–D1041.
41. Wishart, D. S. Introduction to Cheminformatics. *Curr. Protoc. Bioinform.* **2007,** *18* (1), 14.1.
42. Wasley, J. W. *Chemoinformatics, Concepts, Methods, and Tools for Drug Discovery;* Jurgen, B., Ed.; Humana Press Inc., Totowa, NJ, 2004; vol. xiii+, p 524.
43. Bone, R. G.; Firth, M. A.; Sykes, R. A. SMILES Extensions for Pattern Matching and Molecular Transformations: Applications in Chemoinformatics. *J. Chem. Inf. Comput. Sci.* **1999,** *39* (5), 846–860.
44. Begam, B. F.; Kumar, J. S. A Study on Cheminformatics and its Applications on Modern Drug Discovery. *Procedia Eng.* **2012,** *38*, 1264–1275.
45. Qing, X.; Lee, X. Y.; De Raeymaecker, J.; Tame, J. R.; Zhang, K. Y.; De Maeyer, M.; Voet, A. Pharmacophore Modeling: Advances, Limitations, and Current Utility in Drug Discovery. *J. Recept. Ligand Channel Res.* **2014,** *7*, 81–92.
46. Schuster, D.; Spetea, M.; Music, M.; Rief, S.; Fink, M.; Kirchmair, J.; Rollinger, J. M. Morphinans and Isoquinolines: Acetylcholinesterase Inhibition, Pharmacophore Modeling, and Interaction With Opioid Receptors. *Bioorg. Med. Chem.* **2010,** *18* (14), 5071–5080.
47. Langer, T. Pharmacophores in Drug Research. *Mol Inform.* **2010,** *29* (6–7), 470–475.
48. Arcon, J. P.; Modenutti, C. P.; Avendaño, D.; Lopez, E. D.; Defelipe, L. A.; Ambrosio, F. A.; Marti, M. A. AutoDock Bias: Improving Binding Mode Prediction and Virtual Screening Using Known Protein–Ligand Interactions. *Bioinformatics* **2019,** *35* (19), 3836–3838.
49. Schaller, D.; Pach, S.; Wolber, G. PyRod: Tracing Water Molecules in Molecular Dynamics Simulations. *J. Chem. Inf. Model* **2019,** *59* (6), 2818–2829.
50. Yu, W.; Lakkaraju, S. K.; Raman, E. P.; Fang, L.; MacKerell Jr, A. D. Pharmacophore Modeling Using Site-Identification by Ligand Competitive Saturation (SILCS) With Multiple Probe Molecules. *J. Chem. Inf. Model* **2015,** *55* (2), 407–420.
51. Sato, T.; Honma, T.; Yokoyama, S. Combining Machine Learning and Pharmacophore-Based Interaction Fingerprint for in Silico Screening. *J .Chem. Inf. Model* **2010,** *50* (1),170–185.

52. Wang, X.; Shen, Y.; Wang, S.; Li, S.; Zhang, W.; Liu, X.; Li, H. PharmMapper 2017 Update: A Web Server for Potential Drug Target Identification With a Comprehensive Target Pharmacophore Database. *Nucleic Acids Res.* **2017,** *45* (W1), W356–W360.
53. Tian, S.; Wang, J.; Li, Y.; Li, D.; Xu, L.; Hou, T. The Application of in Silico Drug-Likeness Predictions in Pharmaceutical Research. *Adv. Drug Deliv. Rev.* **2015,** *86*, 2–10.
54. Bickerton, G. R.; Paolini, G. V.; Besnard, J.; Muresan, S.; Hopkins, A. L. Quantifying the Chemical Beauty of Drugs. *Nat. Chem.* **2012,** *4* (2), 90–98.
55. Yusof, I.; Segall, M. D. Considering the Impact Drug-Like Properties Have on the Chance of Success. Drug Discov. *Today* **2013,** *18* (13–14), 659–666.
56. Singh, N.; Sun, H.; Chaudhury, S.; AbdulHameed, M. D. M.; Wallqvist, A.; Tawa, G. A Physicochemical Descriptor-Based Scoring Scheme for Effective and Rapid Filtering of Kinase-Like Chemical Space. *J. Cheminform.* **2012,** *4* (1), 1–12.
57. Li, D.; Chen, L.; Li, Y.;Tian, S.; Sun, H.; Hou, T. ADMET Evaluation in Drug Discovery. 13. Development of in Silico Prediction Models for P-glycoprotein Substrates. *Mol. Pharm.* **2014,** *11* (3), 716–726.
58. Wang, S.; Sun, H.; Liu, H.; Li, D.; Li, Y.; Hou, T. ADMET Evaluation in Drug Discovery. Predicting hERG Blockers by Combining Multiple Pharmacophores and Machine Learning Approaches. *Mol. Pharm.* **2016,** *13* (8), 2855–2866.
59. Pathak, R. K.; Gupta, A.; Shukla, R.; Baunthiyal, M. Identification of New Drug-Like Compounds From Millets as Xanthine Oxidoreductase Inhibitors for Treatment of Hyperuricemia: A Molecular Docking and Simulation Study. *Comput. Biol. Chem.* **2018,** *76*, 32–41.
60. Salum, L.; Polikarpov, I.; Andricopulo, A. D. Structure-Based Approach for the Study of Estrogen Receptor Binding Affinity and Subtype Selectivity. *J. Chem. Inf. Model* **2008,** *48*, 2243–2253.
61. Kalyaanamoorthy, S.; Chen, Y. P. Structure-Based Drug Design to Augment Hit Discovery. *Drug Discov. Today* **2011,** *16*, 831–839.
62. Blaney, J. A Very Short History of Structure-Based Design: How Did we Get Here and Where do we Need to go? *J. Comput. Aided Mol. Des.* **2012,** *26*, 13–14.
63. Kumar, A.; Chordia, N. Role of Bioinformatics in Biotechnology. *Res. Rev. Biosci,* **2017,** *12* (1), 116.
64. Pathak, R. K.; Taj, G.; Pandey, D.; Kasana, V. K.; Baunthiyal, M.; Kumar, A. Molecular Modeling and Docking Studies of Phytoalexin (s) With Pathogenic Protein (s) as Molecular Targets for Designing the Derivatives With Anti-Fungal Action on 'Alternaria'spp. of'Brassica'. *Plant Omics.* **2016,** *9* (3), 172–183.
65. Bagaria, A.; Jaravine, V.; Huang, Y. J.; Montelione, G. T.; Güntert, P. Protein Structure Validation by Generalized Linear Model Root-Mean-Square Deviation Prediction. *Protein Sci.* **2012,** *21* (2), 229–238.
66. Singh, D. B.; Tripathi, T. *Frontiers in Protein Structure, Function, and Dynamics*; Springer, 2020.
67. Kryshtafovych, A.; Schwede, T.; Topf, M.; Fidelis, K.; Moult, J. Critical Assessment of Methods of Protein Structure Prediction (CASP)—Round XIII. *Proteins: Struct. Funct. Bioinf.* **2019,** *87* (12), 1011–1020.
68. Mamgain, S.; Dhiman, S.; Pathak, R. K.; Baunthiyal, M. In Silico Identification of Agriculturally Important Molecule (s) for Defense Induction Against Bacterial Blight Disease in Soybean (Glycine max). *Plant Omics.* **2018,** *11* (2), 98–105.

69. Adrian-Scotto, M.; Vasilescu, D. Quantum Molecular Modeling of Glycyl-adenylate. *J. Biomol. Struct. Dyn.* **2008,** *25* (6), 697–708.
70. Yuriev, E.; Ramsland, P. A. Latest Developments in Molecular Docking: 2010–2011 in Review. *J. Mol. Recognit.* **2013,** *26* (5), 215–239.
71. Guedes, I. A.; de Magalhães, C. S.; Dardenne, L. E. Receptor–Ligand Molecular Docking. *Biophys. Rev.* **2014,** *6* (1), 75–87.
72. Feinstein, W. P.; Brylinski, M. Calculating an Optimal Box Size for Ligand Docking and Virtual Screening Against Experimental and Predicted Binding Pockets. *J. Cheminform.* **2015,** *7* (1), 1–10.
73. Durham, E.; Dorr, B.; Woetzel, N.; Staritzbichler, R.; Meiler. J. Solvent Accessible Surface Area Approximations for Rapid and Accurate Protein Structure Prediction. *J. Chem. Inf. Model* **2009,** *15* (9), 1093–1108.
74. Neudert, G.; Klebe, G. DSX: A Knowledge-Based Scoring Function for the Assessment of Protein–Ligand Complexes. *J. Chem. Inf. Model* **2011,** *51* (10), 2731–2745.
75. Singh, D. B.; Dwivedi, S. Structural Insight Into Binding Mode of Inhibitor With SAHH of Plasmodium and Human: Interaction of Curcumin With Anti-Malarial Drug Targets. *J. Chem. Biol.* **2016,** *9* (4), 107–120.
76. Wallace, A. C.; Laskowski, R. A.; Thornton, J. M. LIGPLOT: A Program to Generate Schematic Diagrams of Protein-Ligand Interactions. *Protein Eng. Des. Sel.* **1995,** *8* (2), 127–134.
77. Laskowski, R. A.; Swindells, M. B. LigPlot+: Multiple Ligand–Protein Interaction Diagrams for Drug Discovery. *J. Chem. Inf. Model* **2011,** *51* (10), 2778–2786.
78. DeLano, W. L. Pymol: An Open-Source Molecular Graphics Tool. *CCP4 Newsletter on Protein Crystallography*, 2002, vol *40* (1), pp 82–92.
79. Pettersen, E. F.; Goddard, T. D.; Huang, C. C.; Couch, G. S.; Greenblatt, D. M.; Meng, E. C.; Ferrin, T. E. UCSF Chimera-a Visualization System for Exploratory Research and Analysis. *J. Comput. Chem.* **2004,** *25* (13), 1605–1612.
80. Lindahl, E. R. Molecular Dynamics Simulations. In *Molecular modeling of proteins*; Kukol, A., Ed.; Humana Press, 2008; vol 443, pp 3–23.
81. Khan, F. I.; Wei, D. Q.; Gu, K. R.; Hassan, M. I.; Tabrez, S. Current Updates on Computer Aided Protein Modeling and Designing. *Int. J. Biol. Macromol.* **2016,** *85*, 48–62.
82. Pronk, S.; Pall, S.; Schulz, R.; Larsson, P.; Bjelkmar, P.; Apostolov, R.; Shirts, M. R.; Smith, J. C.; Kasson, P. M.; van der Spoel, D; Hess, B.; Lindahl, E. GROMACS 4.5: A High-Throughput and Highly Parallel Open Source Molecular Simulation Toolkit. *Bioinformatics* **2013,** *29* (7), 845–854.
83. Salomon-Ferrer, R.; Gotz, A. W.; Poole, D.; Le Grand, S.; Walker, R. C. Routine Micro Second Molecular Dynamics Simulations With AMBER on GPUs. 2. Explicit Solvent Particle MeshEwald. *J. Chem. Theory Comput.* **2013,** *9 (*9), 3878–3888.
84. Brooks, B. R.; Brooks III, C. L.; Mackerell Jr, A. D.; Nilsson, L.; Petrella, R. J.; Roux, B.; Karplus, M. CHARMM: The Biomolecular Simulation Program. *J. Comput. Chem.* **2009,** *30* (10), 1545–1614.
85. Phillips, J. C.; Hardy, D. J.; Maia, J. D.; Stone, J. E.; Ribeiro, J. V.; Bernardi, R. C.; Buch, R.; Fiorin, G.; Hénin, J.; Jiang, W.; McGreevy, R. Scalable Molecular Dynamics on CPU and GPU Architectures With NAMD. *J. Chem. Phys.* **2020,** *153* (4), 044130.
86. Humbert, M. T.; Zhang, Y.; Maginn, E. J. PyLAT: Python LAMMPS Analysis Tools. *J. Chem. Inf. Model* **2019,** *59* (4), 1301–1305.

87. Harrold, J. M.; Ramanathan, M.; Mager, D. E. Network-Based Approaches in Drug Discovery and Early Development. *Clin. Pharmacol. Ther.* **2013,** *94* (6), 651–658.
88. Pujol, A.; Mosca, R.; Farres, J.; Aloy, P. Unveiling the Role of Network and Systems Biology in Drug Discovery. *Trends Pharmacol. Sci.* **2010**, *31* (3), 115–123.
89. Hopkins, A. L. Network Pharmacology. *Nat. Biotechnol.* **2007,** *25* (10), 1110–1111.
90. Jia-hu, P. New Paradigm for Drug Discovery Based on Network Pharmacology. *Chinese J. New Drugs Clin. Remed.* **2009,** *10*, 002.
91. Zhang, G. B.; Li, Q. Y.; Chen, Q. L.; Su, S. B. Network Pharmacology: A New Approach for Chinese Herbal Medicine Research. *Evid. Based Complement. Alter. Med.* **2013,** *2013.*
92. Kibble, M.; Saarinen, N.; Tang, J.; Wennerberg, K.; Mäkelä, S.; Aittokallio, T. Network Pharmacology Applications to Map the Unexplored Target Space and Therapeutic Potential of Natural Products. *Nat. Prod. Rep.* **2015,** *32* (8),1249–1266.
93. Muhammad, J.; Khan, A.; Ali, A.; Fang, L.; Yanjing, W.; Xu, Q.; Wei, D. Q. Network Pharmacology: Exploring the Resources and Methodologies. *Curr. Top. Med. Chem.* **2018,** *18* (12), 949–964.
94. Lapinsh, M.; Prusis, P.; Uhlén, S.; Wikberg, J. E. Improved Approach for Proteochemometrics Modeling: Application to Organic Compound-Amine G Protein-Coupled Receptor Interactions. *Bioinformatics* **2005,** *21* (23), 4289–4296.
95. van Westen, G. J.; Hendriks, A.; Wegner, J. K.; IJzerman, A. P.; van Vlijmen, H. W.; Bender, A. Significantly Improved HIV Inhibitor Efficacy Prediction Employing Proteochemometric Models Generated From Antivirogram Data. *PLoS Comput. Biol.* **2013,** *9* (2), e1002899.
96. Strömbergsson, H.; Lapins, M.; Kleywegt. G. J.; Wikberg, J. E. Towards Proteome–Wide Interaction Models Using the Proteochemometrics Approach. *Mol. Inform.* **2010,** *29* (6–7), 499–508.
97. Dakshanamurthy, S.; Issa, N. T.; Assefnia, S.; Seshasayee, A.; Peters, O. J.; Madhavan, S.; Byers, S. W. Predicting New Indications for Approved Drugs Using a Proteochemometric Method. *J. Med. Chem.* **2012,** *55* (15), 6832–6848.
98. Simeon, S.; Spjuth, O.; Lapins, M.; Nabu, S.; Anuwongcharoen, N.; Prachayasittikul, V.; Wikberg, J. E.; Nantasenamat, C. Origin of Aromatase Inhibitory Activity Via Proteochemometric Modeling. *Peer J.* **2016,** *4*, e1979.
99. Nazarshodeh, E.; Sheikhpour, R.; Gharaghani, S.; Sarram, M. A. A Novel Proteochemometrics Model for Predicting the Inhibition of Nine Carbonic Anhydrase Isoforms Based on Supervised Laplacian Score and k-Nearest Neighbour Regression. *SAR QSAR Environ. Res.* **2018,** *29* (6), 419–437.
100. Murrell, D. S.; Cortes-Ciriano, I.; van Westen, G. J. P.; Stott, I. P.; Bender, A.; Malliavin, T. E.; Glen, R. C. Chemically Aware Model Builder (CAMB): An R Package for Property and Bioactivity Modelling of Small Molecules. *J. Cheminformatics.* **2015,** *7*, 45.
101. Prusis, P.; Junaid, M.; Petrovska, R.; Yahorava, S.; Yahorau, A.; Katzenmeier, G.; Wikberg, J. E. Design and Evaluation of Substrate-Based Octapeptide and Non Substrate-Based Tetrapeptide Inhibitors of Dengue Virus NS2b-NS3 Proteases. *Biochem. Biophys. Res. Commun.* **2013,** *434* (4), 767–772.
102. Jaeger, S.; Fulle, S.; Turk, S. Mol2vec: Unsupervised Machine Learning Approach With Chemical Intuition. *J. Chem. Inf. Model* **2018,** *58* (1), 27–35.

103. Qiu, T.; Wu, D.; Qiu, J.; Cao, Z. Finding the Molecular Scaffold of Nuclear Receptor Inhibitors Through Highthroughput Screening Based on Proteochemometric Modelling. *J. Cheminformatics* **2018,** *10*, 21.
104. Subramanian, V.; Ain, Q. U.; Henno, H.; Pietil€a, L. O.; Fuchs, J. E.; Prusis, P.; Wohlfahrt, G. 3D Proteochemometrics: Using Three-Dimensional Information of Proteins and Ligands to Address Aspects of the Selectivity of Serine Proteases. *Med. Chem. Comm.* **2017,** *8* (5), 1037–1045.
105. Mitrea, C.; Taghavi, Z.; Bokanizad, B.; Hanoudi, S.; Tagett, R.; Donato, M.; Voichita, C.; Draghici, S. Methods and Approaches in the Topology-Based Analysis of Biological Pathways. *Front. Physiol.* **2013,** *4*, 278.
106. Ackermann, M.; Strimmer, K. A General Modular Framework for Gene Set Enrichment Analysis. *BMC Bioinform.* **2009,** *10*, 47.
107. Gentleman, R. C.; Carey, V. J.; Bates, D. M.; Bolstad, B.; Dettling, M.; Dudoit, S.; , Ellis, B.; Gautier, L.; Ge, Y.; Gentry, J.; Hornik, K. Bioconductor: Open Software Development for Computational Biology and Bioinformatics. *Genome Biol.* **2004,** *5* (10), 1–6.
108. Dabbish, L.; Stuart, C.; Tsay, J.; Herbsleb, J. In *Social Coding in Github: Transparency and Collaboration in an Open Software Repository*, Proceedings of the ACM Conference on Computer Supported Cooperative Work (New York), 2012; pp 1277–1286.
109. Rath, B.; Qais, F. A.; Patro, R.; Mohapatra, S.; Sharma, T. Design, Synthesis and Molecular Modeling Studies of Novel Mesalamine Linked Coumarin for Treatment of Inflammatory Bowel Disease. *Bioorganic Med. Chem. Lett.* **2021,** *41*, 128029.
110. Klambauer, G.; Sepp, H.; Matthias, R. Machine Learning in Drug Discovery. *J. Chem. Inf. Model* **2019,** *59* (3), 945–946.
111. Patel, L.; Shukla, T.; Huang, X.; Ussery, D. W.; Wang, S. Machine Learning Methods in Drug Discovery. *Molecules* **2020,** *25* (22), 5277.
112. Zhang, L.; Tan, J; Han, D.; Zhu, H. From Machine Learning to Deep Learning: Progress in Machine Intelligence for Rational Drug Discovery. *Drug Discov. Today* **2017,** *22* (11), 1680–1685.
113. Hughes, J. P.; Rees, S.; Kalindjian, S. B.; Philpott, K. L. Principles of Early Drug Discovery. *Br. J. Pharmacol.* **2011,** *162* (6), 1239–1249.
114. Schenone, M.; Dančík, V.; Wagner, B. K.; Clemons, P. A. Target Identification and Mechanism of Action in Chemical Biology and Drug Discovery. *Nat. Chem. Biol.* **2013,** *9* (4), 232–240.
115. Croston, G. E. The Utility of Target-Based Discovery. *Expert Opin. Drug Discov.* **2017,** *12* (5), 427–429.
116. Lionta, E.; Spyrou, G. K.; Vassilatis, D.; Cournia, Z. Structure-Based Virtual Screening for Drug Discovery: Principles, Applications and Recent Advances. *Curr. Top. Med. Chem* . **2014,** *14* (16), 1923–1938.
117. Paul, D.; Sanap, G.; Shenoy, S.; Kalyane, D.; Kalia, K.; Tekade, R. K. Artificial Intelligence in Drug Discovery and Development. *Drug Discov. Today* **2021,** *26* (1), 80.
118. Ghose, A. K.; Herbertz, T.; Pippin, D. A; Salvino, J. M.; Mallamo, J. P. Knowledge Based Prediction of Ligand Binding Modes and Rational Inhibitor Design for Kinase Drug Discovery. *J. Med. Chem.* **2008,** *51* (17), 5149–5171.
119. Ghose, A. K.; Herbertz, T.; Pippin, D. A; Salvino, J. M.; Mallamo, J. P. Knowledge-Based Chemoinformatic Approaches to Drug Discovery. *Drug Discov. Today* **2006,** *11* (23–24), 1107–1114.

120. Szymański, P.; Markowicz, M.; Mikiciuk-Olasik, E. Adaptation of High-Throughput Screening in Drug Discovery—Toxicological Screening Tests. *Int. J. Mol. Sci.* **2012,** *13* (1), 427–452.
121. Atanasov, A. G.; Zotchev, S. B.; Dirsch, V. M.; Supuran, C. T. Natural Products in Drug Discovery: Advances and Opportunities. *Nat. Rev. Drug Discov.* **2021,** *20* (3), 200–216.
122. Hu, L.; Chen, G.; Chau, R. M. A Neural Networks-Based Drug Discovery Approach and its Application for Designing Aldose Reductase Inhibitors. *J. Mol. Graph. Model* **2006,** *24* (4), 244–253.
123. Subramanian, I.; Verma, S.; Kumar, S.; Jere, A.; Anamika, K. Multi-omics Data Integration, Interpretation, and Its Application. *Bioinform. Biol. Insights* **2020,** *14*, 1177932219899051.
124. Fotis, C.; Antoranz, A.; Hatziavramidis, D.; Sakellaropoulos, T.; Alexopoulos, L. G. Network-Based Technologies for Early Drug Discovery. *Drug Discov. Today* **2018,** *23* (3), 626–635.
125. Talevi, A. Multi-Target Pharmacology: Possibilities and Limitations of the "Skeleton Key Approach" From a Medicinal Chemist Perspective. *Front. Pharmacol.* **2015,** *6*, 205.
126. Ramsay, R. R.; Popovic-Nikolic, M. R.; Nikolic, K.; Uliassi, E.; Bolognesi, M. L. A Perspective on Multi-Target Drug Discovery and Design for Complex Diseases. *Clin. Transl. Med.* **2018,** *7* (1), 1–14.
127. Newman, D. J.; Cragg, G. M. Natural Products as Sources of New Drugs Over the Nearly Four Decades From 01/1981 to 09/2019. *J. Nat. Prod.* **2020,** *83* (3), 770–803.
128. Sliwoski, G.; Kothiwale, S.; Meiler, J.; Lowe, E. W. Computational Methods in Drug Discovery. *Pharmacol. Rev.* **2014,** *66* (1), 334–395.
129. Langedijk, J.; Mantel-Teeuwisse, A. K.; Slijkerman, D. S.; Schutjens, M. H. Drug Repositioning and Repurposing: Terminology and Definitions in Literature. *Drug Discov. Today* **2015,** *20* (8), 1027–1034.
130. Adasme, M. F.; Parisi, D.; Sveshnikova, A.; Schroeder, M. Structure-Based Drug Repositioning: Potential and Limits. *Semin. Cancer Biol.* **2021,** *68*, 192–198.
131. Xue, H.; Li, J.; Xie, H.; Wang, Y. Review of Drug Repositioning Approaches and Resources. *Int. J. Biol. Sci* . **2018,** *14* (10), 1232.
132. Saberian, N.; Peyvandipour, A.; Donato, M.; Ansari, S.; Draghici, S. A New Computational Drug Repurposing Method Using Established Disease–Drug Pair Knowledge. *Bioinformatics* **2019,** *35* (19), 3672–3678.
133. Bymaster, F. P.; Katner, J. S.; Nelson, D. L.; Hemrick-Luecke, S. K.; Threlkeld, P. G.; Heiligenstein, J. H.; Morin, S. M.; Gehlert, D. R.; Perry, K. W. Atomoxetine Increases Extracellular Levels of Norepinephrine and Dopamine in Prefrontal Cortex of Rat: A Potential Mechanism for Efficacy in Attention Deficit/Hyperactivity Disorder. *Neuropsychopharmacology* **2002,** *27* (5), 699–711.
134. Ferry, L.; Johnston, J. A. Efficacy and Safety of Bupropion SR for Smoking Cessation: Data From Clinical Trials and Five Years of Postmarketing Experience. *Int. J. Clin. Pract.* **2003,** *57* (3), 224–230.
135. Raskin, J.; Pritchett, Y. L.; Wang, F.; D'Souza, D. N.; Waninger, A. L.; Iyengar, S.; Wernicke, J. F. A Double-Blind, Randomized Multicenter Trial Comparing Duloxetine With Placebo in the Management of Diabetic Peripheral Neuropathic Pain. *Pain Med.* **2005,** *6* (5), 346–356.

136. McMahon, C. G.; Althof, S. E.; Kaufman, J. M.; Buvat, J.; Levine, S. B.; Aquilina, J. W.; Tesfaye, F.; Rothman, M.; Rivas, D. A.; Porst, H. Efficacy and Safety of Dapoxetine for the Treatment of Premature Ejaculation: Integrated Analysis of Results From Five Phase 3 Trials. *J. Sex Med.* **2011,** *8* (2), 524–539.
137. Yao, J. C.; Shah, M. H.; Ito, T.; Bohas, C. L; Wolin, E. M.; Van Cutsem, E.; Hobday, T. J.; Okusaka, T.; Capdevila, J.; De Vries, E. G.; Tomassetti, P. Everolimus for Advanced Pancreatic Neuroendocrine Tumors. *N. Engl. J. Med.* **2011,** *364* (6), 514–523.
138. Cohen, L. S.; Miner, C.; Brown, E.; Freeman, E. W.; Halbreich, U.; Sundell, K., McCray, S. Premenstrual Daily Fluoxetine for Premenstrual Dysphoric Disorder: A Placebo-Controlled, Clinical Trial Using Computerized Diaries. *Obstet. Gynecol.* **2002,** *100* (3), 435–444.
139. Heinrich, M. C.; Corless, C. L.; Demetri, G. D.; Blanke, C.; Von Mehren. M.; Joensuu, H.; McGreevey, L. S.; Chen, C. J.; Van den Abbeele, A. D.; Druker, B. J.; Kiese, B. Kinase Mutations and Imatinib Response in Patients With Metastatic Gastrointestinal Stromal Tumor. *Clin. Oncol.* **2003,** *21* (23), 4342–4349.
140. Fleseriu, M.; Biller, B. M.; Findling, J. W.; Molitch, M. E.; Schteingart, D. E.; Gross, C. SEISMIC Study Investigators, SEISMIC Study Investigators Include. Mifepristone, A glucocorticoid Receptor Antagonist, Produces Clinical and Metabolic Benefits in Patients With Cushing's Syndrome. *J. Clin. Endocrinol.* **2012,** *97* (6), 2039–2049.
141. Vitton, O.; Gendreau, M.; Gendreau, J.; Kranzler, J.; Rao, S. G. A Double-Blind Placebo-Controlled Trial of Milnacipran in the Treatment of Fibromyalgia. *Hum. Psychopharmacol.* **2004,** *19* (S1), S27–S35.
142. Tepe, G.; Zeller, T.; Albrecht, T.; Heller, S.; Schwarzwälder, U.; Beregi, J. P.; Claussen, M.C.; Oldenburg, A.; Scheller, B.; Speck, U. Local Delivery of Paclitaxel to Inhibit Restenosis During Angioplasty of the Leg. *N. Engl. J. Med.* **2008,** *358* (7), 689–699.
143. Diener, H. C.; Tfelt-Hansen, P.; Dahlöf, C.; Láinez, M. J.; Sandrini, G.; Wang, S. J.; Neto, W.; Vijapurkar, U.; Doyle, A.; Jacobs, D. Topiramate in Migraine Prophylaxis. *J. Neurol.* **2004,** *251* (8), 943–950.
144. Tek, C. Naltrexone HCI/Bupropion HCI for Chronic Weight Management in Obese Adults: Patient Selection and Perspectives. *Patient Prefer. Adherence* **2016,** *10*, 751.
145. Jarada, T. N.; Rokne, J. G.; Alhajj, R. A Review of Computational Drug Repositioning: Strategies, Approaches, Opportunities, Challenges, and Directions. *J. Cheminformatics* **2020,** *12* (1), 1–23.
146. Schaduangrat, N.; Lampa, S.; Simeon, S.; Gleeson, M. P.; Spjuth, O. Nantasenamat, C. Towards Reproducible Computational Drug Discovery. *J. Cheminformatics* **2020,** *12* (1), 1–30.

## CHAPTER 3

# Drug Repurposing and Computational Drug Discovery for Viral Infections and Coronavirus Disease-2019 (COVID-19)

SIDDHARTHA MAJI[1], VISHNU NAYAK BADAVATH[2], and SWASTIKA GANGULY[3]

[1]*Department of Chemistry, Oklahoma State University, Stillwater, Oklahoma, USA*

[2]*School of Pharmacy & Technology Management, SVKM's NMIMS University, Hyderabad, India*

[3]*Department of Pharmaceutical Sciences and Technology, Birla Institute of Technology, Mesra, India*

## ABSTRACT

Since the pandemic created by the novel coronavirus (SARS-CoV-2), no exact therapeutic options came to light except some vaccines and one or two drugs which are still under clinical trials. Computational techniques including the computer-aided drug design model (CADD) which has been proven to be a best starting point in the development of drugs for different diseases with a fast and fruitful outcome is also the best possible way to design a potential drug for SARS-CoV-2. Such approaches have been previously used in many viral infectious diseases like Hepatitis C virus (HCV), Ebola virus, influenza, and HIV. Recent clinical studies have revealed that the use of several repurposed drugs as well as multi-targeted drugs can be a successful and promising approach to inhibit virus such as lopinavir, remdesivir, ribavirin,

Drug Repurposing and Computational Drug Discovery: Strategies and Advances.
Mithun Rudrapal, PhD (Ed)

favipiravir, ritonavir, arbidol, darunavir, chloroquine, hydroxychloroquine, and tocilizumab. In this chapter we have focused on the ongoing research on the efficiency of repurposing drugs for the novel coronavirus through experimental studies based on in-vitro, in-vivo analyses. An approach to design new drugs/inhibitors will also be focused through CADD approach.

## 3.1 INTRODUCTION

Viruses are a broad category of microorganisms that cause life-threatening infections. Many antiviral medicines that target viral proteins or host factors have been produced effectively over the last 30 years. Chronic viral infectious disorders such as HIV, influenza, hepatitis C virus (HCV), picorna viruses, and corona viruses (SARS-CoV-2), and the rise in need for novel antiviral medicines are mostly due to the development of resistance to existing antiviral drugs. The increasing understanding of the molecular mechanics of infection has paved the way for the development of novel antiviral medicines that target specific viral proteins or host components. The demand for novel antiviral medications in the treatment of chronic infectious illnesses, as well as the emergence of more efficient new viruses, drives research into new targets and processes for antiviral development.[1] Only 21 novel antiviral medications were approved by the Food and Drug Administration (FDA) in the United States between 2012 and 2021, with eight of them being for the treatment of hepatitis C virus (HCV)-related pathologies and seven being used as anti-HIV drug (www.fda.gov). At the same time, governments, and the World Health Organization (WHO) are grappling with the worldwide danger of a slew of new and re-emerging viruses that have caused worrying outbreaks in recent years. Many new viruses are emerging, such as Zika virus (ZIKV), Ebola virus (EBOV),[2] and SARS corona virus. During the last of couple of years researchers have taken deeper dig into repurposed drugs. The process of finding new uses outside the scope of the original medical indication for existing drugs is also known as redirecting, repurposing, repositioning, and re-profiling. The problem in productivity and worldwide pressure on increasing prices and the growing number of regulatory hurdles one must pass through many drug developers to find new uses and new different targets as redirecting, repurposing, repositioning, and re-profiling are all terms for the process of identifying new applications for existing medications outside their original medical indications. The challenge of productivity, global pressure on rising pricing, and an increasing number of regulatory impediments must be overcome by many drug researchers to identify new applications and new targets as improved versions of current treatments are improvised version of the already existing drugs.[3] Traditional drug development is difficult, expensive, and time-consuming. Drug repurposing decreases the time and cost of drug development for contagious diseases dramatically. The efficiency of developed drugs targeting viral proteins and host components is limited by the resistance viruses and gives unfavorable side effects.[4] The drug repurposing strategy is the process of identifying new indications for already-approved FDA treatments and is a potential way to boost up the drug discovery process for viral diseases and a variety of other disorders.[5] Drug repurposing is critical in the fight against quickly spreading diseases including HIV, influenza, hepatitis C, Ebola, dengue fever, Coronavirus, and a variety of other fatal diseases.[6] Aside from the evident financial benefit, drugs discovery using the Drug Repurposing strategy can swiftly enter the clinical trials, especially for the contiguous diseases having no specific therapy. The drug repurposing technique provides a steady flow of information for studying viral biology and unknown molecular pathways. Present drugs with previously unknown antiviral activities can be used to explore viral mechanisms and pathology.[7] Although there are a few drawbacks to the drug repurposing approach, such as difficulty in identifying the target because the drug may have poly-pharmacology, the effective concentration being higher than what can be achieved in human plasma, and intellectual property rights issues, drug repurposing is still a better approach because it has the potential to reduce research time and costs.[8] In recent research, computational approaches have been widely used to anticipate novel therapeutic targets or drug repurposing prospects (Fig. 3.1). In comparison with wet lab experiment, computational high-throughput screening such as structure-based drug screening, deep-learning (DL)-based drug screening, and artificial-intelligence (AI)-based screening, in silico techniques are faster, less expensive, and can serve as an initial filtering step for thousands of molecules for lead structure identification[9] and farther modification with experimental confirmation. This necessitates the use of appropriate algorithmic tools to explore disease-relevant or disease-specific mechanisms. Antiviral capabilities of several drugs that were originally produced for a disease or disorder are being investigated to combat the worldwide problem of new and re-emerging viral diseases. Table 3.1 presents a list of pharmaceutical products that have been repurposed for a specific ailment, as well as the original indication for which they were produced.[7]

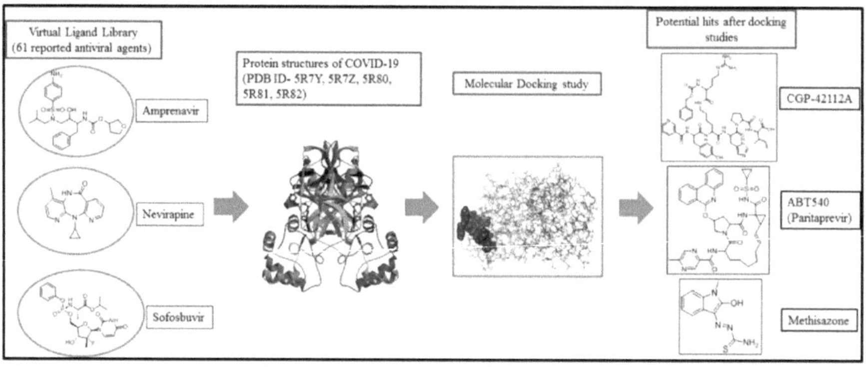


**FIGURE 3.1** The workflows of virus-targeting computational drug repurposing approaches. The input data consist of protein structure information (experimental or predicted) and chemical structure of drugs from public databases. Computational approach for antiviral drug discovery consisting of docking followed by molecular dynamics (MD) simulations. Finally, the output data approaches the potential molecules.

*Source*: Reprinted with permission from Ref. [10]. © 2020 Elsevier.

**TABLE 3.1** Approved and Candidate Drugs with Repurposing Potential as Antiviral Agents.[7]

| Compound | Status/indication | Virus |
|---|---|---|
| Mycophenolic acid | Approved/ immunomodulator | ZIKV |
| Daptomycin | Approved/antibacterial | ZIKV |
| Niclosamide | Approved/antiparasitic | ZIKV |
| Azithromycin | Approved/antibacterial | ZIKV |
| Novobiocin | Approved/antibacterial | ZIKV |
| Nanchangmycin | Investigational | ZIKV |
| Hippeastrine hydrobromide | Investigational | ZIKV |
| Sofosbuvir | Approved/antiviral | ZIKV |
| Ribavirin | Approved/antiviral | ZIKV |
| Chloroquine | Approved/antimalarial | ZIKV, MERS-, and SARS-CoV |
| Memantine | Approved/treatment of Alzheimer's disease | ZIKV |
| Prochlorperazine | Approved/antiemetic | DENV |
| Chlorcyclizine | Approved/antihistamine | HCV |
| Manidipine | Approved/antihypertensive | JEV, ZIKV, and HCMV |
| Favipiravir | Approved/antiviral | EBOV |

**TABLE 3.1** *(Continued)*

| Compound | Status/indication | Virus |
|---|---|---|
| GS-5734 | Investigational/antiviral | MERS and SARS-CoV |
| Imatinib | Approved/anticancer | MERS and SARS-CoV |
| Chlorpromazine | Approved/antipsychotic | MERS and SARS-CoV |
| Chlarithromycin/naproxen + oseltamivir | Approved/antibacterial, anti-inflammatory, antiviral | Influenza |
| Nitazoxanide | Approved/antiparasitic | Influenza, rotavirus, and norovirus |
| Raltegravir | Approved/antiviral | Herpesvirus |
| Lopinavir/ritonavir + interferon b-1b | Approved/antiviral | MERS-CoV |
| Lopinavir/ritonavir | Approved/antiviral | HPV |

## 3.2 COMPUTER-AIDED DRUG DESIGN APPROACH OF REPURPOSED DRUGS FOR ANTIVIRAL THERAPY

Computer-aided drug discovery/design (CADD) methods have been vital in the development of therapeutically important small molecules for more than three decades. There are two types of methods: structure-based and ligand-based. Structural-based approaches are similar to high-throughput screening in that they require both target and ligand structure knowledge. Structure-based techniques include ligand docking, pharmacophore design, and ligand docking. Using just ligand information, pharmacophores, molecular descriptors, and quantitative structure-activity connections, ligand-based approaches predict activity based on its similarity/dissimilarity to previously known active ligands.[11] The article outlines the theory behind the most essential strategies as well as recent successful implementations of repurposed drug screening as COVID-19 caused a large number of deaths in 2020, prompting a global emergency. Vaccines were developed as a result of continuing research and clinical trials. However, due to the evolving coronavirus, the vaccine's long-term effectiveness is still in doubt, which makes drug repositioning a realistic choice.[12] In the aftermath of the Zika virus outbreak a few years ago, one possible path to preventing viral epidemics is to identify broad-spectrum antiviral medications that are effective against entire families of viruses, has been suggested by Dr Anthony Fauci, who is one of the world's foremost authorities on infectious diseases and the longstanding director of the National Institute of Allergy and Infectious Diseases.[13]

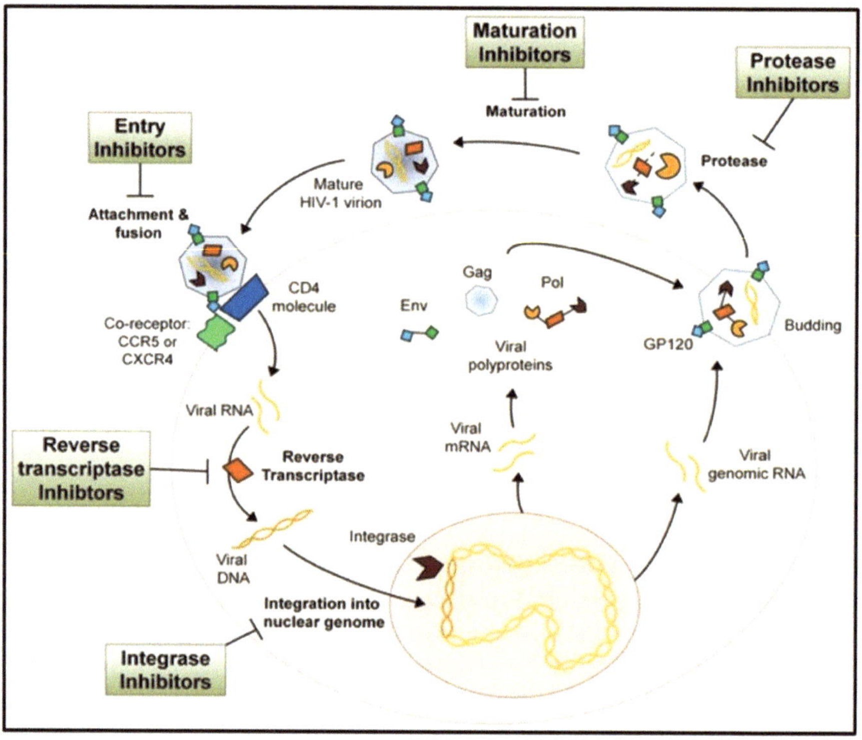


**FIGURE 3.2** Antiviral strategy class viruses rely on infected cells to promote viral genome replication and virus particle synthesis. As a result, infection is a critical stage in the virus's life cycle. Reverse transcriptase inhibitors are antiviral medications that prevent viral genome replication, hence limiting the formation of new virus particles. They operate within infected cells. Entry inhibitors, on the other hand, interact with existing virus particles outside of cells to prevent infection. They aid in viral load reduction and have been shown to improve preventive and therapeutic effects.

*Source*: Reprinted from Ref. [14]. © 2013 Smith, de Boer, Brul, Budovskaya and van der Spek. https://creativecommons.org/licenses/by/3.0/

### *3.2.1 VIRUS-TARGETING APPROACHES*

Each virus has its own structural characteristics, yet many therapeutically significant viruses have essential characteristics that can be used to develop broad-spectrum antiviral drugs (Fig. 3.2). Many viruses, for example, replicate their viral genomes in identical ways within infected cells, leading to the creation of antiviral replication inhibitors.[15] The majority of virus-targeting methods depend on structure-based drug and deep learning screening methods, which use three-dimensional structures of target proteins to estimate affinities or interaction energies of known chemical compounds with the proteins. These procedures are referred to virus-targeting approaches, since they were primarily utilized to find potential medications that target viral proteins; however, they may also be used to host proteins.[16]

Three main methodological workflows in *structure-based drug screening*:

- **Same target–new virus:** The first option is when an antiviral drug that is known to target a specific viral or cellular function/pathway is found to possess activity against other viruses. Antiviral action is based on structural homology and shared enzymatic characteristics of the viral target, as well as shared virus reproduction pathways. Antiviral RNA-polymerase inhibitors like favipiravir and Sofosbuvir (used to treat influenza and HCV infections, respectively) demonstrated its repurposing capabilities against EBOV and ZIKV (Table 3.1). Another example, drugs (e.g., chloroquine) that interfere with the late-stage entrance process of viruses like filo viruses and corona viruses, which employ cellular endocytotic routes to enter the host cell.[2]
- **Same target–new indication**: This occurs when a pharmacological target (i.e., a protein uria pathway that can be modulated by an approved drug) is found to be essential in a pathogenic process associated with a viral infection. In this case, the approved drug can be exploited also as an antiviral therapeutic agent (new indication). The case is exemplified by the anticancer drug imatinib that inhibits cellular ABL-kinase[17] and was found to be also active against pathogenic coronaviruses.[18]
- **New target–new indication:** This occurs when an approved drug with established bioactivity in a specific pathway or mechanism is found to have a new molecular target (i.e., it shows poly-pharmacology, see Glossary) which is essential for virus replication. Examples are antimicrobial agents (e.g., teicoplan in, ivermect in, itraconazole, and nitazoxanide) that were found to have a target also in virus-infected cells, whose inhibition has detrimental effects on viral replication.[19]

#### *3.2.1.1 DEEP LEARNING-BASED REPURPOSING STRATEGIES*

Deep learning (DL) models can predict binding affinities or docking scores and have shown advantages over conventional docking protocols. While standard docking protocols are limited to millions, DL approaches can analyze billions of chemical compounds. This allows them to be applied to whole databases, which increase the diversity of the tested compounds and the likelihood of finding unconventional compounds.[20] Furthermore, they are capable of processing more physico-chemical features (Fig. 3.2)[21,22] and can find features related to a nonfavorable docking.[20] However, most of these methods require datasets for training, which often come from real docking simulations; thus, the performance of many DL-based approaches still relies on the accuracy of the docking software used for training.

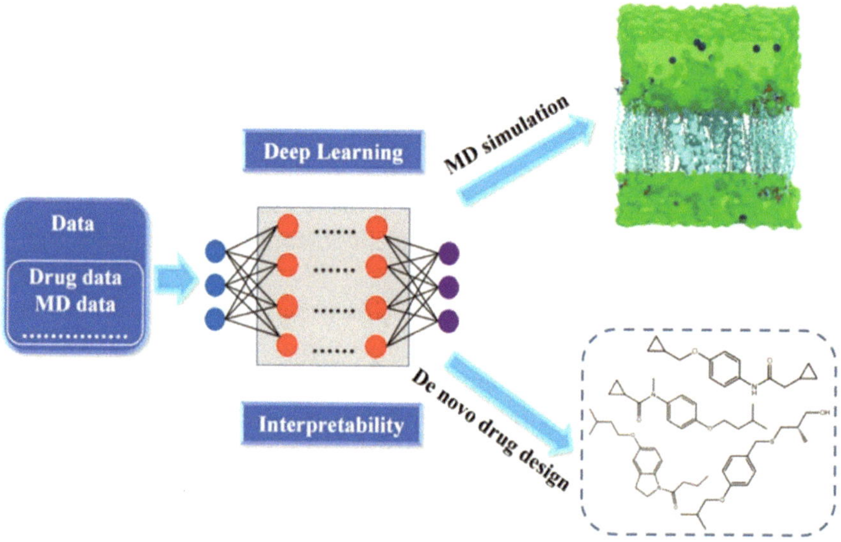


**FIGURE 3.3** Workflow of deep learning docking

*Source*: Reprinted with permission from Ref. [22]. © 2021 John Wiley & Sons.

Deep docking was created by Ton et al., who used quantitative structure–activity relationship models to predict docking scores of drugs targeting the SARS-CoV-2 3CLpro protein.[23] Because it docks specific subsets of compounds, it uses fewer docking processes and can generate a smaller list of compounds that are also rich in possible top hits.

Math DL is a technique created by Nguyen et al.[24] that uses low-dimensional mathematical representations of drug–target protein complex structures, which are then fed into DL algorithms to estimate drug–protein complex binding energies. For SARS-CoV-2, the authors used experimental binding affinity data from SARS-CoV ligand–3CL$^{pro}$ complexes from PDB bind and SARS-CoV protease inhibitors as training data to predict binding energies on DrugBank compounds for SARS-CoV-2 $3CL^{pro}$[25] and do not depend on docking software.

Molecule transformer–drug target interaction is a DL-based drug–target interaction prediction model created by Beck et al.[26] It predicts affinities using simplified molecular-input line-entry system (SMILES) with 51 representations for pharmaceuticals and protein sequences as input. The model was trained on commercially available antiviral drugs as well as viral target proteins for SARS-CoV-2. Among the potential molecules discovered were antiviral agents that had previously been used to treat SARS-CoV-2.

### *3.2.2 HOST-TARGETING APPROACHES*

The goal of host-targeting techniques is to find drugs that interfere with host pathways that contribute to viral pathogenesis, making them less susceptible to drug resistance.[27] This strategy has been driven by research in molecular virology and reached more advanced stages of the drug development progress so far, with compelling potential advantages over existing antiviral strategies. Thus, it provides a successful blueprint for broad-spectrum antiviral strategies developed from a materials science and engineering angle. In general, using small-molecule inhibitors that target a host cell factor that is not under genetic control of the virus can present a more difficult evolutionary task for the virus to escape drug susceptibility. This approach contrasts with direct-acting antivirals that can bind a viral enzyme with high affinity, where a single-point mutation at the drug's binding site can result in loss of drug efficacy. For example, broad-spectrum kinase inhibitors, which have been approved for anticancer therapy, have demonstrated the potential to impair intracellular viral trafficking and thus inhibit a wide range of viruses, such as hepatitis C, dengue, and Ebola, that depend on this particular host cell function.[28,29] In addition, SARS-CoV-2 infections can trigger a hyperreactive immune response characterized by the excessive release of pro-inflammatory cytokines and chemokines.[30] Thus, by targeting specific dysregulated pathways, a molecule that affects the host immune response can help critically sick individuals with COVID-19.[31,32]

- ***Signature-based approaches:*** Signature-based approaches primarily utilize transcriptome datasets from samples infected with viruses to identify candidate drugs through connectivity mapping, a well-established approach that relies on finding drug-induced expression signatures exhibiting reverse profiles to a disease signature.[33,34]

- ***Network-based approaches:*** Multiple data sources, such as virus–host interactions, PPIs, co-expression networks, functional connections, or drug–target interactions, are used in the general network-based method used in drug repurposing research on COVID-19. To discover important host protein targets or sections of the host interactome that can be addressed, network-based techniques or topological measurements are used to the generated networks.[35]

## 3.3 REPURPOSING DRUGS FOR VARIOUS VIRAL INFECTIONS AND COVID-19

### *3.3.1 REPURPOSING IN ZIKA VIRUS INFECTION*

The Zika virus (ZIKV) is a flavivirus that is transmitted by mosquitos and causes severe birth defects and Guillain–Barré syndrome. There are no antiviral drugs or vaccines available to treat ZIKA virus infection.[2] Barrows and colleagues examined a library of 774 FDA-approved drugs for their ability to prevent or inhibit a newly identified ZIKV strain from infecting human HuH-7 hepatocyte cells. Ivermectin, mycophenolic acid (MPA), and daptomycin were among the roughly 24 possible anti-ZIKV molecules discovered in their investigation. The immunosuppressants drug mycophenolic acid and the antibiotic, daptomycin were the promising inhibitors of ZIKA virus replication.[36] Xu et al. screened roughly 6000 compounds using a high-throughput screening approach, including FDA-approved pharmaceuticals, molecules in clinical trials, and pharmacologically active compounds. They detected over 100 chemicals in SNB-19 cells that inhibited ZIKV-induced caspase 3 activation.37 Another study demonstrated that the bacterial polyether nanchangmycin prevented ZIKV infection in a range of cell lines and ex vivo embryonic mouse midbrain neuron-glia mixed cultures[38] Chloroquine, a standard anti-inflammatory and antimalarial drug, has antiviral properties against several viruses. In Vero, human brain microvascular endothelial cells, and neural stem cells, this candidate also has antiviral efficacy against ZIKV. Without causing cytotoxicity, chloroquine lowers viral replication, the number of infected cells, and cell death caused by ZIKV infection. Sofosbuvir ($C_{22}H_{29}FN_3O9P$) has found to be active against ZIKV.[39] The most often used drugs in ZIKV therapy in pregnant women are niclosamide and azithromycin, both of which have a high effective concentration in human plasma.[37] Hippeastrine hydrobromide, a natural substance, has been found to be a powerful inhibitor of ZIKV infection and microcephaly-related consequences. The discovery of new genes and pathways for the creation of new antiviral drug molecules for ZIKV infection will be aided by drug–target network analysis and functional validation. Developing novel high-throughput drug repurposing tests and using current functional genomics methods to viral replication pathways is a possible avenue toward finding efficient antiviral treatments for ZIKV and other infectious agents.[2]

### 3.3.2 REPURPOSING IN EBOLA VIRUS INFECTION

Since the discovery Ebola virus in the late 1970s, it has caused multiple outbreaks, the most recent of which, in 2014–2016, was the most worrying owing to its scale and spread. Because of the urgent need for an effective Ebola virus cure, researchers have been studying current medications as prospective anti-Ebola virus pharmacological therapeutic agents, a process known as drug repurposing or drug repositioning. In vitro and in vivo tests of favipiravir against Ebola virus showed promising results.[40] Chloroquine's has also been found to be potent against Ebola virus in numerous in vitro investigations with various cell types.[41] Selective estrogen reuptake modulators toremifene and clomiphene are widely accessible and licensed for the treatment of breast cancer and infertility, respectively. These drugs were determined to have antiviral properties because they blocked Ebola virus entry by more than 90% in vitro.[42] Amiodarone is a multi-ion channel blocker that is commonly used to treat atrial fibrillation and ventricular tachycardia. It has been found to be an effective Ebola virus inhibitor in a variety of cell lines.[43] Among the many medications evaluated for anti-Ebola virus activity in vitro and in vivo, azithromycin was shown to be a potent in vitro inhibitor of the virus. A targeted drug combination approach led to the discovery of many therapeutic combinations that operate synergistically to prevent Ebola virus entrance.[44]

### 3.3.3 REPURPOSING IN HIV, CMV, HSV, AND HCV INFECTIONS

HIV/AIDS is one of the world's deadliest pandemics. Since 1981, 26 million people have perished, according to the World Health Organization and 1.6 million died only in the year of 2012.[45] Chloroquine and its hydroxyl derivatives, hydroxyl Chloroquine, were found to inhibit HIV-1 replication in various investigations.[46] Human Cytomegalovirus is a prime example of virus host adaptability and the potential of viruses to fully undermine cellular physiological functions in infected cells. Several approved or investigational drugs with an anti-Human Cytomegalovirus mechanism that differs from existing drugs like statins, cardiac glycosides, antiparasitic drugs emetine and nitazoxanide, kinase inhibitors, and the antihypertensive drug manidipine have been identified through various Drug Repurposing campaigns. This drug pharmacologically alters the environment of host proteins; the anti-Human Cytomegalovirus action is most likely due to the pharmaceuticals interfering with host pathways that the virus has taken away.[47–51] Hepatitis C Virus-1 replication is reduced when treated with ciclopirox olamine topically. Zhao et al. discovered that the Histone deacetylase inhibitor suberoylanilide hydroxamic acid, which is utilized in cancer therapy, inhibited Hepatitis C Virus replication in OR6 cells.[52] Several drugs, including anticancer treatments erlotinib and dasatinib, cholesterol drug ezetimibe, and ferroquine, have shown antiviral activity against Hepatitis C Virus. Cyclizine and phenothiazine, two of the most effective H1-antihistamines, were discovered to exhibit anti-HCV efficacy.[53]

### *3.3.4 REPURPOSING IN INFLUENZA AND DENGUE*

The influenza virus, which is a member of the Orthomyxoviridae family, is a pathogen of worldwide public health because it generates pandemic and epidemics. Drug repurposing efforts found anti-influenza drugs such as BAY 81-8781, dapivirine, naproxen, and the antibiotic clarithromycin, which are already approved or in clinical trials.[54] A three-drug combination of Clarithromycin, naproxen, and oseltamivir has been found to be effective in the treatment of severe influenza. The most advanced example of drug repurposing is the antiparasitic drug nitazoxanide, which is currently being repurposed for the treatment of influenza after an in silico screening specifically targeting mutant viral neuraminidase showed efficacy against oseltamivir-resistant influenza for nalidixic acid and dorzolamide, and the most advanced example of drug repurposing is the antiparasitic drug Dinaciclib, flavopiridol, and PIK-75 are kinase inhibitors that have been demonstrated to be highly effective against the H7N9 virus while being relatively safe.[55–57]

Dengue fever is a viral illness spread by mosquitos and caused by four antigenically different serotypes of Dengue Virus (DENV), specifically DENV1–4. Dengue fever is the world's most common arthropod-borne viral disease. Given the fast spread of DENV and the lengthy time it takes to bring a novel medicine to market, repurposing existing drugs appears to be an appealing option for a quick therapeutic intervention.[58]

Nelfinavir and other viral protease inhibitors like lopinavir and ritonavir were repurposed for Dengue virus infection using computer-aided drug design.[59] As a result of multiple research utilizing chloroquine in various drug repositioning studies for dengue virus infection, it was found to be able to suppress Dengue Virus Type 2 replication in Vero cells at a dose of 5 g/mL by plaque assay and qRT-PCR.[60] In vitro, castanospermine is active against influenza virus, CMV, HIV-1, and DENV-1, while in-vivo, it is active against Herpes Simplex Virus and Rauscher Murine Leukemia Virus.[61] Chemotherapeutic agents like dasatinib, bortezomib, and AZD053; prochlorperazine an antipsychotic drug, antiparasitic drugs ivermectin, suramin, nitazoxanide A; dexamethasone, prednisolone (steroids), few antibiotics like geneticin, narasin, and minocycline were found to be effective against DENV.[58]

### 3.3.5 DRUG REPURPOSING FOR SARS COV-2

A novel strain of coronavirus that causes SARS-like symptoms in humans was found in Wuhan, China, in 2019. A phylogenetic study of the entire viral genome was performed to better understand this novel virus (29,903 nucleotides). The findings suggested that the existing virus shares 89.1% nucleotide similarity with the genus beta coronavirus—subgenus Sarbecovirus—which previously caused the SARS pandemic. The new virus is known as COVID-19, and it has essential structural proteins such as the spike (S), envelope (E), membrane (M), and nucleocapsid (N) proteins. There are no particular treatment options for this highly infectious disease at the moment.[62] So, during pandemic scientists from all over the world have been trying to inhibit the SARS-CoV-19 with millions of million known drugs by the process of repurposing. New COVID-19 therapy studies include the use of remdesivir,[63] an antiviral medicine previously licensed to treat the Ebola virus, or a combination of two antivirals, ritonavir + lopinavir, previously approved to treat HIV infection. Additional active clinical trials involve the use of drugs approved for different therapeutic indications. Antimalarial drugs like chloroquine and hydroxychloroquine, as well as monoclonal antibodies targeting the interleukin-6 receptor (anti-IL-6R), are FDA-approved and may help COVID-19 patients by reducing abnormal inflammatory responses during cytokine storms and therefore improving organ function. Drug repurposing is a "recycling" technique based on the reuse of recognized drug that has been shown to be mainly successful, as evidenced by examples of repurposing therapies in cancer and other human illnesses.[64]

## 3.4 CONCLUSION

With the Pharma industry's ever-increasing obstacles, medication repurposing is the finest strategy for decreasing risks in the pipeline of new drug research. Indeed, the economic benefit is considerable, with yearly sales of up to USD20 billion. The lack of antiviral agents or vaccinations is becoming a serious medical problem. Existing FDA-approved medications can be repurposed or repositioned to meet the need. Understanding the possibility of drug repurposing that targets host activities is a quick and inexpensive way to produce broad-spectrum antivirals. Drug repurposing has already shown extremely favorable results with the drugs that have been effectively repurposed, and this technique may potentially open new routes to combat the issues of rising viral threats and antiviral resistance. This technique has previously demonstrated feasibility in the creation of novel anticancer medications (such as the antifungal drug itraconazole and its "second life" as an anticancer drug), but there are currently just a few successful instances in antiviral drug discovery (against influenza, EBOV, and MERS-CoV). The drug repurposing strategy has produced promising prospects for treating a variety of infectious diseases, and it can be expanded to address the drug discovery bottleneck for new and re-emerging viral infectious diseases.

## KEYWORDS

- **viral infections**
- **COVID-19**
- **drug repurposing**
- **computational drug discovery**
- **antiviral therapy**

## REFERENCES

1. Lou, Z.; Sun, Y.; Rao, Z. Current Progress in Antiviral Strategies. *Trends Pharmacol. Sci.* **2014,** *35* (2), 86–102.
2. Mercorelli, B.; Palù, G.; Loregian, A. Drug Repurposing for Viral Infectious Diseases: How Far Are We? *Trends Microbiol.* **2018,** *26* (10), 865–876.
3. Ashburn, T. T.; Thor, K. B. Drug Repositioning: Identifying and Developing New Uses for Existing Drugs. *Nat. Rev. Drug Discov.* **2004,** *3* (8), 673–683.

4. Cheng, F.; Murray, J. L.; Rubin, D. H. Drug Repurposing: New Treatments for Zika Virus Infection? *Trends Mol. Med.* **2016,** *22* (11), 919–921.
5. Nosengo, N. Can You Teach Old Drugs New Tricks? *Nat. News* **2016,** *534* (7607), 314.
6. Sakurai, Y.; Kolokoltsov, A. A.; Chen, C.-C.; Tidwell, M. W.; Bauta, W. E.; Klugbauer, N.; Grimm, C.; Wahl-Schott, C.; Biel, M.; Davey, R. A. Two-Pore Channels Control Ebola Virus Host Cell Entry and Are Drug Targets for Disease Treatment. *Science* **2015,** *347* (6225), 995–998.
7. Mani, D.; Wadhwani, A.; Krishnamurthy, P. T. Drug Repurposing in Antiviral Research: A Current Scenario. *J. Young Pharm.* **2019,** *11* (2), 117.
8. Guha, M. Repositioning Existing Drugs for Cancer Treatment. *Pharm. J.* **2015,** *294*, 7867.
9. Gangadevi, S.; Badavath, V. N.; Thakur, A.; Yin, N.; De Jonghe, S; Acevedo, O,; Jochmans, D.; Leyssen, P.; Wang, K.; Neyts, J.; Yujie, T. Kobophenol a Inhibits Binding of Host Ace2 Receptor With Spike rbd Domain of Sars-cov-2, A Lead Compound for Blocking Covid-19. *J. Phys. Chem. Lett.* **2021,** *12* (7), 1793–802.
10. Shah, B.; Modi, P.; Sagar, S. R. In Silico Studies on Therapeutic Agents for COVID-19: Drug Repurposing Approach. *Life Sci.* **2020,** *252*, 117652.
11. Sliwoski, G.; Kothiwale, S.; Meiler, J.; Lowe, E. W. Computational Methods in Drug Discovery. *Pharmacol. Rev.* **2014,** *66* (1), 334–395.
12. Mongia, A.; Saha, S. K.; Chouzenoux, E.; Majumdar, A. A Computational Approach to Aid Clinicians in Selecting Anti-Viral Drugs for COVID-19 Trials. *Sci. Rep.* **2021,** *11* (1), 1–12.
13. Fauci, A. S.; Morens, D. M. Zika Virus in the Americas—Yet Another Arbovirus Threat. *N. Engl. J. Med.* **2016,** *374* (7), 601–604.
14. Smith, R. L.; de Boer, R.; Brul, S.; Budovskaya, Y. V; van der Spek, H. Premature and Accelerated Aging: HIV or HAART? *Front. Genet.* **2013,** *3*, 328.
15. De Clercq, E. Antivirals and Antiviral Strategies. *Nat. Rev. Microbiol.* **2004,** *2* (9), 704–720.
16. Yoshino, R.; Yasuo, N.; Sekijima, M. Identification of Key Interactions Between SARS-CoV-2 Main Protease and Inhibitor Drug Candidates. *Sci. Rep.* **2020,** *10* (1), 1–8.
17. Aita, S.; Badavath, V. N.; Gundluru, M.; Sudileti, M.; Nemallapudi, B. R.; Gundala, S.; Zyryanov, G. V.; Chamarti, N. R.; Cirandur, S. R. Novel α-Aminophosphonates of Imatinib Intermediate: Synthesis, Anticancer Activity, Human Abl Tyrosine Kinase Inhibition, ADME and Toxicity Prediction. *Bioorg. Chem.* **2021,** *109*, 104718.
18. Coleman, C. M.; Sisk, J. M.; Mingo, R. M.; Nelson, E. A.; White, J. M.; Frieman, M. B. Abelson Kinase Inhibitors Are Potent Inhibitors of Severe Acute Respiratory Syndrome Coronavirus and Middle East Respiratory Syndrome Coronavirus Fusion. *J. Virol.* **2016,** *90* (19), 8924–8933.
19. Strating, J. R. P. M.; van der Linden, L.; Albulescu, L.; Bigay, J.; Arita, M.; Delang, L.; Leyssen, P.; van der Schaar, H. M.; Lanke, K. H. W.; Thibaut, H. J. Itraconazole Inhibits Enterovirus Replication by Targeting the Oxysterol-Binding Protein. *Cell Rep.* **2015,** *10* (4), 600–615.
20. Gentile, F.; Agrawal, V.; Hsing, M.; Ton, A. -T.; Ban, F.; Norinder, U.; Gleave, M. E.; Cherkasov, A. Deep Docking: A Deep Learning Platform for Augmentation of Structure Based Drug Discovery. *ACS Cent. Sci.* **2020,** *6* (6), 939–949.
21. Torres, P. H. M.; Sodero, A. C. R.; Jofily, P.; Silva-Jr, F. P. Key Topics in Molecular Docking for Drug Design. *Int. J. Mol. Sci.* **2019,** *20* (18), 4574.

22. Bai, Q., Liu, S., Tian, Y., Xu, T., Banegas-Luna, A. J., Pérez-Sánchez, H., Huang, J., Liu, H., Yao, X., Application Advances of Deep Learning Methods for De Novo Drug Design and Molecular Dynamics Simulation. Wiley Interdiscip. Rev. Comput. Mol. Sci. **2021,** e1581.
23. Ton, A.; Gentile, F.; Hsing, M.; Ban, F.; Cherkasov, A. Rapid Identification of Potential Inhibitors of SARS-CoV-2 Main Protease by Deep Docking of 1.3 Billion Compounds. *Mol. Inform.* **2020,** *39* (8), 2000028.
24. Nguyen, D. D.; Gao, K.; Wang, M.; Wei, G.-W. MathDL: Mathematical Deep Learning for D3R Grand Challenge 4. *J. Comput. Aided. Mol. Des.* **2020,** *34* (2), 131–147.
25. Nguyen, D. D.; Gao, K.; Chen, J.; Wang, R.; Wei, G.-W. Potentially Highly Potent Drugs for 2019-NCoV. *BioRxiv* **2020.**
26. Beck, B. R.; Shin, B.; Choi, Y.; Park, S.; Kang, K. Predicting Commercially Available Antiviral Drugs That May Act on the Novel Coronavirus (SARS-CoV-2) Through a Drug-Target Interaction Deep Learning Model. *Comput. Struct. Biotechnol. J.* **2020,** *18*, 784–790.
27. Lee, S. M.-Y.; Yen, H.-L. Targeting the Host or the Virus: Current and Novel Concepts for Antiviral Approaches Against Influenza Virus Infection. *Antiviral Res.* **2012,** *96* (3), 391–404.
28. Bekerman, E.; Neveu, G.; Shulla, A.; Brannan, J.; Pu, S.-Y.; Wang, S.; Xiao, F.; Barouch-Bentov, R.; Bakken, R. R.; Mateo, R. Anticancer Kinase Inhibitors Impair Intracellular Viral Trafficking and Exert Broad-Spectrum Antiviral Effects. *J. Clin. Invest.* **2017,** *127* (4), 1338–1352.
29. Schor, S.; Einav, S. Repurposing of Kinase Inhibitors as Broad-Spectrum Antiviral Drugs. *DNA Cell Biol.* **2018,** *37* (2), 63–69.
30. Catanzaro, M.; Fagiani, F.; Racchi, M.; Corsini, E.; Govoni, S.; Lanni, C. Immune Response in COVID-19: Addressing a Pharmacological Challenge by Targeting Pathways Triggered by SARS-CoV-2. *Signal Transduct. Target. Ther.* **2020,** *5* (1), 1–10.
31. Liao, J.; Way, G.; Madahar, V. Target Virus or Target Ourselves for COVID-19 Drugs Discovery?—Lessons Learned From Anti-Influenza Virus Therapies; Elsevier, 2020.
32. Chen, L.; Li, Q.; Zheng, D.; Jiang, H.; Wei, Y.; Zou, L.; Feng, L.; Xiong, G.; Sun, G.; Wang, H. Clinical Characteristics of Pregnant Women with Covid-19 in Wuhan, China. *N. Engl. J. Med.* **2020,** *382* (25), e100.
33. Lamb, J.; Crawford, E. D.; Peck, D.; Modell, J. W.; Blat, I. C.; Wrobel, M. J.; Lerner, J.; Brunet, J.-P.; Subramanian, A.; Ross, K. N. The Connectivity Map: Using Gene-Expression Signatures to Connect Small Molecules, Genes, and Disease. *Science* **2006,** *313* (5795), 1929–1935.
34. Subramanian, A.; Narayan, R.; Corsello, S. M.; Peck, D. D.; Natoli, T. E.; Lu, X.; Gould, J.; Davis, J. F.; Tubelli, A. A.; Asiedu, J. K. A Next Generation Connectivity Map: L1000 Platform and the First 1,000,000 Profiles. *Cell* **2017,** *171* (6), 1437–1452.
35. Cheng, F.; Rao, S.; Mehra, R. COVID-19 Treatment: Combining Anti-Inflammatory and Antiviral Therapeutics Using a Network-Based Approach. *Cleve. Clin. J. Med.* **2020.**
36. Barrows, N. J.; Campos, R. K.; Powell, S. T.; Prasanth, K. R.; Schott-Lerner, G.; Soto-Acosta, R.; Galarza-Muñoz, G.; McGrath, E. L.; Urrabaz-Garza, R.; Gao, J. A Screen of FDA-Approved Drugs for Inhibitors of Zika Virus Infection. *Cell Host Microbe* **2016,** *20* (2), 259–270.
37. Xu, M.; Lee, E. M.; Wen, Z.; Cheng, Y.; Huang, W.-K.; Qian, X.; Julia, T. C. W.; Kouznetsova, J.; Ogden, S. C.; Hammack, C. Identification of Small-Molecule Inhibitors

of Zika Virus Infection and Induced Neural Cell Death via a Drug Repurposing Screen. *Nat. Med.* **2016,** *22* (10), 1101–1107.
38. Yuan, S.; Chan, J. F.-W.; den-Haan, H.; Chik, K. K.-H.; Zhang, A. J.; Chan, C. C.-S.; Poon, V. K.-M.; Yip, C. C.-Y.; Mak, W. W.-N.; Zhu, Z. Structure-Based Discovery of Clinically Approved Drugs as Zika Virus NS2B-NS3 Protease Inhibitors That Potently Inhibit Zika Virus Infection in Vitro and in Vivo. *Antiviral Res.* **2017,** *145*, 33–43.
39. Alam, A.; Imam, N.; Ali, S.; Malik, M. Z.; Ishrat, R. Recent Trends in ZikV Research: A Step Away From Cure. *Biomed. Pharmacother.* **2017,** *91*, 1152–1159.
40. Sweiti, H.; Ekwunife, O.; Jaschinski, T.; Lhachimi, S. K. Repurposed Therapeutic Agents Targeting the Ebola Virus: A Systematic Review. *Curr. Ther. Res.* **2017,** *84*, 10–21.
41. Dowall, S. D.; Bosworth, A.; Watson, R.; Bewley, K.; Taylor, I.; Rayner, E.; Hunter, L.; Pearson, G.; Easterbrook, L.; Pitman, J. Chloroquine Inhibited Ebola Virus Replication in Vitro but Failed to Protect Against Infection and Disease in the in Vivo Guinea Pig Model. *J. Gen. Virol.* **2015,** *96* (Pt 12), 3484.
42. Madrid, P. B.; Chopra, S.; Manger, I. D.; Gilfillan, L.; Keepers, T. R.; Shurtleff, A. C.; Green, C. E.; Iyer, L. V; Dilks, H. H.; Davey, R. A. A Systematic Screen of FDA-Approved Drugs for Inhibitors of Biological Threat Agents. *PLoS One* **2013,** *8* (4), e60579.
43. Salata, C.; Baritussio, A.; Munegato, D.; Calistri, A.; Ha, H. R.; Bigler, L.; Fabris, F.; Parolin, C.; Palu, G.; Mirazimi, A. Amiodarone and Metabolite MDEA Inhibit Ebola Virus Infection by Interfering With the Viral Entry Process. *Pathog. Dis.* **2015,** *73* (5), ftv032.
44. Gielen, V.; Johnston, S. L.; Edwards, M. R. Azithromycin Induces Anti-Viral Responses in Bronchial Epithelial Cells. *Eur. Respir. J.* **2010,** *36* (3), 646–654.
45. Kinch, M. S.; Patridge, E. An Analysis of FDA-Approved Drugs for Infectious Disease: HIV/AIDS Drugs. *Drug Discov. Today* **2014,** *19* (10), 1510–1513.
46. Savarino, A.; Shytaj, I. L. Chloroquine and Beyond: Exploring Anti-Rheumatic Drugs to Reduce Immune Hyperactivation in HIV/AIDS. *Retrovirology* **2015,** *12* (1), 1–10.
47. Mercorelli, B.; Luganini, A.; Nannetti, G.; Tabarrini, O.; Palù, G.; Gribaudo, G.; Loregian, A. Drug Repurposing Approach Identifies Inhibitors of the Prototypic Viral Transcription Factor IE2 That Block Human Cytomegalovirus Replication. *Cell Chem. Biol.* **2016,** *23* (3), 340–351.
48. Gardner, T. J.; Cohen, T.; Redmann, V.; Lau, Z.; Felsenfeld, D.; Tortorella, D. Development of a High-Content Screen for the Identification of Inhibitors Directed Against the Early Steps of the Cytomegalovirus Infectious Cycle. *Antiviral Res.* **2015,** *113*, 49–61.
49. Ponroy, N.; Taveira, A.; Mueller, N. J.; Millard, A. Statins Demonstrate a Broad Anti-cytomegalovirus Activity in Vitro in Ganciclovir-Susceptible and Resistant Strains. *J. Med. Virol.* **2015,** *87* (1), 141–153.
50. Kapoor, A.; Cai, H.; Forman, M.; He, R.; Shamay, M.; Arav-Boger, R. Human Cytomegalovirus Inhibition by Cardiac Glycosides: Evidence for Involvement of the HERG Gene. *Antimicrob. Agents Chemother.* **2012,** *56* (9), 4891–4899.
51. Mukhopadhyay, R.; Roy, S.; Venkatadri, R.; Su, Y.-P.; Ye, W.; Barnaeva, E.; Mathews Griner, L.; Southall, N.; Hu, X.; Wang, A. Q. Efficacy and Mechanism of Action of Low Dose Emetine Against Human Cytomegalovirus. *PLoS Pathog.* **2016,** *12* (6), e1005717.

52. Zhao, F.; Liu, N.; Zhan, P.; Liu, X. Repurposing of HDAC Inhibitors Toward Anti-Hepatitis C Virus Drug Discovery: Teaching an Old Dog New Tricks. *Future Med. Chem.* **2015,** *7* (11), 1367–1371.
53. He, S.; Lin, B.; Chu, V.; Hu, Z.; Hu, X.; Xiao, J.; Wang, A. Q.; Schweitzer, C. J.; Li, Q.; Imamura, M. Repurposing of the Antihistamine Chlorcyclizine and Related Compounds for Treatment of Hepatitis C Virus Infection. *Sci. Transl. Med.* **2015,** *7* (282), 282ra49–282ra49.
54. Hung, I. F. N.; To, K. K. W.; Chan, J. F. W.; Cheng, V. C. C.; Liu, K. S. H.; Tam, A.; Chan, T.-C.; Zhang, A. J.; Li, P.; Wong, T.-L. Efficacy of Clarithromycin-Naproxen-Oseltamivir Combination in the Treatment of Patients Hospitalized for Influenza A (H3N2) Infection: An Open-Label Randomized, Controlled, Phase IIb/III Trial. *Chest* **2017,** *151* (5), 1069–1080.
55. Bao, J.; Marathe, B.; Govorkova, E. A.; Zheng, J. J. Drug Repurposing Identifies Inhibitors of Oseltamivir-resistant Influenza Viruses. *Angew. Chemie Int. Ed.* **2016,** *55* (10), 3438–3441.
56. Haffizulla, J.; Hartman, A.; Hoppers, M.; Resnick, H.; Samudrala, S.; Ginocchio, C.; Bardin, M.; Rossignol, J.-F.; Group, U. S. N. I. C. S. Effect of Nitazoxanide in Adults and Adolescents With Acute Uncomplicated Influenza: A Double-Blind, Randomised, Placebo-Controlled, Phase 2b/3 Trial. *Lancet Infect. Dis.* **2014,** *14* (7), 609–618.
57. Perwitasari, O.; Yan, X.; O'Donnell, J.; Johnson, S.; Tripp, R. A. Repurposing Kinase Inhibitors as Antiviral Agents to Control Influenza A Virus Replication. *Assay Drug Dev. Technol.* **2015,** *13* (10), 638–649.
58. Botta, L.; Rivara, M.; Zuliani, V.; Radi, M. Drug Repurposing Approaches to Fight Dengue Virus Infection and Related Diseases. *Front Biosci.* **2018,** *23*, 997–1019.
59. Bhakat, S.; Delang, L.; Kaptein, S.; Neyts, J.; Leyssen, P.; Jayaprakash, V. Reaching Beyond HIV/HCV: Nelfinavir as a Potential Starting Point for Broad-Spectrum Protease Inhibitors Against Dengue and Chikungunya Virus. *RSC Adv.* **2015,** *5* (104), 85938–85949.
60. Rolain, J.-M.; Colson, P.; Raoult, D. Recycling of Chloroquine and Its Hydroxyl Analogue to Face Bacterial, Fungal and Viral Infections in the 21st Century. *Int. J. Antimicrob. Agents* **2007,** *30* (4), 297–308.
61. Whitby, K.; Pierson, T. C.; Geiss, B.; Lane, K.; Engle, M.; Zhou, Y.; Doms, R. W.; Diamond, M. S. Castanospermine, a Potent Inhibitor of Dengue Virus Infection in Vitro and in Vivo. *J. Virol.* **2005,** *79* (14), 8698–8706.
62. Wu, C.; Liu, Y.; Yang, Y.; Zhang, P.; Zhong, W.; Wang, Y.; Wang, Q.; Xu, Y.; Li, M.; Li, X. Analysis of Therapeutic Targets for SARS-CoV-2 and Discovery of Potential Drugs by Computational Methods. *Acta Pharm. Sin. B* **2020,** *10* (5), 766–788.
63. Badavath, V. N.; Kumar, A.; Samanta, P. K.; Maji, S.; Das, A.; Blum, G.; Jha, A.; Sen, A. Determination of Potential Inhibitors Based on Isatin Derivatives Against SARS-CoV-2 Main Protease (Mpro): A Molecular Docking, Molecular Dynamics and Structure-Activity Relationship Studies. *J. Biomol. Struct. Dyn.* **2020,** 1–19.
64. Pushpakom, S.; Iorio, F.; Eyers, P. A.; Escott, K. J.; Hopper, S.; Wells, A.; Doig, A.; Guilliams, T.; Latimer, J.; McNamee, C. Drug Repurposing: Progress, Challenges and Recommendations. *Nat. Rev. Drug Discov.* **2019,** *18* (1), 41–58.

# CHAPTER 4

# Drug Repurposing and Computational Drug Discovery for Parasitic Diseases and Neglected Tropical Diseases (NTDs)

JAMES H. ZOTHANTLUANGA[1], ARPITA PAUL[1], ABD. KAKHAR UMAR[2], and DIPAK CHETIA[1]

[1]*Department of Pharmaceutical Sciences, Faculty of Science and Engineering, Dibrugarh University, Dibrugarh, Assam, India*

[2]*Department of Pharmaceutics and Pharmaceutical Technology, Faculty of Pharmacy, Universitas Padjadjaran, Jatinangor, Indonesia*

## ABSTRACT

Neglected tropical diseases (NTDs) (Chagas disease, lymphatic filariasis, leprosy, Buruli ulcer, trypanosomiasis, cysticercosis, fascioliasis, dracunculiasis, mycetoma, schistosomiasis, trachoma, and onchocerciasis) burden the low-income or poverty-embedded populations of the tropical region. Many drugs are currently being used to treat parasitic diseases and NTDs. However, drug resistance and toxicity have limited the efficacy of these drugs. An alternative to the traditional drug discovery process is the technique of drug repurposing or repositioning, wherein an existing Food and Drug Administration (FDA)-approved drug used for the treatment of a particular disease was repurposed/repositioned to treat another disease. In this chapter, drug repurposing techniques and computational techniques will be discussed. Specific drugs that have been repurposed for parasitic diseases and NTDs will be covered. Potential leads identified for parasitic diseases and NTDs through computational techniques will also be covered. Many

Drug Repurposing and Computational Drug Discovery: Strategies and Advances.
Mithun Rudrapal, PhD (Ed)

existing FDA-approved drugs showed remarkable potential to be repurposed for the treatment of parasitic diseases and NTDs. Different computational techniques such as virtual screening, 3D-QSAR, homology modeling, molecular docking, MD simulations, target fishing, etc. have played a key role in the identification of new compounds for the treatment of parasitic diseases and NTDs.

## 4.1 INTRODUCTION

Parasitic diseases (malaria, chikungunya, dengue, Zika virus, soil-transmitted helminths, and leishmaniasis) are spread by parasites such as protozoans, helminths, viruses, and ectoparasites through contaminated water, food, or by insect vectors.[1,2] Neglected tropical diseases (NTDs) (Chagas disease, lymphatic filariasis, leprosy, Buruli ulcer, trypanosomiasis, cysticercosis, fascioliasis, dracunculiasis, mycetoma, schistosomiasis, trachoma, and onchocerciasis) burden the low-income or poverty-embedded populations of the tropical region.[3,4] Parasites, viruses, and bacteria that are responsible for causing NTDs are transmitted by insect vectors, along with contaminated food and water.[3,5]Several drugs are currently being used to treat parasitic diseases and NTDs.[6,7] However, drug resistance and toxicity have limited the efficacy of these drugs.[5,6,8–12] Millions of people are being infected and thousands are killed every year due to parasitic diseases and NTDs.[1,4] To reduce the disease burden and fatality rate, newer drugs with improved efficacy having a good safety profile are the need of the hour.

The conventional method for the discovery and development of a new drug is a complex, risky, costly, tedious, and time-consuming process.[13] An alternative to the traditional drug discovery process is the technique of drug repurposing or repositioning, wherein an existing Food and drug administration (FDA)-approved drug used for the treatment of a particular disease was repurposed/repositioned to treat another disease.[14] Discovering new therapies for diseases with existing drugs offers several advantages. While the traditional drug discovery process generally takes 10–16 years, the time required to find a new therapy for a disease with an existing drug requires 3–12 years. The traditional process costs USD12 billion while drug repurposing costs only ~USD2 billion. It takes 1–2 years to find new drug targets for repurposed drugs and around 8 years to develop a repurposed drug. Toxicity and bioavailability issues are significantly reduced as drugs intended to be repurposed for other diseases have already been approved by

FDA.[15] For example, thalidomide originally intended for morning sickness has been repurposed for multiple myeloma, sildenafil initially used for angina and hypertension was repurposed for erectile dysfunction, and amantadine used for influenza have been repurposed for Parkinson's disease.[16]

Computer-aided drug design (CADD) is broadly classified into ligand-based drug design such as virtual screening, structure–activity relationship (SAR), 2D/3D quantitative-SAR, pharmacophore modeling; structure-based drug design such as molecular docking, molecular dynamics; and other computational techniques such as binding free energy (MM-PBSA, MM-GBSA) calculations.[17] Also, many online web servers had eased the process of toxicity prediction, physicochemical analysis, pharmacokinetic studies, along bioavailability assessment of compounds.[18,19] Major contributions of CADD in the field of drug discovery are carbonic anhydrase inhibitor (dorzolamide),[20] angiotensin-converting enzyme inhibitor (captopril),[21] anti-HIV drugs (saquinavir, ritonavir, indinavir),[22] and fibrinogen antagonist (tirofiban).[23] An example of the efficiency and reliability of CADD in drug design, discovery, and development can be observed when two groups of independent researchers, that worked separately to identify novel inhibitors for transforming growth factor-$\beta 1$ receptor kinase, reported a strikingly similar results in the lead compound they had identified.[17] To name a few, many potential leads had been identified with CADD for parasitic diseases such as malaria,[24] dengue,[25] Zika virus,[26] and chikungunya.[27] CADD can provide promising results while reducing the workload and cost.[28]

In this chapter, drug repurposing techniques and computational techniques will be discussed. Specific drugs that have been repurposed for parasitic diseases and NTDs will be covered. Potential leads identified for parasitic diseases and NTDs through computational techniques will also be covered.

## 4.2 REPURPOSED DRUGS FOR PARASITIC DISEASES AND NTDS

Different approaches exist for repurposing drugs from one disease for another. In the drug-based approach, several parameters of the drug such as structural features, pharmacological activity, toxicities, and adverse effects are taken into consideration. In a drug-based approach, the biological activity of a molecule is evaluated with prior information on a target protein and thus, the traditional drug discovery process is followed for this approach.[29] On the other hand, if there is an ample amount of information regarding the disease, then the disease-based approach becomes relevant for repurposing drugs. Specific target proteins that are involved in a disease pathway (proteomics), specific genomic data associated with a disease (genomics), and metabolic pathways of a disease (metabolomics) are taken into consideration for disease-based drug repurposing.[30] A flow chart of the drug repurposing process is given in Figure 4.1. Some examples of drugs that have been repurposed for parasitic diseases and NTDs are given in Tables 4.1 and 4.2, respectively.

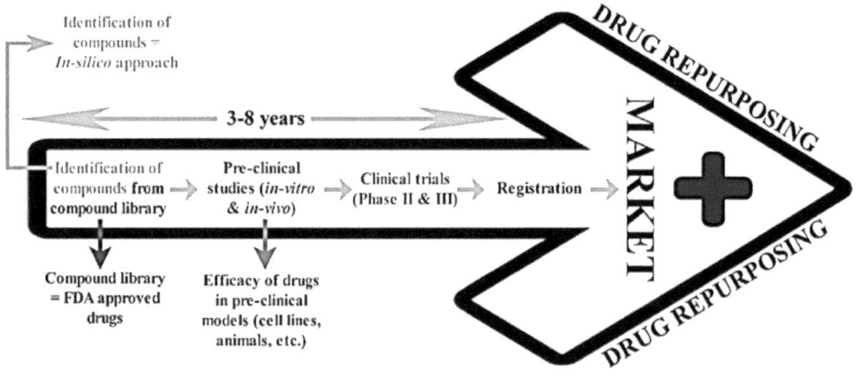


**FIGURE 4.1** Drug repurposing process.

**TABLE 4.1** Repurposed Drugs for Parasitic Diseases.

| Drug | Original use | New indication | Possible mechanism against newly indicated disease | References |
|---|---|---|---|---|
| **Idelalisib** | Anticancer | Malaria | Inhibition of plasmodium enzymes such as kinases | [31] |
| **Regorafenib** | Anticancer | | Inhibition of plasmodium enzymes such as kinases | [31] |
| **Bleomycin** | Anticancer | | Induction of oxidative stress by producing free radicals | [31] |
| **Roxithromycin** | Antibiotic | | Inhibition of essential enzymes | [31] |
| **Erythromycin** | Antibiotic | | Inhibition of essential enzymes | [31] |

**TABLE 4.1** *(Continued)*

| Drug | Original use | New indication | Possible mechanism against newly indicated disease | References |
|---|---|---|---|---|
| Prochlorperazine | Antipsychotic | Dengue | Inhibits viral binding and viral entry | [32] |
| Quinine | Antimalarial | | Inhibits viral protein synthesis, induces production of viral genes for improved immunity against dengue infection | [33] |
| Minocycline | Antibiotic | | Inhibition of viral protein synthesis, and upregulation of antiviral genes | [34] |
| Metoclopramide | Antiemetic | | Inhibition of viral replication | [35] |
| Memantine hydrochloride | Anti-Alzheimer | Chikungunya | – | [36] |
| Novobiocin | Antibiotic | | Inhibition of viral replication by inhibiting nsP2 protease | [37] |
| Telmisartan | Antihypertensive | | Inhibition of viral replication by inhibiting nsP2 protease | [37] |
| Suramin | Antiparasitic | | Inhibition of cellular entry | [38] |
| Fluconazole | Antifungal | Leishmaniasis | Inhibition of lanosterol-14-α-demethylase | [39] |
| Itroconazole | Antifungal | | Inhibition of lanosterol-14-α-demethylase | [39] |

**TABLE 4.1** *(Continued)*

| Drug | Original use | New indication | Possible mechanism against newly indicated disease | References |
|---|---|---|---|---|
| Ketoconazole | Antifungal | | Inhibition of lanosterol-14-α-demethylase | [39] |
| Amphotericin B | Antifungal | | Inhibition of lanosterol-14-α-demethylase | [39] |
| Posaconazole | Antifungal | | Inhibition of lanosterol-14-α-demethylase | [39] |
| Chloroquine | Antimalarial | Zika virus | Inhibition of viral protein synthesis, or inhibition of cellular entry | [40] |
| Niclosamide | Anthelmintic | | Inhibition of essential enzymes such as kinases | [40] |
| Suramin | African-sleeping sickness | | Inhibition of viral protein synthesis | [40] |
| Nitazoxanide | Antiprotozoal | | Inhibition of viral replication | [40] |
| Imatinib | Anticancer | Soil-transmitted helminths | – | [41] |
| Artemether | Antimalarial | | – | [41] |
| Artesunate | Antimalarial | | – | [41] |
| Dihydroartemisinin | Antimalarial | | – | [41] |
| Nilutamide | Anticancer | | – | [41] |

**TABLE 4.2** Repurposed Drugs for NTDs.

| Drug | Original use | New indication | Possible mechanism against newly indicated disease | References |
|---|---|---|---|---|
| Rifampin | Used in combination therapy for tuberculosis | Buruli ulcer | Interferes with β-subunit of bacterial RNA-polymerase and prevents the synthesis of RNA | [42] |

**TABLE 4.2** *(Continued)*

| Drug | Original use | New indication | Possible mechanism against newly indicated disease | References |
|---|---|---|---|---|
| Streptomycin | Antitubercular | | Binds to 30s subunit of the bacterial ribosome and inhibit protein synthesis | [43] |
| Clarithromycin | Antibiotic | | Inhibits polypeptide synthesis by binding with 23S rRNA on 50S ribosomal subunit leading to a bacteriostatic effect | [44] |
| Sparfloxacin | Antibacterial | | Interfere with DNA replication and transcription by inhibiting bacterial DNA gyrase enzyme | [45] |
| Clofazimine | Antileprotic | | Reduced to a reactive oxygen species by mycobacterial type 2 NADH:quinone oxidoreductase causing toxic effects to the bacteria | [46] |
| Clomipramine | Antidepressant | Chagas disease | Irreversible inhibition of trypanothione reductase | [47] |
| Thioridazine | Antipsychotic | | Irreversible inhibition of trypanothione reductase | [48] |
| Ketoconazole | Antifungal | | Impairment in the function of cytochrome P-450 sterol 14 alpha-demethylase, retarding parasitic growth | [49] |
| Itraconazole | Antifungal | | Impairment in the function of cytochrome P-450 sterol 14 alpha-demethylase, retarding parasitic growth | [49] |

**TABLE 4.2** *(Continued)*

| Drug | Original use | New indication | Possible mechanism against newly indicated disease | References |
|---|---|---|---|---|
| Fluconazole | Antifungal | | Impairment in the function of cytochrome P-450 sterol 14 alpha-demethylase, retarding parasitic growth | [49] |
| Artesunate and artemether | Antimalarial | Fascioliasis | Disruption of spermatogenesis | [50] |
| Bithionol | Anthelminthic | | Poorly understood, causes morphological changes | [51] |
| Emetine | Amebicidal | | Disrupts protein synthesis | [52] |
| Praziquantel | Anthelmintic | | Activates a transient receptor potential melastatin ion channel leading to paralysis of the parasite | [53] |
| Metronidazole | Antibiotic | | Distortion of the helical structure of the DNA | [54] |
| Eflornithine | Anticancer | Human African trypanosomiasis | Retards cell proliferation by inhibiting ornithine decarboxylase which in turn depletes putrescine and spermidine | [55] |
| Nifurtimox | Chagas disease | | Generation of reactive oxygen species causes detrimental effects to the cellular components | [56] |
| Pafuramidine | Pneumocystis pneumonia | | Interrupts with DNA synthesis | [57] |
| Rifampin | Antitubercular | *Onchocerciasis* | Inhibits RNA synthesis | [58] |

**TABLE 4.2** *(Continued)*

| Drug | Original use | New indication | Possible mechanism against newly indicated disease | References |
|---|---|---|---|---|
| Ivermectin | Antiparasitic | | Increases the influx of chloride ions by glutamate-gated chloride channels as a consequence dysfunction of the excretory pore, flaccid paralysis, and death of the parasite takes place | [59] |
| Moxidectin | Anthelmintic (for animals) | | Increases the influx of chloride ions by glutamate-gated chloride channels as a consequence dysfunction of the excretory pore, flaccid paralysis, and death of the parasite takes place | [60] |
| Emodepside | Anthelmintic | | Interacts with calcium-gated and potassium-gated voltage channels | [61] |
| Albendazole | Anthelmintic | | Prevents microtubule elongation which interferes with chromosome segregation and cell division ultimately leading to defective embryogenesis | [62] |
| Rifampicin | Antitubercular | Leprosy | Interrupts the binding of β subunit with DNA which inhibits mRNA production leading to the death of the bacteria | [63] |
| Ofloxacin | Antibiotic | | Inhibits DNA gyrase, DNA replication, and transcription | [64] |

**TABLE 4.2** *(Continued)*

| Drug | Original use | New indication | Possible mechanism against newly indicated disease | References |
|---|---|---|---|---|
| Dapsone | Antibiotic | | Inhibits folate biosynthesis in the bacterial cells | [63] |
| Minocycline | Antibiotic | | Inhibits protein synthesis by binding with 30S subunit of the ribosome | [63] |
| Thalidomide | Morning sickness and insomnia | | Inhibits pro-inflammatory cytokine TNF-alpha | [65] |
| Nitazoxanide | Antibiotic | Lymphatic filariasis | Interferes with anaerobic electron transport channel | [66] |
| Doxycycline | Antibiotic | | Blocks embryogenesis, inhibits inflammation, angiogenesis, proteolysis, and apoptosis | [67] |
| Tizoxanide | Antibiotic | | Interferes with anaerobic electron transport channel | [66] |
| Ivermectin | Antiparasitic | | Increases the influx of chloride ions by glutamate-gated chloride channels as a consequence dysfunction of the excretory pore, flaccid paralysis, and death of the parasite takes place | [68] |
| Albendazole | Anthelmintic | | Prevents microtubule elongation which interferes with chromosome segregation and cell division ultimately leading to defective embryogenesis | [62] |

**TABLE 4.2** *(Continued)*

| Drug | Original use | New indication | Possible mechanism against newly indicated disease | References |
|---|---|---|---|---|
| Azithromycin | Antibiotic | *Trachoma* | Inhibits polypeptide synthesis | [69] |
| Artemether | Antimalarial | *Schistosomiasis* | Alters glycogen content in the parasite | [70] |
| Artesunate | Antimalarial | | Impairs fecundity of adult female | [71] |
| Mefloquine | Antimalarial | | Interferes with hemozoin formation | [72] |
| Synriam | Antimalarial | | Interferes with hemozoin formation | [73] |
| Edelfosine | Anticancer | | Downregulates the function of T helper 1 and T helper 2 response, thereby reducing granuloma formation | [74] |
| Ravuconazole | Chagas disease | *Mycetoma* | – | [75] |
| Sulphamethoxazole | Antibiotic | | Inhibits folic acid synthesis | [76] |
| Trimethoprim | Antibiotic | | Inhibits the activity of dihydrofolate-reductase | [76] |
| Linezolid | Antibiotic | | – | [77] |

## 4.3 COMPUTATIONAL APPROACHES AND TECHNIQUES

Drug design, discovery, or development using computational techniques have emerged as a cost-effective and efficient approach in the field of pharmaceutical research.[78] Captopril (antihypertensive), dorzolamide (treatment of glaucoma), saquinavir (anti-HIV), zanamivir (anti-influenza), oseltamivir (anti-influenza), aliskiren (antihypertensive), boceprevir (treatment of hepatitis), nolatrexed (anticancer), TMI-005 (anti-inflammatory), LY-517717 (prevention of thrombosis), rupintrivir (antiviral), and NVP-AUY922 (anticancer) are few examples of drugs that were discovered or optimized with CADD.[21] A flowchart of the computational drug discovery process is given in Figure 4.2. In the following section, different techniques of CADD will be discussed in brief with special reference given to structure-based drug design (molecular docking, MD simulations) and ligand-based drug design (similarity searching, virtual screening, SAR, QSAR, pharmacophore modeling). In addition, newer approaches such as the application of artificial intelligence will also be briefly discussed.

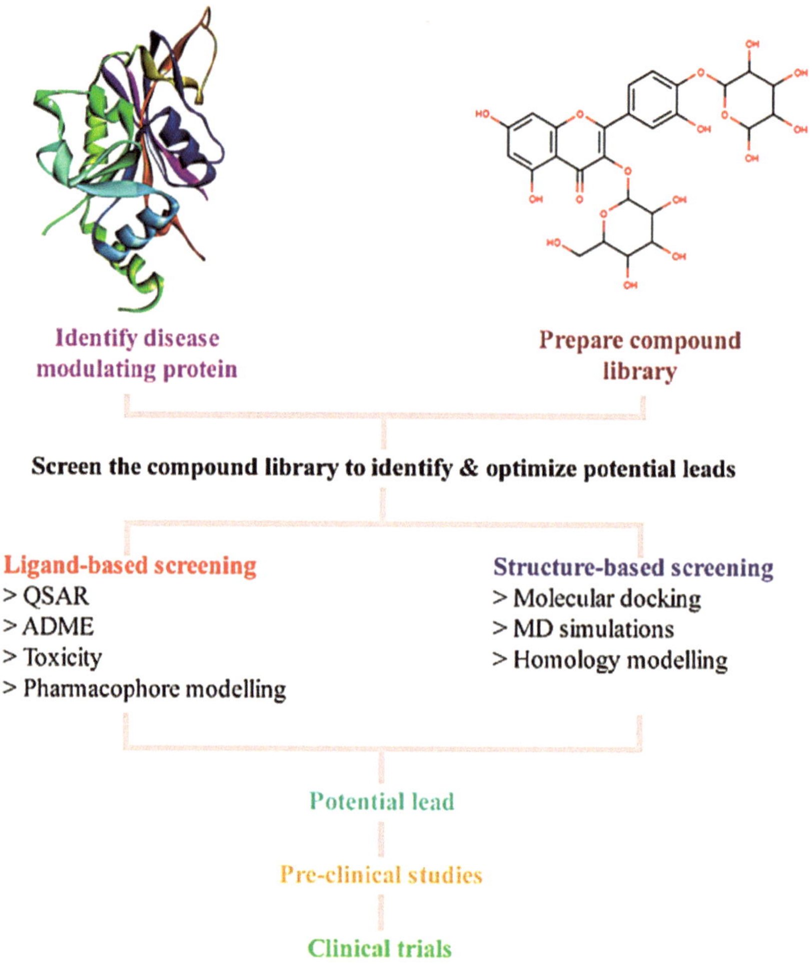


**FIGURE 4.2** Flowchart of drug discovery process involving computational approach.

## 4.3.1 STRUCTURE-BASED DRUG DESIGN

### 4.3.1.1 Molecular Docking Simulation Studies

Since the development of the first algorithm back in the 1980s,[79] MDSS has been the most widely and consistently used CADD technique.[80–82] Even in the ongoing coronavirus disease 2019 (COVID-19) pandemic, molecular docking is widely used by researchers to identify potential inhibitors of essential enzymes of severe acute respiratory syndrome coronavirus 2 (SARS-CoV-2).[83–86] MDSS can predict the binding affinity and binding pose of a molecular at the active binding pocket of a known protein of interest.[87] Following are a few key pieces of information that are necessary to understand the basics of MDSS:

- MDSS is used when the target proteins are known and the protein structures are available or can be prepared.
- The X-ray 3D structure of proteins can be downloaded from the Research Collaboratory for Structural Bioinformatics-Protein Data Bank website (RCSB-PDB) (https://www.rcsb.org/).
- When the desired protein structures are not available on the RCSB-PDB website, the structures of proteins are prepared manually using the homology modeling technique.[88]
- The ligands that are to be docked toward the active binding pocket of the target protein can be prepared manually, or their structures can also be downloaded from an online database such as PubChem (https://pubchem.ncbi.nlm.nih.gov/), COCONUT (https://coconut.naturalproducts.net/), etc.
- Once a ligand is docked toward the active binding site of a protein, the algorithms of the docking software generate different binding poses of the ligand with the first pose having the best binding affinity (lowest binding energy) toward the target protein and so on.[84]
- When multiple ligands are docked simultaneously, the software algorithms rank the binding affinity of the ligands by giving each ligand a numerical score. For example, "ligand A" with a binding energy of −10.0 kcal/mol has the best binding affinity toward a target protein while "ligand J" with a binding energy of −1.0 kcal/mol has the worst binding affinity toward a target protein.[83] Generally, "ligand A" will be selected for further studies.

- Ligand interaction studies reveal what amino acid residues interact with a ligand. This is important because a ligand must interact with the crucial amino acids of a target protein. For example, in the case of SARS-CoV-2, ligands (potential inhibitors) must interact with the HIS41-CYS145 catalytic dyad and other amino acids at the catalytic cavity of the main protease.[86]

#### *4.3.1.2 MOLECULAR DYNAMICS SIMULATION*

Molecular dynamics (MD) simulation is a computational technique that is widely used to validate the results obtained from the MDSS.[89,90] MDSS is a preliminary computational study to investigate the binding affinity of a ligand toward the active site of a target protein.[84] On the other hand, MD simulation is used to evaluate the stability of a protein–ligand complex for a specific period (nanoseconds) in a computationally simulated environment to mimic a real-life biological environment.[91] When MDSS results are supported by in vitro data, MD simulations are generally not carried out.[92,93] However, in the absence of in vitro data, MD simulations are conducted to validate the MDSS results. [91] Following are a few key pieces of information that are necessary to understand the basics of MD simulations:

- MD simulation is used to validate MDSS. From MDSS, we can get information regarding the binding affinity and protein–ligand interactions.[94] However, the stability of a ligand at the active binding pocket of a protein is not provided by MDSS. Therefore, MD simulations are used to check and evaluate the stability of a protein–ligand complex. MD simulations are carried out in conditions that mimic real-life biological environments where a ligand will supposedly interact with a protein.[91]
- Root mean square deviation (RMSD) of a protein and ligand is the most common property evaluated to check the stability of protein–ligand interaction. The RMSD trajectory of a protein and ligand is evaluated for a specified period. The original docked pose of ligand at the active binding pocket of a protein is taken as the reference structure. From this, the stability of the protein and the ligand is determined by observing their movements in the form of an RMSD trajectory plot. For small proteins, an RMSD fluctuation between 1 and 3 Å is considered acceptable. The RMSD of a ligand should not

differ significantly from the RMSD equilibrium of the protein that it is complexed with.[91]

- Root mean square fluctuation (RMSF) is an important parameter generated by MD simulations for each amino acid residue of a protein. The RMSF of a protein can be used to determine specific changes occurring at a particular amino acid residue during the MD simulations. By taking the RMSF of a protein as a reference, the changes occurring to a particular amino acid in a protein that is complexed with a ligand can be found out. The smaller the RMSF value, the more stable is the protein.[95]
- The radius of gyration (rGyr) is another important parameter generated by MD simulations. rGyr determines the compactness of a protein–ligand complex during the MD simulations. For a protein–ligand complex, it is desirable to have a small value of rGyr.[95]

## 4.3.2 LIGAND-BASED DRUG DESIGN

### 4.3.2.1 VIRTUAL SCREENING

In the late 1990s, the term "virtual screening" (VS) was coined wherein VS refers to the use of computational models and algorithms to screen out potential lead compounds.[96] Following are a few key pieces of information that are necessary to understand the basics of VS:

- VS can be used to identify potentially toxic compounds from a small to a large compound library. For example, the median lethal dose ($LD_{50}$), toxicity class, and organ toxicity of a compound can be determined by an online web tool "ProTox-II[19]" Offline software such as DataWarrior may also be used for toxicity studies.[97] In this way, from a library of 1000 compounds, all compounds with potential signs of toxicities can be identified and discarded from a study.
- VS can be used to identify compounds with desired drug-like properties. For example, the physicochemical properties, lipophilicity, water-solubility, pharmacokinetics, bioavailability, and synthetic accessibility of a test compound can be studied with the help of an online tool "SwissADME.[18]" When we have hundreds or thousands of compounds, it is not possible to synthesize or carry out MDSS for all those compounds. Also, it is not rational to carry out MDSS or MD simulations for a compound with toxicities and undesirable drug-like

properties. To avoid such types of issues, VS can be used to remove toxic compounds with undesirable drug-like properties from a study. In this way, the chances of failure at a preclinical or clinical level will be reduced. In addition to SwissADME, DataWarrior[97] and Molinspiration Chemoinformatics (https://www.molinspiration.com/) may also be used for VS.

- Among VS, quantitative structure–activity relationship (QSAR) is a powerful computational technique that is used to study the relationship between the chemical structure of a compound and its biological activity. The presence or absence of certain functional groups or the attachment of a functional group at a particular position can alter the biological activity of a compound.[98] QSAR models can be created, validated, and subsequently be used to predict the inhibitory concentration ($IC_{50}$) value of compounds against target proteins or specific pathogens.[99] QSAR models are mathematical equations that are built using molecular descriptors (i.e., properties of a compound such as lipophilicity, polar surface area, etc.). With the help of machine learning techniques, the molecular descriptors of a compound are correlated with its biological activity using the developed and validated QSAR models. In this way, the biological activity of novel compounds can also be tested. The ultimate aim of QSAR is not to replace in vitro or in vivo studies, but rather, to rationalize the process of compound selection that will be tested in vitro or in vivo.[100]

#### *4.3.2.2 PHARMACOPHORE MODELING*

According to the International Union of Pure and Applied Chemistry, the pharmacophore model is defined as "an ensemble of steric and electronic features that is necessary to ensure the optimal supramolecular interactions with a specific biological target and to trigger (or block) its biological response."[101] Pharmacophore modeling is a powerful computational technique that is used for virtual screening, de novo drug design, lead optimization, multi-target drug design, activity profiling, and target identification.[102] The electronic and steric features of a compound that are most responsible for eliciting interactions and biological activity are identified using pharmacophore modeling.[103] When a ligand structure and target protein are not known, pharmacophore modeling is impossible. When a ligand structure is known but the target protein is not known, ligand-based pharmacophore modeling can be carried out. When a ligand structure is not known but the target protein is known, structure-based pharmacophore modeling can be carried out.[104] Pharmacophore modeling can be used with other computational techniques such as MDSS and QSAR.[104,105] Similar to some VS techniques that are used to identify nontoxic compounds with desirable drug-like properties, pharmacophore modeling may be used to identify potential leads before they are subjected to further studies.

#### *4.3.2.3 ARTIFICIAL INTELLIGENCE*

Artificial intelligence (AI) has made significant progress in big data analysis for screening and discovery of potent active compounds for various types of diseases. Several deep learning and machine learning methods have been discussed in several reviews regarding their usefulness in drug discovery and optimization.[106,107] Simulation modeling, computation, and prediction methods that apply AI have become more powerful and accurate, with excellent accessibility. These methods have been used for repurposing existing drugs, approved natural products, and clinical trial candidates. Here are some applications of AI in the discovery of antiparasitic drugs and NTDs drugs:

- One of the first uses of the machine learning (ML) method was to troubleshoot the formulation of benznidazole to treat Chagas disease (American trypanosomiasis).[106] In this research, they used chitosan microparticles to improve its pharmacokinetic properties. Particle size, encapsulation efficiency, and dissolution rate were modeled using JST to optimize their oral absorption.
- The use of ML models was also reported by Guera et al. in which 72 compounds were tested in vitro against *Trypanosoma cruzi*, with descriptors generated by the CODES software.[108] The model was successful in predicting the activity of compounds in the test set with the standard error (SE) prediction and root-mean-square error (RMSE) of 0.17 and the value of the area under the receiving operator curve (AROC) of 0.7 (value 0.5 is random).
- Research conducted by de Souza et al. showed that the use of artificial neural networks (ANN) and kernel-based PLS (KPLS) in modeling 363 compounds as anti-*T. cruzi* has the good predictive ability by producing r square and RMSE values of 0.85 and 0.75, respectively.[109] The KPLS model was used to provide a comprehensive analysis of the

effect of molecular features on increasing or decreasing the activity of antitrypanocidal compounds.

- A good predictive model of antitrypanosomal drug activity has been developed by Kryshchyshyn et al. using stochastic gradient boosting (SGB), random forests (RF), Gaussian process regression (GP), and multivariate adaptive regression splines (MARS).[110] The RF model has the highest predictive power followed by SGB and GP, while MARS gives the worst prediction. Logit transform can improve the predictive power of the model.
- Analysis of the activity of anti-*T. cruzi* compounds. cruzi was also studied by Luchi et al. by adopting different types of molecular descriptors.[111] They used the support vector machines (SVM) method to understand the basic molecular activity by selecting 87 features. However, the SVM method tends to overestimate the data so the use of a large number of features is not recommended.[112]

In the process of drug discovery and development, computational techniques serve multiple purposes. They are used to identify compounds for web-lab testing, or are used to predict the potential molecular mechanism, or are used to optimize the properties or biological activity of lead compounds. Different computational techniques have been used to identify potential drug candidates against many parasitic diseases and NTDs (Table 4.3). Molecular docking, MD simulations, EIIP filtering, 3D-QSAR, pharmacophore modeling, virtual screening, homology modeling, MM-PBSA binding free energy calculations, and target fishing are a few examples of computational techniques that are used for evaluating compounds against parasitic diseases and NTDs.

## 4.4 FUTURE PERSPECTIVES

Many drugs have been repurposed for parasitic diseases and NTDs (Tables 4.1 and 4.2). Also, several compounds that showed potent activity against parasitic diseases and NTDs are highlighted in Table 4.3. For all the compounds that are highlighted in the paper, the majority of them have been repurposed or tested only at a preclinical level. Even at the preclinical stage, the majority had been tested at the in vitro level only. For those compounds that have the potential to be repurposed for a different disease, or those newly identified compounds that showed potential against parasitic diseases and NTDs, exhaustive in vivo studies are recommended so that these potential

**TABLE 4.3** Computational Techniques Used for the Identification and Development of Potential Drug Candidates Against Parasitic Diseases and NTDs.

| Disease | Compound | Possible mechanism of action | Computational technique involved | Stage | Reference |
|---|---|---|---|---|---|
| Malaria | 1-(heteroaryl)-2-((5-nitroheteroaryl)methylene) hydrazine derivatives | Inhibition of plasmodium falciparum lactate dehydrogenase | Molecular docking | Preclinical; in vivo; Peter's test, Compound 2, 99.09% parasitemia inhibition at 125 mg/kg/day | [113] |
| | 18β-glycyrrhetinic acid | Inhibition of plasmodium falciparum lactate dehydrogenase | Molecular docking | Preclinical; in vitro, $IC_{50}$ = 1.69 µg/mL; in vivo, 68–100% suppression at 62.5–250 mg/kg | [114] |
| Dengue | Quercetin | Inhibition of NS5 protein | Molecular docking | Preclinical; in vitro; $EC_{50}$ = 19.2 µg/mL | [115] |
| | Gallic acid | Inhibition of NS5 protein | Molecular docking | Preclinical; in vitro; $EC_{50}$ = 25.8 µg/mL | [115] |
| Chikungunya | Aminopiperidine analogue (compound 11) | Inhibition of envelope protein | Molecular dynamics simulations | Preclinical; in vitro; $EC_{50}$ = 1.6 µM | [116] |
| | 1,3-thiazolodin-4-one derivatives (compound 8) | Inhibition of nsP2 | Molecular docking | Preclinical; in vitro; $EC_{50}$ = 1.5 µM | [117] |
| Leishmaniasis | *N'-[(5-Bromo-2-thiophenyl)methylene] thiophene-2-carbohydrazide* | Inhibition of arginase (*L. donovani*) | EIIP filtering, 3D-QSAR, molecular docking | Preclinical; in vitro; $IC_{50}$ = 2.18 µM | [118] |

**TABLE 4.3** *(Continued)*

| Disease | Compound | Possible mechanism of action | Computational technique involved | Stage | Reference |
|---|---|---|---|---|---|
| | 5-acetyl-6-((4-(5-(4-hydroxyphenyl)-4,5-dihydro-1H-pyrazol-3-yl) phenylamino)methyl)-1-methyl35 4-phenyl-3,4-dihydropyrimidin-2(1H)-one | Inhibition of pteridine reductase 1 (*L. donovani*) | Molecular docking | Preclinical; in vitro; $IC_{50}$ = 1.5 µM | [119] |
| Zika virus | Asunaprevir | Inhibition of NS2B/NS3 protease | Pharmacophore anchor model, molecular docking | Preclinical; in vitro; enzymatic assay; $IC_{50}$ = 6.0 µM | [120] |
| | Simeprevir | Inhibition of NS2B/NS3 protease | Pharmacophore anchor model, molecular docking | Preclinical; in vitro; enzymatic assay; $IC_{50}$ = 2.6 µM | [120] |
| Buruli ulcer | Euscaphic acid | Potential inhibitor of isocitrate lyase | Molecular docking, virtual screening | Preclinical; in silico; −8.6 kcal/mol | [121] |
| | ZINC95485880 | Potential inhibitor of isocitrate lyase | Molecular docking, virtual screening | Preclinical; in silico; −8.6 kcal/mol | [121] |
| Chagas disease | 28SMB032 | Inhibition of DNA | Homology modeling, molecular docking | Preclinical; in vitro, $IC_{50}$ = 0.54 µM; in vivo, 40% parasitemia reduction, 0.1 mg/kg for 5 days, intraperitoneal | [122] |

**TABLE 4.3** *(Continued)*

| Disease | Compound | Possible mechanism of action | Computational technique involved | Stage | Reference |
|---|---|---|---|---|---|
| | 4-[4-(2-Chloro-benzyloxy)-phenyl]-3,6-dimethyl-4,8-dihydro-1H-pyrazolo[3,4-e][1,4]thiazepin-7-one | Inhibition of CYP51TC | Molecular docking | Preclinical; in vitro, intracellular strain ($EC_{50}$ = 3.86 μM), bloodstream strain ($EC_{50}$ = 4.00 μM); in vivo, 43% parasitaemia peak reduction | [123] |
| Fascioliasis | HTS12701 | Inhibition of cathepsin L3 | Virtual screening, molecular docking, MD simulations, MM-PBSA binding free energy calculations | Preclinical; in silico; −10.68 kcal/mol | [124] |
| | BTB03219 | Inhibition of cathepsin L3 | Virtual screening, molecular docking, MD simulations, MM-PBSA binding free energy calculations | Preclinical; in silico; −7.16 kcal/mol | [124] |
| Human African Trypanosomiasis | 2,3,4,5-tetrahydrobenzo[F][1,4]oxazepines derivatives (compound 7a) | Inhibition of peroxisomal import matrix 14 | Molecular docking, pharmacophore modelling | Preclinical; in vitro; $IC_{50}$ = 4.0 μM | [125] |
| Leprosy | Neobavaisoflavone | Inhibition of dihydropteroate synthase | Molecular docking | Preclinical; in silico | [126] |

**TABLE 4.3** *(Continued)*

| Disease | Compound | Possible mechanism of action | Computational technique involved | Stage | Reference |
|---|---|---|---|---|---|
| | Dapsone-thymol conjugate | Inhibition of dihydropteroate synthase | Molecular docking, MD simulations, MM-PBSA binding free energy calculations | Preclinical; in vivo; Bacilli cell count before treatment $= 5 \times 10^5$; Bacilli cell count after treatment $= 2.6 \times 10^4$ | [127] |
| Lymphatic filariasis | Ursolic acid | Inhibition of glutathione-s-transferase | Molecular docking | Preclinical; in vitro, $IC_{50} = 8.84$ µM; in vivo, 54% macrofilaricidal activity, 56% female worm sterility | [128] |
| | Moxidectin | Inhibition of glutamate-gated chloride channels | Molecular docking | Preclinical; in vitro, $IC_{50} =$ 0.242 µM (female parasite), 0.186 µM (male worm), 0.813 µM (microfilaria); in vivo, 49% macrofilaricidal activity, 54% female worm sterility | [129] |
| Schistosomiasis | GPQF-108 | Inhibition of carbonic anhydrase and arginase | Target fishing, molecular docking | Preclinical; in vitro, $IC_{50} =$ 29.4 µM; in vivo, 400 mg/kg single dose, 54% reduction in total worm burden, 38.72% reduction in number of immature eggs, 56.38% reduction of eggs | [130] |

**TABLE 4.3** *(Continued)*

| Disease | Compound | Possible mechanism of action | Computational technique involved | Stage | Reference |
|---|---|---|---|---|---|
| | (E)-2'-hydroxy-4-methoxychalcone | Inhibition of ATP diphosphohydrolase | Molecular docking | Preclinical; in vitro enzymatic assay (ATPase), $IC_{50}$ = 30.62 μM; in vivo, 32.8% reduction in total worm burden | [131] |
| Mycetoma | Olorofim (Phase II in clinical trials for the treatment of invasive fungal infections) | Inhibition of dihydroorotate dehydrogenase | Molecular docking | Preclinical; in vitro, MIC = 0.004–0.125 mg/L | [132] |

drug candidates may have sufficient safety and efficacy data to enter clinical trials. The possible mechanisms by which drug repurposing and computational approaches may help solve the problems that are associated with parasitic diseases and NTDs are given in Figure 4.3.

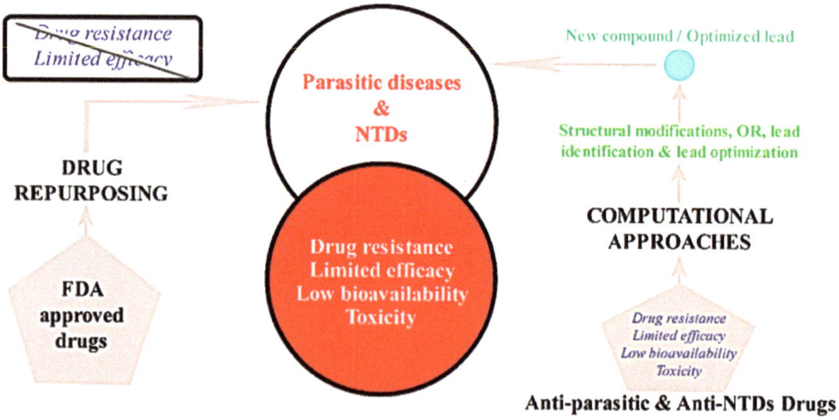


**FIGURE 4.3** Possible mechanisms by which problems associated with parasitic diseases and NTDs may be solved with drug repurposing and computational approaches.

## 4.5 CONCLUSION

Several existing FDA-approved drugs showed remarkable potential to be repurposed for the treatment of parasitic diseases and NTDs. Different computational techniques such as virtual screening, 3D-QSAR, homology modeling, molecular docking, MD simulations, and target fishing have played a key role in the identification of new compounds for the treatment of parasitic diseases and NTDs.

## KEYWORDS

- **computer-aided drug design**
- **drug repurposing**
- **neglected tropical diseases**
- **parasitic diseases**

## REFERENCES

1. CDC. Parasites [Online]. https://www.cdc.gov/parasites/az/index.html (accessed Oct 6, 2021).
2. WHO. Vector-Borne and Parasitic Diseases [Online]. https://www.euro.who.int/en/health-topics/communicable-diseases/vector-borne-and-parasitic-diseases (accessed Oct 6, 2021).
3. Engels, D.; Zhou, X.-N. Neglected Tropical Diseases: An Effective Global Response to Local Poverty-Related Disease Priorities. *Infect. Dis. Poverty* **2020,** *9*, 10.
4. CDC. Neglected Tropical Diseases (NTDs) [Online]. https://www.cdc.gov/globalhealth/ntd/diseases/index.html (accessed Oct 6, 2021).
5. Molyneux, D. H.; Savioli, L.; Engels, D. Neglected Tropical Diseases: Progress Towards Addressing the Chronic Pandemic. *Lancet* **2017,** *389*, 312–325.
6. Cortez-Maya, S.; Moreno-Herrera, A.; Palos, I.; Rivera, G. Old Antiprotozoal Drugs: Are They Still Viable Options for Parasitic Infections or New Options for Other Diseases? *Curr. Med. Chem.* **2020,** *27*, 5403–5428.
7. Scotti, L.; Filho, F. J. B. M.; de Moura, R. O.; Ribeiro, F. F.; Ishiki, H.; da Silva, M. S.; Filho, J. M. B.; Scotti, M. T. Multi-Target Drugs for Neglected Diseases. *Curr. Pharm. Des.* **2016,** *22*, 3135–3163.
8. Pink, R.; Hudson, A.; Mouriès, M. A.; Bendig, M. Opportunities and Challenges in Antiparasitic Drug Discovery. *Nat. Rev. Drug Discov.* **2005,** *4*, 727–740.
9. Capela, R.; Moreira, R.; Lopes, F. An Overview of Drug Resistance in Protozoal Diseases. *Int. J. Mol. Sci.* **2019,** *20*, 5748.
10. Christaki, E.; Marcou, M.; Tofarides, A. Antimicrobial Resistance in Bacteria: Mechanisms, Evolution, and Persistence. *J. Mol. Evol.* **2020,** *88*, 26–40.
11. Lampejo, T. Influenza and Antiviral Resistance: An Overview. *Eur. J. Clin. Microbiol. Infect. Dis.* **2020,** *39*, 1201–1208.
12. Pramanik, P. K.; Alam, M. N.; Roy Chowdhury, D.; Chakraborti, T. Drug Resistance in Protozoan Parasites: An Incessant Wrestle for Survival. *J. Glob. Antimicrob. Resist.* **2019,** *18*, 1–11.
13. Xue, H.; Li, J.; Xie, H.; Wang, Y. Review of Drug Repositioning Approaches and Resources. *Int. J. Biol. Sci.* **2018,** *14*, 1232–1244.
14. Parvathaneni, V.; Kulkarni, N. S.; Muth, A.; Gupta, V. Drug Repurposing: A Promising Tool to Accelerate the Drug Discovery Process. *Drug Discov. Today* **2019,** *24*, 2076–2085.
15. Rudrapal, M.; Khairnar, S.; Jadhav, A. Drug Repurposing (DR): An Emerging Approach in Drug Discovery. In *Drug Repurposing - Hypothesis, Molecular Aspects and Therapeutic Applications*; IntechOpen, 2020.
16. Sultana, J.; Crisafulli, S.; Gabbay, F.; Lynn, E.; Shakir, S.; Trifirò, G. Challenges for Drug Repurposing in the COVID-19 Pandemic Era. *Front. Pharmacol.* **2020**. DOI: 10.3389/fphar.2020.588654.
17. Sliwoski, G.; Kothiwale, S.; Meiler, J.; Lowe, E. W. Computational Methods in Drug Discovery. *Pharmacol. Rev.* **2014,** *66*, 334–395.
18. Daina, A.; Michielin, O.; Zoete, V. SwissADME: A Free Web Tool to Evaluate Pharmacokinetics, Drug-Likeness and Medicinal Chemistry Friendliness of Small Molecules. *Sci. Rep.* **2017,** *7*, 42717.

19. Banerjee, P.; Eckert, A. O.; Schrey, A. K.; Preissner, R. ProTox-II: A Webserver for the Prediction of Toxicity of Chemicals. *Nucleic Acids Res.* **2018,** *46*, W257–W263.
20. Vijayakrishnan, R. Structure-Based Drug Design and Modern Medicine. *J. Postgrad. Med.* **2009,** *55*, 301.
21. Talele, T.; Khedkar, S.; Rigby, A. Successful Applications of Computer Aided Drug Discovery: Moving Drugs from Concept to the Clinic. *Curr. Top. Med. Chem.* **2010,** *10*, 127–141.
22. Van Drie, J. H. Computer-Aided Drug Design: The Next 20 Years. *J. Comput. Aided Mol. Des.* **2007,** *21*, 591–601.
23. Hartman, G. D.; Egbertson, M. S.; Halczenko, W.; Laswell, W. L.; Duggan, M. E.; Smith, R. L.; Naylor, A. M.; Manno, P. D.; Lynch, R. J. Non-Peptide Fibrinogen Receptor Antagonists. 1. Discovery and Design of Exosite Inhibitors. *J. Med. Chem.* **1992,** *35*, 4640–4642.
24. Rajguru, T.; Bora, D.; Modi, M. K. Combined CADD and Virtual Screening to Identify Novel Nonpeptidic Falcipain-2 Inhibitors. *Curr. Comput. Aided Drug Des.* **2021,** *17*, 579–588.
25. Maddipati, V. C.; Mittal, L.; Mantipally, M.; Asthana, S.; Bhattacharyya, S.; Gundla, R. A Review on the Progress and Prospects of Dengue Drug Discovery Targeting NS5 RNA- Dependent RNA Polymerase. *Curr. Pharm. Des.* **2020,** *26*, 4386–4409.
26. Bowen, L. R.; Li, D. J.; Nola, D. T.; Anderson, M. O.; Heying, M.; Groves, A. T.; Eagon, S. Identification of Potential Zika Virus NS2B-NS3 Protease Inhibitors via Docking, Molecular Dynamics and Consensus Scoring-Based Virtual Screening. *J. Mol. Model* **2019,** *25*, 194.
27. Nguyen, P. T. V.; Yu, H.; Keller, P. A. Molecular Docking Studies to Explore Potential Binding Pockets and Inhibitors for Chikungunya Virus Envelope Glycoproteins. *Interdiscip. Sci. Comput. Life Sci.* **2018,** *10*, 515–524.
28. Sawyer, J. S.; Anderson, B. D.; Beight, D. W.; Campbell, R. M.; Jones, M. L.; Herron, D. K.; Lampe, J. W.; McCowan, J. R.; McMillen, W. T.; Mort, N.; Parsons, S.; Smith, E. C.; Vieth, M.; Weir, L. C.; Yan, L.; Zhang, F.; Yingling, J. M. Synthesis and Activity of New Aryl- and Heteroaryl-Substituted Pyrazole Inhibitors of the Transforming Growth Factor-β Type I Receptor Kinase Domain. *J. Med. Chem.* **2003,** *46*, 3953–3956.
29. Koch, U.; Hamacher, M.; Nussbaumer, P. Cheminformatics at the Interface of Medicinal Chemistry and Proteomics. *Biochim. Biophys. Acta Proteins Proteom.* **2014,** *1844*, 156–161.
30. Chong, C. R.; Chen, X.; Shi, L.; Liu, J. O.; Sullivan, D. J. A Clinical Drug Library Screen Identifies Astemizole as an Antimalarial Agent. *Nat. Chem. Biol.* **2006,** *2*, 415–416.
31. Yadav, K.; Shivahare, R.; Shaham, S. H.; Joshi, P.; Sharma, A.; Tripathi, R. Repurposing of Existing Therapeutics to Combat Drug-Resistant Malaria. *Biomed. Pharmacother.* **2021,** *136*, 111275.
32. Simanjuntak, Y.; Liang, J.-J.; Lee, Y.-L.; Lin, Y.-L. Repurposing of Prochlorperazine for Use Against Dengue Virus Infection. *J. Infect. Dis.* **2015,** *211*, 394–404.
33. Malakar, S.; Sreelatha, L.; Dechtawewat, T.; Noisakran, S.; Yenchitsomanus, P.; Chu, J. J. H.; Limjindaporn, T. Drug Repurposing of Quinine as Antiviral Against Dengue Virus Infection. *Virus Res.* **2018,** *255*, 171–178.
34. Leela, S. L.; Srisawat, C.; Sreekanth, G. P.; Noisakran, S.; Yenchitsomanus, P.; Limjindaporn, T. Drug Repurposing of Minocycline Against Dengue Virus Infection. *Biochem. Biophys. Res. Commun.* **2016,** *478*, 410–416.

35. Shen, T.-J.; Hanh, V. T.; Nguyen, T. Q.; Jhan, M.-K.; Ho, M.-R.; Lin, C.-F. Repurposing the Antiemetic Metoclopramide as an Antiviral Against Dengue Virus Infection in Neuronal Cells. *Front Cell Infect. Microbiol.* **2021,** *10.*
36. Pereira, A. K. dos S.; Santos, I. A.; da Silva, W. W.; Nogueira, F. A. R.; Bergamini, F. R. G.; Jardim, A. C. G.; Corbi, P. P. Memantine Hydrochloride: A Drug to Be Repurposed Against Chikungunya Virus? *Pharmacol. Rep.* **2021,** *73,* 954–961.
37. Tripathi, P. K.; Soni, A.; Singh Yadav, S. P.; Kumar, A.; Gaurav, N.; Raghavendhar, S.; Sharma, P.; Sunil, S.; Ashish; Jayaram, B.; Patel, A. K. Evaluation of Novobiocin and Telmisartan for Anti-CHIKV Activity. *Virology* **2020,** *548,* 250–260.
38. Albulescu, I. C.; White-Scholten, L.; Tas, A.; Hoornweg, T. E.; Ferla, S.; Kovacikova, K.; Smit, J. M.; Brancale, A.; Snijder, E. J.; van Hemert, M. J. Suramin Inhibits Chikungunya Virus Replication by Interacting With Virions and Blocking the Early Steps of Infection. *Viruses* **2020,** *12,* 314.
39. Braga, S. S. Multi-Target Drugs Active Against Leishmaniasis: A Paradigm of Drug Repurposing. *Eur. J. Med. Chem.* **2019,** *183,* 111660.
40. Devillers, J. Repurposing Drugs for Use Against Zika Virus Infection. *SAR QSAR Environ. Res.* **2018,** *29,* 103–115.
41. Panic, G.; Duthaler, U.; Speich, B.; Keiser, J. Repurposing Drugs for the Treatment and Control of Helminth Infections. *Int. J. Parasitol. Drugs Drug Resist.* **2014,** *4,* 185–200.
42. Goldstein, B. P. Resistance to Rifampicin: A Review. *J. Antibiot. (Tokyo)* **2014,** *67,* 625–630.
43. Etuaful, S.; Carbonnelle, B.; Grosset, J.; Lucas, S.; Horsfield, C.; Phillips, R.; Evans, M.; Ofori-Adjei, D.; Klustse, E.; Owusu-Boateng, J.; Amedofu, G. K.; Awuah, P.; Ampadu, E.; Amofah, G.; Asiedu, K.; Wansbrough-Jones, M. Efficacy of the Combination Rifampin-Streptomycin in Preventing Growth of Mycobacterium Ulcerans in Early Lesions of Buruli Ulcer in Humans. *Antimicrob. Agents Chemother.* **2005,** *49,* 3182–3186.
44. Portaels, F.; Traore, H.; De Ridder, K.; Meyers, W. M. In Vitro Susceptibility of Mycobacterium Ulcerans to Clarithromycin. *Antimicrob. Agents Chemother.* **1998,** *42,* 2070–2073.
45. Thangaraj, H. S.; Adjei, O.; Allen, B. W.; Portaels, F.; Evans, M. R. W.; Banerjee, D. K.; Wansbrough-Jones, M. H. In Vitro Activity of Ciprofloxacin, Sparfloxacin, Ofloxacin, Amikacin and Rifampicin Against Ghanaian Isolates of Mycobacterium Ulcerans. *J. Antimicrob. Chemother.* **2000,** *45,* 231–233.
46. Yano, T.; Kassovska-Bratinova, S.; Teh, J. S.; Winkler, J.; Sullivan, K.; Isaacs, A.; Schechter, N. M.; Rubin, H. Reduction of Clofazimine by Mycobacterial Type 2 NADH:Quinone Oxidoreductase. *J. Biol. Chem.* **2011,** *286,* 10276–10287.
47. Rivarola, H. W.; Fernández, A. R.; Enders, J. E.; Fretes, R.; Gea, S.; Paglini-Oliva, P. Effects of Clomipramine on Trypanosoma Cruzi Infection in Mice. *Trans. R Soc. Trop. Med. Hyg.* **2001,** *95,* 529–533.
48. Gutierrez-Correa, J.; Fairlamb, A. H.; Stoppani, A. O. M. Trypanosoma Cruzi Trypanothione Reductase is Inactivated by Peroxidase-Generated Phenothiazine Cationic Radicals. *Free Radic. Res.* **2001,** *34,* 363–378.
49. Goad, L. J.; Berens, R. L.; Marr, J. J.; Beach, D. H.; Holz, G. G. The Activity of Ketoconazole and Other Azoles Against Trypanosoma Cruzi: Biochemistry and Chemotherapeutic Action in Vitro. *Mol. Biochem. Parasitol.* **1989,** *32,* 179–189.
50. O'Neill, J. F.; Johnston, R. C.; Halferty, L.; Hanna, R. E. B.; Brennan, G. P.; Fairweather, I. Disruption of Spermatogenesis in the Liver Fluke, Fasciola Hepatica by Two Artemisinin Derivatives, Artemether and Artesunate. *J. Helminthol.* **2017,** *91,* 55–71.

51. Dawes, B. Some Apparent Effects of Bithionol ('Actamer') on Fasciola Hepatica. *Nature* **1966,** *209*, 424–425.
52. Mera y Sierra, R.; Agramunt, V. H.; Cuervo, P.; Mas-Coma, S. Human Fascioliasis in Argentina: Retrospective Overview, Critical Analysis and Baseline for Future Research. *Parasit Vectors*. **2011,** *4*, 104.
53. Park, S.-K.; Friedrich, L.; Yahya, N.; Rohr, C.; Maillard, D.; Rippmann, F.; Spangenberg, T.; Marchant, J. S. Mechanism of Praziquantel Action at a Transient Receptor Potential Channel. *Biophys. J.* **2021,** *120*, 336a.
54. Mansour-Ghanaei, F.; Shafaghi, A.; Fallah, M. The Effect of Metronidazole in Treating Human Fascioliasis. *Med. Sci. Monit.* **2003,** *9*, 127–130.
55. Fairlamb, A. H. Chemotherapy of Human African Trypanosomiasis: Current and Future Prospects. *Trends Parasitol.* **2003,** *19*, 488–494.
56. Docampo, R.; Moreno, S. N. J.; Stoppani, A. O. M.; Leon, W.; Cruz, F. S.; Villalta, F.; Muniz, R. F. A. Mechanism of Nifurtimox Toxicity in Different Forms of Trypanosoma Cruzi. *Biochem. Pharmacol.* **1981,** *30*, 1947–1951.
57. Pohlig, G.; Bernhard, S. C.; Blum, J.; Burri, C.; Mpanya, A.; Lubaki, J.-P. F.; Mpoto, A. M.; Munungu, B. F.; N'tombe, P. M.; Deo, G. K.;Mutantu, P. N.; Kuikumbi, F. M.; Mintwo, A. F.; Munungi, A. K.; Dala, A.; Macharia, S.; Bilenge, C. M.; Mesu, V. K.; Franco, J. R.; Dituvanga, N. D.; Tidwell, R. R.; Olson, C. A. Efficacy and Safety of Pafuramidine Versus Pentamidine Maleate for Treatment of First Stage Sleeping Sickness in a Randomized, Comparator-Controlled, International Phase 3 Clinical Trial. *PLoS Negl. Trop. Dis.* **2016,** *10*, e0004363.
58. Bah, G. S.; Ward, E. L.; Srivastava, A.; Trees, A. J.; Tanya, V. N.; Makepeace, B. L.; Efficacy of Three-Week Oxytetracycline or Rifampin Monotherapy Compared With a Combination Regimen Against the Filarial Nematode Onchocerca Ochengi. *Antimicrob. Agents Chemother.* **2014,** *58*, 801–810.
59. Wolstenholme, A. J.; Maclean, M. J.; Coates, R.; McCoy, C. J.; Reaves, B. J. How Do the Macrocyclic Lactones Kill Filarial Nematode Larvae? *Invertebr. Neurosci.* **2016,** *16*, 7.
60. Milton, P.; Hamley, J. I. D.; Walker, M.; Basáñez, M.-G. Moxidectin: An Oral Treatment for Human Onchocerciasis. *Expert Rev. Anti. Infect. Ther.* **2020,** *18*, 1067–1081.
61. Krücken, J.; Holden-Dye, L.; Keiser, J.; Prichard, R. K.; Townson, S.; Makepeace, B. L.; Hübner, M. P.; Hahnel, S. R.; Scandale, I.; Harder, A.; Kulke, D. Development of Emodepside as a Possible Adulticidal Treatment for Human Onchocerciasis—The Fruit of a Successful Industrial–Academic Collaboration. *PLOS Pathog.* **2021,** *17*, e1009682.
62. Rodulfo, S.; Convit, J.; Bartholomew, R.; Eberhard, M. L.; Cline, B. L.; Mather, F. J.; Hernandez, J. L.; Welborn, C. A.; De Maza, S. N. Albendazole in the Treatment of Onchocerciasis: Double-Blind Clinical Trial in Venezuela. *Am. J. Trop. Med. Hyg.* **1992,** *47*, 512–520.
63. Cambau, E.; Williams, D. L. Anti-Leprosy Drugs: Modes of Action and Mechanisms of Resistance in Mycobacterium Leprae. In *The International Textbook of Leprosy*, 2019; vol 7, pp 1–33.
64. Demet Akpolat, N.; Akkus, A.; Kaynak, E. An Update on the Epidemiology, Diagnosis and Treatment of Leprosy. In *Hansen's Disease - The Forgotten and Neglected Disease*; IntechOpen, 2019.
65. Tadesse, A.; Shannon, E. J. Effects of Thalidomide on Intracellular Mycobacterium Leprae in Normal and Activated Macrophages. *Clin. Vaccine Immunol.* **2005,** *12*, 130–134.

66. Rao, R. U.; Huang, Y.; Fischer, K.; Fischer, P. U.; Weil, G. J. Brugia Malayi: Effects of Nitazoxanide and Tizoxanide on Adult Worms and Microfilariae of Filarial Nematodes. *Exp. Parasitol.* **2009,** *121*, 38–45.
67. Mand, S.; Debrah, A. Y.; Klarmann, U.; Batsa, L.; Marfo-Debrekyei, Y.; Kwarteng, A.; Specht, S.; Belda-Domene, A.; Fimmers, R.; Taylor, M.; Adjei, O.; Hoerauf, A. Doxycycline Improves Filarial Lymphedema Independent of Active Filarial Infection: A Randomized Controlled Trial. *Clin. Infect. Dis.* **2012,** *55*, 621–630.
68. Bakowski, M. A.; McNamara, C. W. Advances in Antiwolbachial Drug Discovery for Treatment of Parasitic Filarial Worm Infections. *Trop. Med. Infect. Dis.* **2019,** *4*, 108.
69. Gunter, K. K.; Miller, L. M.; Aschner, M.; Eliseev, R.; Depuis, D.; Gavin, C. E.; Gunter, T. E. XANES Spectroscopy: A Promising Tool for Toxicology. *Neurotoxicology* **2002,** *23*, 127–146.
70. Utzinger, J.; Shuhua, X.; N'Goran, E. K.; Bergquist, R.; Tanner, M. The Potential of Artemether for the Control of Schistosomiasis. *Int. J. Parasitol.* **2001,** *31*, 1549–1562.
71. Abdin, A. A.; Ashour, D. S.; Shoheib, Z. S. Artesunate Effect on Schistosome Thioredoxin Glutathione Reductase and Cytochrome c Peroxidase as New Molecular Targets in Schistosoma Mansoni-Infected Mice. *Biomed. Environ. Sci.* **2013,** *26*, 953–961.
72. Keiser, J.; Chollet, J.; Xiao, S. H.; Mei, J. Y.; Jiao, P. Y.; Utzinger, J.; Tanner, M. Mefloquine—An Aminoalcohol With Promising Antischistosomal Properties in Mice. *PLoS Negl. Trop. Dis.* **2009,** *3*, e350.
73. Mossallam, S. F.; Amer, E. I.; El-Faham, M. H.; Efficacy of Synriam™, a New Antimalarial Combination of OZ277 and Piperaquine, Against Different Developmental Stages of Schistosoma Mansoni. *Acta Trop.* **2015,** *143*, 36–46.
74. Gouveia, M.; Brindley, P.; Gärtner, F.; Costa, J.; Vale, N. Drug Repurposing for Schistosomiasis: Combinations of Drugs or Biomolecules. *Pharmaceuticals* **2018,** *11*, 15.
75. Elkheir, L. Y. M.; Haroun, R.; Mohamed, M. A.; Fahal, A. H. Madurella Mycetomatis Causing Eumycetoma Medical Treatment: The Challenges and Prospects. *PLoS Negl. Trop. Dis.* **2020,** *14*, e0008307.
76. Nitidandhaprabhas, P. Treatment of Nocardial Mycetoma With Trimethoprim and Sulfamethoxazole. *Arch. Dermatol.* **1975,** *111*, 1345.
77. Patra, S.; Senthilnathan, G.; Ramam, M.; Arava, S.; Bhari, N. Linezolid: A Novel Treatment Option for the Treatment of a Non-Responsive Case of Actinomycotic Mycetoma. *Indian J. Dermatol. Venereol. Leprol.* **2021,** *87*, 473–475.
78. Hassan Baig, M.; Ahmad, K.; Roy, S.; Mohammad Ashraf, J.; Adil, M.; Haris Siddiqui, M.; Khan, S.; Amjad Kamal, M.; Provazník, I.; Choi, I. Computer Aided Drug Design: Success and Limitations. *Curr. Pharm. Des.* **2016,** *22*, 572–581.
79. Kuntz, I. D.; Blaney, J. M.; Oatley, S. J.; Langridge, R.; Ferrin, T. E. A Geometric Approach to Macromolecule-Ligand Interactions. *J. Mol. Biol.* **1982,** *161*, 269–288.
80. Thillainayagam, M.; Ramaiah, S.; Anbarasu, A. Molecular Docking and Dynamics Studies on Novel Benzene Sulfonamide Substituted Pyrazole-Pyrazoline Analogues as Potent Inhibitors of Plasmodium Falciparum Histo Aspartic Protease. *J. Biomol. Struct. Dyn.* **2020,** *38*, 3235–3245.
81. Ibrahim, M. A. A.; Abdelrahman, A. H. M.; Hassan, A. M. A. Identification of Novel Plasmodium Falciparum PI4KB Inhibitors as Potential Anti-Malarial Drugs: Homology Modeling, Molecular Docking and Molecular Dynamics Simulations. *Comput. Biol. Chem.* **2019,** *80*, 79–89.

82. Rout, S.; Mahapatra, R. K. In Silico Analysis of Plasmodium Falciparum CDPK5 Protein Through Molecular Modeling, Docking and Dynamics. *J. Theor. Biol.* **2019,** *461*, 254–267.
83. Zothantluanga, J. H.; Gogoi, N.; Shakya, A.; Chetia, D.; Lalthanzara, H. Computational Guided Identification of Potential Leads From Acacia Pennata (L.) Willd. as Inhibitors for Cellular Entry and Viral Replication of SARS-CoV-2. *Futur. J. Pharm. Sci.* **2021,** *7*, 201.
84. Zothantluanga, J. H. Molecular Docking Simulation Studies, Toxicity Study, Bioactivity Prediction, and Structure-Activity Relationship Reveals Rutin as a Potential Inhibitor of SARS-CoV-2 3CL Pro. *J. Sci. Res.* **2021,** *65*, 96–104.
85. da Silva, J. K. R.; Figueiredo, P. L. B.; Byler, K. G.; Setzer, W. N. Essential Oils as Antiviral Agents, Potential of Essential Oils to Treat SARS-CoV-2 Infection: An In-Silico Investigation. *Int. J. Mol. Sci.* **2020,** *21*, 3426.
86. Ghosh, R.; Chakraborty, A.; Biswas, A.; Chowdhuri, S. Evaluation of Green Tea Polyphenols as Novel Corona Virus (SARS CoV-2) Main Protease (Mpro) Inhibitors – an in Silico Docking and Molecular Dynamics Simulation Study. *J. Biomol. Struct. Dyn.* **2021,** *39*, 4362–4374.
87. Ferreira, L.; dos Santos, R.; Oliva, G.; Andricopulo, A. Molecular Docking and Structure-Based Drug Design Strategies. *Molecules* **2015,** *20*, 13384–13421.
88. Arwansyah, A.; Arif, A. R.; Ramli, I.; Kurniawan, I.; Sukarti, S.; Nur Alam, M.; Illing, I.; Farid Lewa, A.; Manguntungi, B. Molecular Modelling on SARS-CoV-2 Papain-Like Protease: An Integrated Study With Homology Modelling, Molecular Docking, and Molecular Dynamics Simulations. *SAR QSAR Environ. Res.* **2021,** *32*, 699–718.
89. Khelfaoui, H.; Harkati, D.; Saleh, B. A. Molecular Docking, Molecular Dynamics Simulations and Reactivity, Studies on Approved Drugs Library Targeting ACE2 and SARS-CoV-2 Binding With ACE2. *J. Biomol. Struct. Dyn.* **2021,** *39*, 7246–7262.
90. Vora, J.; Patel, S.; Athar, M.; Sinha, S.; Chhabria, M. T.; Jha, P. C.; Shrivastava, N. Pharmacophore Modeling, Molecular Docking and Molecular Dynamics Simulation for Screening and Identifying Anti-Dengue Phytocompounds. *J. Biomol. Struct. Dyn.* **2019,** 1–15.
91. Prajapati, J.; Patel, R.; Goswami, D.; Saraf, M.; Rawal, R. M. Sterenin M as a Potential Inhibitor of SARS-CoV-2 Main Protease Identified From MeFSAT Database Using Molecular Docking, Molecular Dynamics Simulation and Binding Free Energy Calculation. *Comput. Biol. Med.* **2021,** *135*, 104568.
92. Nayab, R. S.; Maddila, S.; Krishna, M. P.; Titinchi, S. J. J.; Thaslim, B. S.; Chintha, V.; Wudayagiri, R.; Nagam, V.; Tartte, V.; Chinnam, S.; Chamarthi, N. R. In Silico Molecular Docking and in Vitro Antioxidant Activity Studies of Novel α -Aminophosphonates Bearing 6-Amino-1,3-Dimethyl Uracil. *J. Recept. Signal Transduct.* **2020,** *40*, 166–172.
93. Ugwu, D. I.; Okoro, U. C.; Ukoha, P. O.; Okafor, S.; Ibezim, A.; Kumar, N. M. Synthesis, Characterization, Molecular Docking and in Vitro Antimalarial Properties of New Carboxamides Bearing Sulphonamide. *Eur. J. Med. Chem.* **2017,** *135*, 349–369.
94. Enmozhi, S. K.; Raja, K.; Sebastine, I.; Joseph, J. Andrographolide as a Potential Inhibitor of SARS-CoV-2 Main Protease: An in Silico Approach. *J. Biomol. Struct. Dyn.* **2020,** 1–7.
95. Junejo, J. A.; Zaman, K.; Rudrapal, M.; Celik, I.; Attah, E. I. Antidiabetic Bioactive Compounds From Tetrastigma Angustifolia (Roxb.) Deb and Oxalis Debilis Kunth.:

Validation of Ethnomedicinal Claim by in Vitro and in Silico Studies. *South African J. Bot.* **2021,** *143*, 164–175.
96. Schneider, G. Virtual Screening: An Endless Staircase? *Nat. Rev. Drug Discov.* **2010,** *9*, 273–276.
97. Sander, T.; Freyss, J.; von Korff, M.; Rufener, C. DataWarrior: An Open-Source Program For Chemistry Aware Data Visualization And Analysis. *J. Chem. Inf. Model* **2015,** *55*, 460–473.
98. Kwon, S.; Bae, H.; Jo, J.; Yoon, S. Comprehensive Ensemble in QSAR Prediction for Drug Discovery. *BMC Bioinform.* **2019,** *20*, 521.
99. Faidallah, H. M.; Panda, S. S.; Serrano, J. C.; Girgis, A. S.; Khan, K. A.; Alamry, K. A.; Therathanakorn, T.; Meyers, M. J.; Sverdrup, F. M.; Eickhoff, C. S.; Getchell, S. G.; Katritzky, A. R. Synthesis, Antimalarial Properties and 2D-QSAR Studies of Novel Triazole-Quinine Conjugates. *Bioorg. Med. Chem.* **2016,** *24*, 3527–3539.
100. Neves, B. J.; Braga, R. C.; Melo-Filho, C. C.; Moreira-Filho, J. T.; Muratov, E. N.; Andrade, C. H. QSAR-Based Virtual Screening: Advances and Applications in Drug Discovery. *Front. Pharmacol.* **2018.** DOI: 10.3389/fphar.2018.01275
101. Daisy, P.; Singh, S. K.; Vijayalakshmi, P.; Selvaraj, C.; Rajalakshmi, M.; Suveena, S. A Database for the Predicted Pharmacophoric Features of Medicinal Compounds. *Bioinformation* **2011,** *6*, 167–168.
102. Khedkar, S.; Malde, A.; Coutinho, E.; Srivastava, S. Pharmacophore Modeling in Drug Discovery and Development: An Overview. *Med. Chem. (Los. Angeles)* **2007,** *3*, 187–197.
103. Yang, S.-Y. Pharmacophore Modeling and Applications in Drug Discovery: Challenges and Recent Advances. *Drug Discov. Today* **2010,** *15*, 444–450.
104. Voet, A.; Qing, X.; Lee, X. Y.; De Raeymaecker, J.; Tame, J.; Zhang, K.; De Maeyer, M. Pharmacophore Modeling: Advances, Limitations, and Current Utility in Drug Discovery. *J. Receptor Ligand Channel Res.* **2014,** *2014* (7), 81–92.
105. Geethaavacini, G.; Poh, G. P.; Yan, L. Y.; Deepashini, R.; Shalini, S.; Harish, R.; Sureshkumar, K.; Ravichandran, V. QSAR and Pharmacophore Mapping Studies on Benzothiazinimines to Relate Their Structural Features With Anti-HIV Activity. *Med. Chem.* **2018,** *14*, 733–740.
106. Leonardi, D.; Salomón, C. J.; Lamas, M. C.; Olivieri, A. C. Development of Novel Formulations for Chagas' Disease: Optimization of Benznidazole Chitosan Microparticles Based on Artificial Neural Networks. *Int. J. Pharm.* **2009,** *367*, 140–147.
107. Le, T.; Epa, V. C.; Burden, F. R.; Winkler, D. A. Quantitative Structure–Property Relationship Modeling of Diverse Materials Properties. *Chem. Rev.* **2012,** *112*, 2889–2919.
108. Guerra, A.; Gonzalez-Naranjo, P.; E. Campillo, N.; Cerecetto, H.; Gonzalez, M.; A. Paez, J. Artificial Neural Networks Based on CODES Descriptors in Pharmacology: Identification of Novel Trypanocidal Drugs against Chagas Disease. *Curr. Comput. Aided Drug Des.* **2013,** *9*, 130–140.
109. de Souza, A. S.; Ferreira, L. L. G.; de Oliveira, A. S.; Andricopulo, A. D. Quantitative Structure–Activity Relationships for Structurally Diverse Chemotypes Having Anti-Trypanosoma Cruzi Activity. *Int. J. Mol. Sci.* **2019,** *20*, 2801.
110. Kryshchyshyn, A.; Devinyak, O.; Kaminskyy, D.; Grellier, P.; Lesyk, R. Development of Predictive QSAR Models of 4-Thiazolidinones Antitrypanosomal Activity Using Modern Machine Learning Algorithms. *Mol. Inform.* **2018,** *37*, 1700078.

111. Luchi, A. M.; Villafañe, R. N.; Gómez Chávez, J. L.; Bogado, M. L.; Angelina, E. L.; Peruchena, N. M. Combining Charge Density Analysis With Machine Learning Tools to Investigate the Cruzain Inhibition Mechanism. *ACS Omega* **2019,** *4*, 19582–19594.
112. Burden, F. R.; Winkler, D. A. Relevance Vector Machines: Sparse Classification Methods for QSAR. *J. Chem. Inf. Model* **2015,** *55*, 1529–1534.
113. Tahghighi, A.; Mohamadi-Zarch, S.-M.; Rahimi, H.; Marashiyan, M.; Maleki-Ravasan, N.; Eslamifar, A. In Silico and in Vivo Anti-Malarial Investigation on 1-(Heteroaryl)-2-((5-Nitroheteroaryl)Methylene) Hydrazine Derivatives. *Malar. J.* **2020,** *19*, 231.
114. Kalani, K.; Agarwal, J.; Alam, S.; Khan, F.; Pal, A.; Srivastava, S. K. In Silico and In Vivo Anti-Malarial Studies of 18β Glycyrrhetinic Acid From Glycyrrhiza Glabra. *PLoS One* **2013,** *8*, e74761.
115. Trujillo-Correa, A. I.; Quintero-Gil, D. C.; Diaz-Castillo, F.; Quiñones, W.; Robledo, S. M.; Martinez-Gutierrez, M. In Vitro and in Silico Anti-Dengue Activity of Compounds Obtained From Psidium Guajava Through Bioprospecting. *BMC Complement. Altern. Med.* **2019,** *19*, 298.
116. Battini, L.; Fidalgo, D. M.; Álvarez, D. E.; Bollini, M.; Discovery of a Potent and Selective Chikungunya Virus Envelope Protein Inhibitor Through Computer-Aided Drug Design. *ACS Infect. Dis.* **2021,** *7*, 1503–1518.
117. Das, P. K.; Puusepp, L.; Varghese, F. S.; Utt, A.; Ahola, T.; Kananovich, D. G.; Lopp, M.; Merits, A.; Karelson, M. Design and Validation of Novel Chikungunya Virus Protease Inhibitors. *Antimicrob. Agents Chemother.* **2016,** *60*, 7382–7395.
118. Stevanovic, S.; Sencanski, M.; Danel, M.; Menendez, C.; Belguedj, R.; Bouraiou, A.; Nikolic, K.; Cojean, S.; Loiseau, P.; Glisic, S.; Baltas, M.; Garcia-Sosa, A. T. Synthesis, In Silico, and In Vitro Evaluation of Anti-Leishmanial Activity of Oxadiazoles and Indolizine Containing Compounds Flagged Against Anti-Targets. *Molecules* **2019,** *24*, 1282.
119. Rashid, U.; Sultana, R.; Shaheen, N.; Hassan, S. F.; Yaqoob, F.; Ahmad, M. J.; Iftikhar, F.; Sultana, N.; Asghar, S.; Yasinzai, M.; Ansari, F. L.; Qureshi, N. A. Structure Based Medicinal Chemistry-Driven Strategy to Design Substituted Dihydropyrimidines as Potential Antileishmanial Agents. *Eur. J. Med. Chem.* **2016,** *115*, 230–244.
120. Pathak, N.; Kuo, Y. P.; Chang, T. Y.; Huang, C. T.; Hung, H. C.; Hsu, J. T. A.; Yu, G. Y.; Yang, J. M. Zika Virus NS3 Protease Pharmacophore Anchor Model and Drug Discovery. *Sci. Rep.* **2020,** *10*, 8929.
121. Kwofie, S.; Dankwa, B.; Odame, E.; Agamah, F.; Doe, Lady; Teye, J.; Agyapong, O.; Miller, W.; Mosi, L.; Wilson, M. In Silico Screening of Isocitrate Lyase for Novel Anti-Buruli Ulcer Natural Products Originating From Africa. *Molecules* **2018,** *23*, 1550.
122. Santos, C. C.; Lionel, J. R.; Peres, R. B.; Batista, M. M.; da Silva, P. B.; de Oliveira, G. M.; da Silva, C. F.; Batista, D. G. J.; Souza, S. M. O.; Andrade, C. H.; Neves, B. J.; Braga, R. C.; Patrick, D. A.; Bakunova, S. M.; Tidwell, R. R.; Soeiro, M. In Vitro, In Silico, and In Vivo Analyses of Novel Aromatic Amidines Against Trypanosoma Cruzi. *Antimicrob. Agents Chemother.* **2018,** *62* (2), e02205–e02217.
123. Ferreira de Almeida Fiuza, L.; Peres, R. B.; Simões-Silva, M. R.; da Silva, P. B.; Batista, D. da G. J.; da Silva, C. F.; Nefertiti Silva da Gama, A.; Krishna Reddy, T. R.; Soeiro, M. de N. C. Identification of Pyrazolo[3,4-e][1,4]Thiazepin Based CYP51 Inhibitors as Potential Chagas Disease Therapeutic Alternative: In Vitro and in Vivo Evaluation, Binding Mode Prediction and SAR Exploration. *Eur. J. Med. Chem.* **2018,** *149*, 257–268.

124. Hernández Alvarez, L.; Naranjo Feliciano, D.; Hernández González, J. E.; de Oliveira Soares, R.; Barreto Gomes, D. E.; Pascutti, P. G. Insights Into the Interactions of Fasciola Hepatica Cathepsin L3 With a Substrate and Potential Novel Inhibitors Through In Silico Approaches. *PLoS Negl. Trop. Dis.* **2015,** *9*, e0003759.
125. Fino, R.; Lenhart, D.; Kalel, V. C.; Softley, C. A.; Napolitano, V.; Byrne, R.; Schliebs, W.; Dawidowski, M.; Erdmann, R.; Sattler, M.;Schneider, G.; Plettenburg, O.; Popowicz, G. M. Computer-Aided Design and Synthesis of a New Class of PEX14 Inhibitors: Substituted 2,3,4,5-Tetrahydrobenzo[F][1,4]Oxazepines as Potential New Trypanocidal Agents. *J. Chem. Inf. Model.* **2021,** *61*, 5256–5268.
126. Halder, S.; Dhorajiwala, T.; Samant, L. Multiple Docking Analysis and In Silico Absorption, Distribution, Metabolism, Excretion, and Toxicity Screening of Anti-Leprosy Phytochemicals and Dapsone Against Dihydropteroate Synthase of Mycobacterium Leprae. *Int. J. Mycobacteriol.***2019,** *8*, 229.
127. Swain, S. S.; Paidesetty, S. K.; Dehury, B.; Das, M.; Vedithi, S. C.; Padhy, R. N. Computer-Aided Synthesis of Dapsone-Phytochemical Conjugates Against Dapsone-Resistant Mycobacterium Leprae. *Sci. Rep.* **2020,** *10*, 6839.
128. Kalani, K.; Kushwaha, V.; Sharma, P.; Verma, R.; Srivastava, M.; Khan, F.; Murthy, P. K.; Srivastava, S. K. In Vitro, In Silico and In Vivo Studies of Ursolic Acid as an Anti-Filarial Agent. *PLoS One* **2014,** *9*, e111244.
129. Verma, M.; Pathak, M.; Shahab, M.; Singh, K.; Mitra, K.; Misra-Bhattacharya, S. Moxidectin Causes Adult Worm Mortality of Human Lymphatic Filarial Parasite Brugia Malayi in Rodent Models. *Folia Parasitol.* **2014,** *61*, 561–570.
130. Amorim, C. R.; Pavani, T. F. A.; Lopes, A. F. S.; Duque, M. D.; Mengarda, A. C. A.; Silva, M. P.; de Moraes, J.; Rando, D. G. G. Schiff Bases of 4-Phenyl-2-Aminothiazoles as Hits to New Antischistosomals: Synthesis, in Vitro, in Vivo and in Silico Studies. *Eur. J. Pharm. Sci.* **2020,** *150*, 105371.
131. Pereira, V. R. D.; Junior, I. J. A.; da Silveira, L. S.; Geraldo, R. B.; de F. Pinto, P.; Teixeira, F. S.; Salvadori, M. C.; Silva, M. P.; Alves, L. A.; Capriles, P.V. S. Z.; Almeida, A. das C.; Coimbra, E. S.; Pinto, P. L. S.; Couri, M. R. C.; de Moraes, J.; Da Silva Filho, A. A. In Vitro and in Vivo Antischistosomal Activities of Chalcones. *Chem. Biodivers.* **2018,** *15*, e1800398.
132. Lim, W.; Eadie, K.; Konings, M.; Rijnders, B.; Fahal, A. H.; Oliver, J. D.; Birch, M.; Verbon, A.; van de Sande, W. Madurella Mycetomatis, the Main Causative Agent of Eumycetoma, Is Highly Susceptible to Olorofim. *J. Antimicrob. Chemother.* **2020,** *75*, 936–941.

# CHAPTER 5

# Drug Repurposing and Computational Drug Discovery for Malignant Diseases

ASHISH SHAH[1], GHANSHYAM PARMAR[1], and ASHISH PATEL[2]

[1]*Department of Pharmacy, Sumandeep Vidyapeeth, Vadodara, Gujarat, India*

[2]*Ramanbhai Patel College of Pharmacy, Charusat University, Changa, Gujarat, India*

## ABSTRACT

The concept of drug repurposing excludes any structural modification of the drug. Instead, repositioning makes advantage of either the biological qualities for which the medication has already been licensed. Many of the drugs being studied for oncological repurposing, on the other hand, are either generic or low-cost. One of the most important aspects of drug repurposing is the use of in silico tools (data mining, machine learning, ligand-based, and structure-based approaches) to describe the factors connected with the complex interplay between diseases, drugs, and targets. Considering heterogeneity of cancer, drug development process for cancer is even more complicated. Undruggable target, chemo-resistance, tumor heterogeneity are major barriers for safe and effective cancer chemotherapy. Traditional drug discovery approach fails to provide effective solution. Drug repurposing has provided safe and cost-effective solution for cancer therapy. *In silico* methods also have the capability to repurpose the old and off-target compounds to increase the likelihood in selected patients and predict better response in tumours. In 2011, the FDA approved drugs vemurafenib and crizotinib were repurposed in metastatic melanoma and lung cancer respectively. In this

Drug Repurposing and Computational Drug Discovery: Strategies and Advances.
Mithun Rudrapal, PhD (Ed)

chapter we had discussed about various drug repurposing approaches for cancer therapy along with its limitations.

## 5.1 INTRODUCTION

Cancer is a group of diseases characterized by uncontrollable growth of abnormal cells that have the ability to spread. Normally cancer in body cells grows and multiply as per the need of the body. By the time cell grows old or become damaged or dies and new cells again replace the older cells. Sometimes, this normal response process breaks down and instead of normal new cells, damaged or abnormal cells grow and multiply which may result in the formation of tumor. Tumor may be benign or metastatic. Tumor cells have the ability to spread and invade nearby tissues. Benign tumor does not spread or invade into nearby tissues while malignant can spread. Oncogenic classification of cancer can be done in 100 different types and nomenclature of cancer done based on organ or tissues where the cancer forms. The major classes of cancer include carcinoma, sarcoma, and lymphoma.[1]

Cancer is a leading cause of death. According to the global cancer statistics, there were about 19.3 million new cancer cases and 10 million cancer deaths reported in the year 2020 globally. The pharmacological treatment of cancer includes hormone therapy, immunotherapy, chemotherapy, or combination of all. Among these, the major obstacle on chemotherapy is multi drug resistance (MDR). Continuous doses of chemotherapeutic agent reduce efflux outside the cell which hampers permeation across the cell membranes. To understand permeation of cell across the cell, various nanocarriers are used in the treatment of cancer.[2]

## 5.2 DRUG REPURPOSING: AN APPROACH TO BOOST ANTICANCER DRUG DEVELOPMENT RESEARCH

The concept of drug repurposing thus excludes any structural modification of the drug. Instead, repositioning makes advantage of either the biological qualities for which the medication has already been licensed (perhaps in a new formulation, at a new dose, or via a new method of administration) or the drug's side features that are responsible for its undesirable effects.[3] Drug repurposing is based on two fundamental scientific foundations: (1) the finding, via the elucidation of the human genome, that various diseases have biological targets that are sometimes shared, and (2) the idea of pleiotropic medications. One of the most important aspects of drug repurposing is the use of in silico tools (data mining, machine learning, ligand-based and structure-based approaches) to describe the factors connected with the complex interplay between diseases, drugs, and targets.[4] It is now possible to classify diseases based on their molecular profile (e.g., the genes, biomarkers, signaling pathways, environmental factors, etc.) and to compare diseases that share a number of these molecular traits using computational methods, particularly data mining.[4] Before going on to drug repurposing as a way to break through the stalemate, it is worth noting the limitations of pharmaceutical industry for the development of novel oncology treatments in time being where understanding of cancer at the molecular level is continually improving. The targeted therapeutics paradigm is increasingly driving drug development, paralleling the increased understanding of cancer at the molecular and genetic levels.[5]

A "soft" type of drug repurposing, for example, is the application of current oncology treatments to novel cancer indications. It tries to circumvent many of the problems associated with medication development and testing by repurposing existing pharmaceuticals for new purposes or in novel ways for established indications.[6] Due to this, the drug repurposing approach can be considered a response to oncological drug research's diminishing productivity, a tactic to shorten development periods, and a source of low-cost medicines to fulfil the rising demands and unmet requirements of cancer patients. It is a strategy that is fundamentally different from the predominant model that directs the development of targeted medicines, but it could be a largely unexplored source of new therapies.[7]

Many of the drugs being studied for oncological repurposing, on the other hand, are either generic or low-cost. The incremental costs of these medications, when used in combination protocols with standard therapy, are projected to be negligible. While cost alone should not be used to determine whether therapies are acceptable for patient care, it is a significant consideration for health systems and insurers, as well as a key role in health policy formation. Initiatives with proven efficacy repurposed medications will score higher in any cost-utility analysis than interventions using more expensive targeted therapies. Randomized clinical studies with repurposed pharmaceuticals are therefore critical to proving their usefulness and, as a result, to reducing the financial load on pressured health systems, particularly in poorer economies.

## 5.3 TARGET-BASED CANCER THERAPY

Researchers have identified the biomarkers (in comparison with normal cell) that help cancer cells to grow. The special activity of that biomarker is considered a target for cancer therapy. The researcher develops specific drug that targets a specific biomarker responsible for cancer. This wonder is called *targeted therapy*. Targeted therapy is a special type of chemotherapy which has an advantage to difference normal and cancer cells. Targeted therapy is sometimes used alone or with combination therapy. Targeted therapy drugs can be useful in the treatment of different types of cancers, so these drugs are considered chemotherapeutic drugs, but the way of working is different than standard chemotherapeutic drugs.[8]

### 5.3.1 HOW DOES TARGETED CANCER THERAPY WORK?

Most of the standard chemotherapeutic drugs kill cells in the body which grow and divide fast as the cancer cell grows and divide quickly these drugs that are effective. The problem of standard chemo drugs is that they also kill normal body cells that grow and divide quickly which produce sometimes serious side effects. Target therapy has overcome this problem as the way of working is different. Cancer cells have many changes in their genes which are not observed in normal cells. These types of genetic changes are responsible for development of cancer cells. The mechanisms for the formation of cancer are different and due to this all cancer are not the same. For example, colon cancer and breast cancer have different genes that help them to grow. Sometimes, different genes are responsible for the development of the same type of cancer.

Principles of targeted therapies include blockage of chemical signals, mutation of protein that reduces or stops the growth of cancer cells, decrease in the rate of the angiogenesis, and improvement of immune system. All these mechanisms help to reduce the growth as well as spread of cancer cells. The targeted therapy is more focused than convention therapy and may have the ability to target single or multiple proteins that help cancer cells to grow, divide, and spread. The target therapy drugs are also used to boost the immunity of the body to fight against the cancer cells.[9]

### 5.3.2 ADVANTAGES OF TARGET-BASED ANTICANCER THERAPY[10]

- Depending upon types of cancer, target therapy offers different benefits. Target therapy may be useful to
- Block the signaling pathways that help cancer cells to grow and multiply.
- Mutation of the proteins within cancer cell that results in cell death.
- Prevent the new blood vessel formation to cut the blood supply to tumor cells.
- Activate immune system to attack cancer cells.
- Deliver toxins to kill cancer cells without harming normal cells.

## 5.4 COMPUTATIONAL TOOLS AND RESOURCES FOR IN SILICO DRUG REPURPOSING

### 5.4.1 ROLE OF ARTIFICIAL INTELLIGENCE IN ANTICANCER DRUG DISCOVERY

Artificial intelligence is simulation of human intelligence by computer. It contains subfield that is known as machine learning methods. Various computational tools can be useful for the discovery of new drug and may provide important clue to the researchers.[11] The technique called as computer-aided drug design can be very useful for the discovery of new molecules against this disease. CADD is divided mainly into two categories: Structure-based drug design (SBDD) and ligand-based drug design (LBDD). SBDD use the information of 3D structure of disease protein while LBDD applies when the 3D structure of disease protein is not available. It uses knowledge of existing molecules to design a new molecule with belief that new molecule may have higher potency and less side effects as compared with the previous one. The SBDD can perform using docking and de novo drug design methods. LBDD can be performed using QSAR, virtual screening using pharmacophore.[12]

### 5.4.2 ANTICANCER DRUG TARGET PREDICTION

Approximately 30,000 genes present in human and out of those 6000 to 8000 sites are considered pharmacological targets. However only 400 proteins are evaluated for drug development until now.[13] The traditional drug discovery method follows the principle of one molecule-one target-one disease. This approach does not consider drug–target interactions; however, the disease that is complicated developed through multiple target proteins.[14] Another point is ploy-pharmacological properties of certain drug and due to this they may interact with off-targets that result in undesirable side effects. Also, there are certain examples in which off target effects are beneficial. For example, sildenafil, the drug, was originally developed to treat angina but now it is repurposed for erectile dysfunction therapy.[15] There are several anticancer drugs whose drug targets are still unknown or unidentified. In some cases, targets are known and effective but they remain exterior from the opportunity of pharmacological regulation. These targets include transcription factors, phosphates, and RAS family which are considered undruggable targets due to lack of enzyme active sites.[16]

Characterization of active sites or ligand-binding sites is one of the key aspects of drug repurposing. Bioinformatics methods are useful for the target prediction and ultimately help in an accurate prediction of the drug target. Until now a variety of computational database and prediction tools (Table 5.1) are established that provide important information regarding ligand-binding sites. Various computational tools are also useful to study potential interaction between protein and drugs. Network-based models and Ml-based models are important tools.[17]

**TABLE 5.1** Drug Target Database and Computational Tools for Target Prediction.

| Sl. No. | Database/computational tool | Website |
|---|---|---|
| 1 | DrugBank | https://go.drugbank.com/ |
| 2 | TTD | http://db.idrblab.net/ttd/ |
| 3 | MATADOR | http://matador.embl.de/ |
| 4 | Super target | http://bigd.big.ac.cn/databasecommons/database/id/564 |
| 5 | TDR targets | https://tdrtargets.org/ |
| 6 | BindingDB | https://www.bindingdb.org/bind/index.jsp |
| 7 | CancerDR | https://webs.iiitd.edu.in/raghava/cancerdr/ |
| 8 | DCDB | https://www.dcdeckbuilding.info/ |
| 9 | SEA | https://sea.bkslab.org/ |
| 10 | Pharmmapper | http://www.lilab-ecust.cn/pharmmapper/ |
| 11 | SuperPred | https://prediction.charite.de/ |
| 12 | Swiss target prediction | http://www.swisstargetprediction.ch/ |

### 5.4.3 STRUCTURE-BASED DRUG DISCOVERY APPROACH IN DRUG REPURPOSING

In comparison with the old method, structure-based drug design is becoming a crucial tool for faster and more cost-effective lead discovery. Hundreds of new targets and prospects in drug repurposing have been discovered thanks to genomic, proteomic, and structural research. Protein comparison is utilized in the field of drug repurposing to find secondary targets of an approved drug.[18] Proteins may be compared on a global scale using sequence similarity, which has been used to construct phylogenetic trees, the most common of which is the kinome.[19] Modern methods for doing multiple sequence alignments, such as BLAST, are, nonetheless, extensively used and accessible via online servers. Small variations in critical places, such as those occurring in correspondence to the gatekeeper residue of protein kinases or other oncogenic alterations, can have a significant impact on ligand binding.[20] Furthermore, comparable ligands were found to be capable of binding proteins with distantly related sequences in a study based on the similarity searching approach.[21] In terms of polypharmacology and drug repurposing, local binding site similarities may be more essential than global similarities.[22] Sequence alignments perform well in finding novel targets of known ligands when proteins share a high degree of sequence identity, whereas local protein comparisons perform better when proteins share a low degree of sequence identity.[23] Furthermore, scanning the protein surface for cavities[24] and then calculating descriptors of various types to produce a similarity score are standard methods for identifying and comparing binding sites. It is worth noting that while numerous methodologies and algorithms for comparing binding sites have been proposed, none of them appear to be without flaws or restrictions.

Binding site similarities and other molecular modeling methods were employed together to discover new targets for drugs like Pemigatinib and Capmatinib.[25] The research began with a large number of comparable binding sites, which were then refined by docking to simulate the binding mode of pemigatinib and capmatinib. The proteins with the highest docking scores were ranked for priority and further experimentally confirmed. It is worth noting that, when available, ligand-binding modalities are a valuable asset in the search for new targets. Focusing on target–ligand interactions is one technique to describe molecular recognition.

Various approaches, such as structure-based pharmacophores or interaction fingerprints, can be used to accomplish this. When a protein–ligand complex's structure is unavailable, computational approaches can be used in advance to find out hot spots in the binding site.[26] Another method for connecting ligand information to protein habitats is to employ the idea of chemoisosterism,[27] which refers to the ability of two protein environments to bind the same chemical fragment and can reveal the inherent cross-pharmacology across protein targets. The polypharmacology of chemical fragments was discovered to be related to the degree of chemoisosterism. This method permits interaction networks to be built between chemical fragments and chemoisosteric protein environments. These networks, when combined with target disease correlations, make potentially enticing beginning points for drug repurposing.

Structure-based approaches, on the other hand, are obviously reliant on crystallographic structures of protein–ligand complexes being available. The level of information that may be used to represent a binding site is influenced by the resolution and sensitivity to atomic coordinates. While crystallographic structures give a static model of a protein, conformational changes might cause new pockets to develop. Detecting those cryptic locations has become a growing subject of study, as it may open up new possibilities for medication repurposing. In fact, beyond the more extensively investigated orthosteric site, cryptic allosteric sites may be valuable for gaining selectivity, exploring new chemical regions for drug design, and establishing drug–target relationships. Overall, discovering new allosteric sites in proteins may open up considerably more possibilities for therapeutic repurposing than previously thought.

### *5.4.4 LIGAND-BASED DRUG DISCOVERY APPROACH IN DRUG REPURPOSING*

The idea behind ligand-based techniques is that related molecules have similar biological characteristics. These methods have been widely utilized in drug repurposing to examine and predict the activity of ligands for novel targets. PubChem, ChEMBL, and DrugBank are public databases of bioactive compounds that comprise information extracted and manually selected from literature sources.[28] These databases house a vast and ever-expanding amount of chemical and biological data, including binding affinity, cellular activity, functional, and ADMET data. The availability of databases focused on repurposed medications, failed pharmaceuticals, therapeutic indications, and bioactivity data[29] is one of the most recent advancements in drug repurposing. One advantage of using these methods for drug repurposing is that the number of publicly accessible compound records (more than a hundred million, according to PubChem) greatly outnumbers the number of deposited protein crystal structures (less than 150,000 in the Protein Data Bank as of today).[30] However, ligand-based approaches rely on the chemical space coverage of previously identified compounds. Furthermore, because small structural divergences in chemical scaffolds can lead to "activity cliffs," a high overall similarity does not always imply activity on a secondary target.

Using the similarity ensemble methodology, another ligand-based method correctly identified 23 novel drug–target relationships. The drug repurposing has also benefited from pharmacophore screening.[31] A drug can be represented as a set of pharmacophoric properties in this manner, which can then be used to query chemical compound databases for molecules with various scaffolds. Combining multiple levels of ligand description raises the likelihood of discovering novel repurposing opportunities. Predictive models based on disease feature descriptors, large-scale drug–target, and target–disease relationships all showed improvements in predicting novel drug–disease links for various reasons. Chemical and phenotypic similarities, in particular, have been proven to be complementary to one another, and that combining predictions from both methods is helpful.[32]

### *5.4.5 USE OF COMPUTER-AIDED DRUG REPURPOSING APPROACH TO IDENTIFY ANTICANCER AGENTS*

Targeted gene expression profiles can demonstrate the activation of typical intracellular pathways in cancer which allows predicting the oncogenic signaling pathways that are active in tumor environment.[33–35] These activated oncogenes or dependent pathways control the proliferation and maintenance of tumors which can be considered a molecular target. In this case, the drug repurposing approach can be implicated for the discovery of novel drugs that revert these genetic signatures and exhibit as cancer inhibitors that mainly affect the proliferation of tumor cells.[36] Using MANTRA 2.0 tools Carrella et al identify the anthelmintic drugs as inhibitors of PI3K-dependent oncogene in cancer. The rigorous in vitro and in vivo experimental research allowed us to validate the effectiveness of the inhibition of pathways. Mottini et al. extensively validate the K-RAS oncogene-dependent gene targets using the in-silico approach to predict decitabine FDA-approved drug for myelodysplastic syndrome, as a potential inhibitor for pancreatic cancer.[37]

## 5.5 DRUG-REPURPOSING APPROACH IN CANCER

According to the report of the International Agency for Research on cancer of 2018, there are 9.6 million deaths due to cancer. The most predominant type of cancer is the pulmonary, memory gland, and colorectal.[38] The high prevalence rate of mortality is associated with pancreatic and stomach cancer might be due to limitations in treatment options. The high prevalence of death in cancer is dependent upon certain factors such as chemoresistance or tumor heterogeneity as well as metastases. Due to high demand drug and production crises emerged in the research and development of pharmaceuticals which lead to the formulation of novel drug delivery therapeutics.[39] However, the discovery and development of new drug molecules represent time-consuming and costly processes with low success rates. This might be due to less understanding of the association between dose, exposure, and its effect on the target. The interesting approach in the discovery of novel anticancer therapeutics is to the repurposing of old FDA-approved molecules for new indication termed as a repurposing of drug.[40] The main advantages of drug repurposing are to minimize risk and cost for the development and shorten the time gap in the discovery process due to the availability of pharmacokinetic and pharmacodynamics and clinical data. Additionally, drug repurposing has an opportunity to find new molecules for the treatment of rare cancer that is often neglected by the R&D department of pharmaceutical companies due to less marketing values.[41] In silico drug repurposing approach has typical benefit to transform biological data into a prediction of druggable targets. Computational methods facilitate the use of data generated through different omics tools, that is, genomic, proteomic, transcriptomic, and metabolomics into the understanding of the biology of old and new targets along with the mechanism of action of drugs (Fig. 5.1).

Prediction of druggable targets is crucial for repurposing. Hence, the acquisition of biological data on targets can be collected from in vivo and in vitro studies or from computational in silico methods.[42] Even though, targeted therapy has significant potential to improve patient life span. Despite this in silico drug repurposing is based on the hypothetical approach that uses huge data to predict the drug molecules against the cancer targets. This approach has the unique ability to transform biological data of cancer phenotyping and identification of druggable targets. For a successful computational pipeline, the algorithms are necessary for the integration of huge biological data. In this process, one of the essential steps is the appropriate collection and cross-examination of available omics data about cancer biology and mechanism.

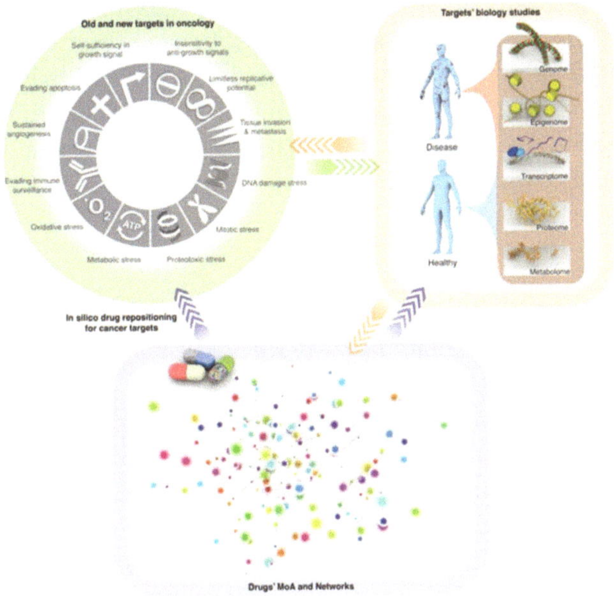


**FIGURE 5.1** A general overview of computer-aided drug repositioning approaches in oncology.

*Source*: Reprinted with permission from Ref. [64]. © 2021 Elsevier.

## 5.5.1 ROLE OF IN SILICO DRUG REPURPOSING IN THE PREDICTION OF DRUG THERAPIES

In vitro and in vivo experiment-based drug repurposing is often the result of chance discovery and not hypothesis-based. This might result from an experimental drug screening or by the identification of target similarities among different diseases.[42] Disulfiram was approved for the treatment of alcoholism in repurposing effectively in the treatment of cancer.[43–45] Anti-convulsant drug valproic acid was effective as an anticonvulsant drug. However, valproic acid is proven as an anticancer drug in multiple clinical trials.[46] A similar example such as nelfinavir was originally employed in the treatment of acquired immune deficiency virus (AIDS) infection that is currently in pipeline to treat various cancers such as lung, breast, and melanoma.[47] Brivudine was first indicated in the treatment of herpes viral infection. Brivudine is a thymidine analogue that stops viral replication in humans. Later, based on the structural interaction brivudine also binds to the human heat shock protein Hsp27 and inhibits its antiapoptotic activity.[48] Ibrutinib was identified as a typical Bruton's tyrosine kinase inhibitor via

BTK inhibition, but later it was proved to be a VEGFR2 inhibitor in cancer therapy. So it is an example of a drug having the same drug acting on different targets.[49] Similarly, nilotinib was validated as a potent MAPK14 inhibitor; additionally, it was proved as a potential anti-inflammatory drug.[50] Based on the structural system pharmacology platform levosimendan a PDE inhibitor for heart failure was developed as a serine/threonine-protein kinase inhibitor in cancer.[51]

### 5.5.2 IMPORTANCE OF IN SILICO DRUG REPURPOSING IN PERSONALIZED MEDICINE IN CANCER

Computational driven drug repurposing might help to improve the efficacy of personalized drug therapy in cancer which offers novel therapeutic indications. The key challenge that might be faced in oncology is the development of targeted and personalized therapy to treat cancer with minimal side effects with a high response rate in each patient.[52] The targeted drug therapies result from either the use of a drug promisingly effective against defined molecular targets identified in a cancer cell or in the tumor microenvironment or more ideally based on the selection of patients who could likely benefit from a specific treatment (Table 5.2). However, the concept of targeting typical mutated tumor protein such as oncogene appeared as a successful pharmacological approach in preclinical models.[53] Other challenging factors such as interindividual variation in patient's physiological parameters, the bioavailability of drug, tumor microenvironment, the aggressiveness of the tumor, and metastatic stage of tumor are also limiting factors in the development of personalized therapy. To overcome these limiting factors computational approaches for drug repurposing might be the appropriate solution to develop targeted therapy in cancer. Computation approaches can take advantage of various biological data considering tumor biology, the clinical outcome of patient's biomarkers along with drug pharmacokinetics and pharmacodynamics modeling for the prediction of a drug in cancer therapy.[54–56] Moreover, in silico methods have also the capability to repurpose the old and off-target compounds to increase the likelihood in selected patients and predict better response in tumors.[37] In 2011, the FDA-approved drugs vemurafenib and crizotinib were repurposed in metastatic melanoma and lung cancer, respectively.[57,58] Everolimus was repurposed in pancreatic tumors with the mutation in mTOR pathway genes.[59]

**TABLE 5.2** Identification of Drug Candidates Targeting Hallmarks of Cancer.[60]

| Sl. no. | Cancer hallmarks | Types of therapy | Example |
|---|---|---|---|
| 1 | Sustaining proliferative signalling | Mono | Rapamycin, prazosin, indomethacin |
| 2 | Evading growth suppressors | Combinatorial | Quinacrine, ritonavir |
| 3 | Resisting cell death | Mono | Artemisinin, chloroquine |
| 4 | Enabling replicative mortality | Combinatorial | Curcumin, genistein |
| 5 | Genome instability and mutation | Combinatorial | Spironolactone, mebendazole |
| 6 | Reprogramming energy metabolism | Mono | Metformin, disulfiram |
| 7 | Inducing angiogenesis | Combinatorial | Thalidomide, itraconazole |
| 8 | Activating invasion and metastasis | Combinatorial | Berberine, niclosamide |
| 9 | Tumor-promoting inflammation | Combinatorial | Aspirin, thiocolchicoside |
| 10 | Evading immune destruction | Mono | Infectious disease vaccines |

## 5.6 ROLE OF OMICS STUDIES AND ITS IMPORTANCE IN COMPUTER-AIDED DRUG REPURPOSING

Development and progression of cancer involves multiple factors, genetic alterations, and mutation of various cellular components. Various types of cancer hallmark already have been discovered that activates positive regulators for cancer cell proliferation, survival, and genetic mutations. Some of the hallmarks are useful to predict disease outcome and prognosis. Deeper understanding of these hallmarks is essential to implement target-based anticancer therapy. Multiple layers of omics studies on specific targets provide disease-related data that helps in in-silico repurposing.[61,62]

### *5.6.1 THE CANCER GENOME ATLAS (TCGA) STUDY*

TCGA is a public project that has a collection of over 11,000 tumors, representing 33 most common types of cancer. The goal of this project is to create a comprehensive atlas of molecular alterations that occur in cancer cells and to discover genetic alterations using multidimensional platforms. The multidimensional platform integrates genomic, epigenomic, and transcriptomic data from various types of cancer to define their molecular subtypes. TCGA studies also compare multiple cancer types to identify common molecular features that help to do accurate molecular classification of tumors. Multi-dimensional studies of TCGA provide detailed information about cancer biology that can be applied for various computational approaches for drug repurposing. The availability of this large publicly available data will boost translational cancer research and could be powerful resource to support drug repurposing.[63]

### 5.6.2 *CANCER TARGETS RELATED OMICS DATA*

Activation of oncogene and inactivation of tumor suppressor genes are one of the fundamental reasons for development and progression of cancer. Currently, targeted anticancer agents that are approved or under clinical trial are targeting oncogenic activation pathways. However, development of resistance is one of the major drawbacks of targeted therapy. Certain mutated oncogenes like K-RAS, β-catenin are irresponsive to targeted cancer therapy that turns into development of resistance. Due to this reason current targeted therapy still shows limited clinical success, thus opening the opportunities of drug repurposing.

Omics studies provide deeper insights into understanding of oncogenic signaling activation or tumor suppressor genes that support cancer cell proliferation and survival. In certain studies, multiple omics approach had been implemented that focused on specific oncogenes (like C-MYC, K-RAS, and β-catenin) and tumor suppressor genes (like PTEN and p53) for implementation of modulators or inhibitors in a clinical setting. For example, multiple omics studies on K-RAs help to define biology of K-RAS in different types of tumors. Various in vivo and in vitro models are developed for better understanding.[64]

## 5.7 CONCLUSION AND FUTURE PERSPECTIVES

Cancer is one of the major threats for human health. Every year around 10 million people die from various forms of cancer. Cancer is the second leading disease that causes human death. Drug discovery and development process take around 12 years with around 2.7 billion USD cost for the development of new molecules. Considering heterogeneity of cancer, the drug development process for cancer is even more complicated. Undruggable target, chemo-resistance, and tumor heterogeneity are major barriers for safe and effective cancer chemotherapy. Traditional drug discovery approach fails to provide an effective solution. Computer-aided drug repurposing has a potential to improve cancer therapy due to major three reasons: (1) each cancer hallmark is deeply investigated by Omics approaches. (2) Use of traditional algorithm with computational approaches can expand the ability of data integration even in the absence of clinical hypothesis. (3) Prediction of synthetic lethality using computational tools can be useful for effective clinical outcomes.

The era of computer-aided drug repurposing is just started. In the future, the drug repurposing approach can solve therapeutic limitations of current cancer therapy. The combination of useful predictions generated by various computational tools with experimental validation can speed up the anticancer drug development process.

## KEYWORDS

- ***in silico* repurposing**
- **malignant diseases**
- **computational tools**
- **SBDD approach**
- **omics**

## REFERENCES


1. Shewach, D. S.; Kuchta, R. D. Introduction to Cancer Chemotherapeutics. *Chem. Rev.* **2009**, *109* (7), 2859–2861.
2. Sung, H.; Ferlay, J.; Siegel, R. L.; Laversanne, M.; Soerjomataram, I.; Jemal, A.; Bray, F. Global Cancer Statistics 2020: GLOBOCAN Estimates of Incidence and Mortality Worldwide for 36 Cancers in 185 Countries. **2021,** *71* (3), 209–249.
3. Pushpakom, S.; Iorio, F.; Eyers, P. A.; Escott, K. J.; Hopper, S.; Wells, A.; Doig, A.; Guilliams, T.; Latimer, J.; McNamee, C.; Norris, A.; Sanseau, P.; Cavalla, D.; Pirmohamed, M. Drug Repurposing: Progress, Challenges and Recommendations. *Nat. Rev. Drug Discov.* **2019**, *18* (1), 41–58.
4. March-Vila, E.; Pinzi, L.; Sturm, N.; Tinivella, A.; Engkvist, O.; Chen, H.; Rastelli, G. On the Integration of In Silico Drug Design Methods for Drug Repurposing. *Front. Pharmacol.* **2017,** *8*, 298.
5. Fojo, T.; Parkinson, D. R. Biologically Targeted Cancer Therapy and Marginal Benefits: Are we Making too Much of too Little or are we Achieving too Little by Giving too Much? *Clin. Cancer Res.* **2010,** *16* (24), 5972–5980.

6. Huang, M.; Shen, A.; Ding, J.; Geng, M. Molecularly Targeted Cancer Therapy: Some Lessons From the Past Decade. *Trends Pharmacol. Sci.* **2014,** *35* (1), 41–50.
7. Ashburn, T. T.; Thor, K. B. Drug Repositioning: Identifying and Developing New Uses for Existing Drugs. *Nat. Rev. Drug Discov.* **2004,** *3* (8), 673–683.
8. Yan, L.; Rosen, N.; Arteaga, C. Targeted Cancer Therapies. *Chin. J. Cancer* **2011,** *30* (1), 1–4.
9. Baudino, T. A. Targeted Cancer Therapy: The Next Generation of Cancer Treatment. *Curr. Drug Discov. Technol.* **2015,** *12* (1), 3–20.
10. Padma, V. V. An Overview of Targeted Cancer Therapy. *Biomedicine (Taipei)* **2015,** *5* (4), 19.
11. Mak, K. K.; Pichika, M. R. Artificial Intelligence in Drug Development: Present Status and Future Prospects. *Drug Discov. Today* **2019,** *24* (3), 773–780.
12. Baig, M. H.; Ahmad, K.; Roy, S.; Ashraf, J. M.; Adil, M.; Siddiqui, M. H.; Khan, S.; Kamal, M. A.; Provazník, I.; Choi, I. Computer Aided Drug Design: Success and Limitations. *Curr. Pharm. Des.* **2016,** *22* (5), 572–581.
13. Chen, X.; Yan, C. C.; Zhang, X.; Zhang, X.; Dai, F.; Yin, J.; Zhang, Y. Drug-Target Interaction Prediction: Databases, Web Servers and Computational Models. *Brief Bioinform.* **2016,** *17* (4), 696–712.
14. Chen, X.; Liu, M. X.; Yan, G. Y. Drug-Target Interaction Prediction by Random Walk on the Heterogeneous Network. *Mol. Biosyst.* **2012,** *8* (7), 1970–1978.
15. Takarabe, M.; Kotera, M.; Nishimura, Y.; Goto, S.; Yamanishi, Y. Drug Target Prediction Using Adverse Event Report Systems: A Pharmacogenomic Approach. *Bioinformatics* **2012,** *28* (18), i611–i618.
16. Lazo, J. S.; Sharlow, E. R. Drugging Undruggable Molecular Cancer Targets. *Annu. Rev. Pharmacol. Toxicol.* **2016,** *56*, 23–40.
17. Cui, W.; Aouidate, A.; Wang, S.; Yu, Q.; Li, Y.; Yuan, S. Discovering Anti-Cancer Drugs via Computational Methods. *Front. Pharmacol.* **2020,** *11*, 733.
18. Ehrt, C.; Brinkjost, T.; Koch, O. Impact of Binding Site Comparisons on Medicinal Chemistry and Rational Molecular Design. *J. Med. Chem.* **2016,** *59* (9), 4121–4151.
19. Manning, G.; Whyte, D. B.; Martinez, R.; Hunter, T.; Sudarsanam, S. The Protein Kinase Complement of the Human Genome. *Science* **2002,** *298* (5600), 1912–1934.
20. Huang, L.; Fu, L. Mechanisms of Resistance to EGFR Tyrosine Kinase Inhibitors. *Acta Pharm. Sin. B* **2015,** *5* (5), 390–401.
21. Keiser, M. J.; Roth, B. L.; Armbruster, B. N.; Ernsberger, P.; Irwin, J. J.; Shoichet, B. K. Relating Protein Pharmacology by Ligand Chemistry. *Nat. Biotechnol.* **2007,** *25* (2), 197–206.
22. Anighoro, A.; Stumpfe, D.; Heikamp, K.; Beebe, K.; Neckers, L. M.; Bajorath, J.; Rastelli, G. Computational Polypharmacology Analysis of the Heat Shock Protein 90 Interactome. *J. Chem. Inf. Model* **2015,** *55* (3), 676–686.
23. Chen, Y. C.; Tolbert, R.; Aronov, A. M.; McGaughey, G.; Walters, W. P.; Meireles, L. Prediction of Protein Pairs Sharing Common Active Ligands Using Protein Sequence, Structure, and Ligand Similarity. *J. Chem. Inf. Model* **2016,** *56* (9), 1734–1745.
24. Laurie, A. T.; Jackson, R. M. Methods for the Prediction of Protein-Ligand Binding Sites for Structure-Based Drug Design and Virtual Ligand Screening. *Curr. Protein Pept. Sci.* **2006,** *7* (5), 395–406.
25. Kinnings, S. L.; Liu, N.; Buchmeier, N.; Tonge, P. J.; Xie, L.; Bourne, P. E. Drug Discovery Using Chemical Systems Biology: Repositioning the Safe Medicine Comtan

to Treat Multi-Drug and Extensively Drug Resistant Tuberculosis. *PLoS Comput. Biol.* **2009,** *5* (7), e1000423.
26. Hall, D. R.; Kozakov, D.; Whitty, A.; Vajda, S. Lessons From Hot Spot Analysis for Fragment-Based Drug Discovery. *Trends Pharmacol. Sci.* **2015,** *36* (11), 724–736.
27. Jalencas, X.; Mestres, J. Chemoisosterism in the Proteome. *J. Chem. Inf. Model* **2013,** *53* (2), 279–292.
28. Wang, Y.; Bryant, S. H.; Cheng, T.; Wang, J.; Gindulyte, A.; Shoemaker, B. A.; Thiessen, P. A.; He, S.; Zhang, J. PubChem BioAssay: 2017 Update. *Nucleic Acids Res.* **2017,** *45* (D1), D955–D963.
29. Brown, A. S.; Patel, C. J. A Standard Database for Drug Repositioning. *Sci. Data* **2017,** *4*, 170029.
30. Berman, H. M.; Westbrook, J.; Feng, Z.; Gilliland, G.; Bhat, T. N.; Weissig, H.; Shindyalov, I. N.; Bourne, P. E. The Protein Data Bank. *Nucleic Acids Res.* **2000,** *28* (1), 235–242.
31. Liu, X.; Ouyang, S.; Yu, B.; Liu, Y.; Huang, K.; Gong, J.; Zheng, S.; Li, Z.; Li, H.; Jiang, H. PharmMapper Server: A Web Server for Potential Drug Target Identification Using Pharmacophore Mapping Approach. *Nucleic Acids Res.* **2010,** *38* (Web Server issue), W609–W614.
32. Sawada, R.; Iwata, H.; Mizutani, S.; Yamanishi, Y. Target-Based Drug Repositioning Using Large-Scale Chemical-Protein Interactome Data. *J. Chem. Inf. Model* **2015,** *55* (12), 2717–2730.
33. Bild, A. H.; Yao, G.; Chang, J. T.; Wang, Q.; Potti, A.; Chasse, D.; Joshi, M.-B.; Harpole, D.; Lancaster, J. M.; Berchuck, A.; Olson, J. A.; Marks, J. R.; Dressman, H. K.; West, M.; Nevins, J. R. Oncogenic Pathway Signatures in Human Cancers as a Guide to Targeted Therapies. *Nature* **2006,** *439* (7074), 353–357.
34. Furge, K. A.; Tan, M. H.; Dykema, K.; Kort, E.; Stadler, W.; Yao, X.; Zhou, M.; Teh, B. T. Identification of Deregulated Oncogenic Pathways in Renal Cell Carcinoma: An Integrated Oncogenomic Approach Based on Gene Expression Profiling. *Oncogene* **2007,** *26* (9), 1346–1350.
35. Nevins, J. R.; Potti, A. Mining Gene Expression Profiles: Expression Signatures as Cancer Phenotypes. *Nat. Rev. Genet.* **2007,** *8* (8), 601–609.
36. Carrella, D.; Manni, I.; Tumaini, B.; Dattilo, R.; Papaccio, F.; Mutarelli, M.; Sirci, F.; Amoreo, C. A.; Mottolese, M.; Iezzi, M.; Ciolli, L.; Aria, V.; Bosotti, R.; Isacchi, A.; Loreni, F.; Bardelli, A.; Avvedimento, V. E.; di Bernardo, D.; Cardone, L. Computational Drugs Repositioning Identifies Inhibitors of Oncogenic PI3K/AKT/P70S6K-Dependent Pathways Among FDA-Approved Compounds. *Oncotarget* **2016,** *7* (37), 58743–58758.
37. Mottini, C.; Tomihara, H.; Carrella, D.; Lamolinara, A.; Iezzi, M.; Huang, J. K.; Amoreo, C. A.; Buglioni, S.; Manni, I.; Robinson, F. S.; Minelli, R.; Kang, Y.; Fleming, J. B.; Kim, M. P.; Bristow, C. A.; Trisciuoglio, D.; Iuliano, A.; Del Bufalo, D.; Di Bernardo, D.; Melisi, D.; Draetta, G. F.; Ciliberto, G.; Carugo, A.; Cardone, L. Predictive Signatures Inform the Effective Repurposing of Decitabine to Treat KRAS–Dependent Pancreatic Ductal Adenocarcinoma. *Cancer Res.* **2019,***79* (21), 5612–5625
38. World Health Organization. Latest Global Cancer Data: Cancer Burden Rises to 18.1 Million New Cases and 9.6 Million Cancer Deaths in 2018. International Agency for Research on Cancer; World Health Organization: Geneva, 2018.

39. Booth, B.; Zemmel, R. Prospects for Productivity. *Nat. Rev. Drug Discov.* **2004,** *3* (5), 451–456.
40. Waring, M. J.; Arrowsmith, J.; Leach, A. R.; Leeson, P. D.; Mandrell, S.; Owen, R. M.; Pairaudeau, G.; Pennie, W. D.; Pickett, S. D.; Wang, J.; Wallace, O.; Weir, A. An Analysis of the Attrition of Drug Candidates From Four Major Pharmaceutical Companies. *Nat. Rev. Drug Discov.* **2015,** *14* (7), 475–486.
41. Pammolli, F.; Magazzini, L.; Riccaboni, M. The Productivity Crisis in Pharmaceutical R&D. *Nat. Rev. Drug Discov.* **2011,** *10* (6), 428–438.
42. Wilkinson, G. F.; Pritchard, K. In Vitro Screening for Drug Repositioning. *J. Biomol. Screen.* **2014,** *20* (2), 167–179.
43. Iljin, K.; Ketola, K.; Vainio, P.; Halonen, P.; Kohonen, P.; Fey, V.; Grafström, R. C.; Perälä, M.; Kallioniemi, O. High-Throughput Cell-Based Screening of 4910 Known Drugs and Drug-Like Small Molecules Identifies Disulfiram as an Inhibitor of Prostate Cancer Cell Growth. *Clin. Cancer Res.* **2009,** *15* (19), 6070–6078.
44. Skrott, Z.; Mistrik, M.; Andersen, K. K.; Friis, S.; Majera, D.; Gursky, J.; Ozdian, T.; Bartkova, J.; Turi, Z.; Moudry, P.; Kraus, M.; Michalova, M.; Vaclavkova, J.; Dzubak, P.; Vrobel, I.; Pouckova, P.; Sedlacek, J.; Miklovicova, A.; Kutt, A.; Li, J.; Mattova, J.; Driessen, C.; Dou, Q. P.; Olsen, J.; Hajduch, M.; Cvek, B.; Deshaies, R. J.; Bartek, J. Alcohol-Abuse Drug Disulfiram Targets Cancer via p97 Segregase Adaptor NPL4. *Nature* **2017,** *552* (7684), 194–199.
45. Huang, J.; Campian, J. L.; Gujar, A. D.; Tran, D. D.; Lockhart, A. C.; DeWees, T. A.; Tsien, C. I.; Kim, A. H. A Phase I Study to Repurpose Disulfiram in Combination With Temozolomide to Treat Newly Diagnosed Glioblastoma After Chemoradiotherapy. *J. Neuro-Oncol.* **2016,** *128* (2), 259-266.
46. Chateauvieux, S.; Morceau, F.; Dicato, M.; Diederich, M. Molecular and Therapeutic Potential and Toxicity of Valproic Acid. *J. Biomed. Biotechnol.* **2010,** *2010*, 479364.
47. Shim, J. S.; Liu, J. O. Recent Advances in Drug Repositioning for the Discovery of New Anticancer Drugs. *Int. J. Biol. Sci.* **2014,** *10* (7), 654–663.
48. Salentin, S.; Adasme, M. F.; Heinrich, J. C.; Haupt, V. J.; Daminelli, S.; Zhang, Y.; Schroeder, M. From Malaria to Cancer: Computational Drug Repositioning of Amodiaquine Using PLIP Interaction Patterns. *Sci. Rep.* **2017,** *7* (1), 11401.
49. Adasme, M. F.; Parisi, D.; Van Belle, K.; Salentin, S.; Haupt, V. J.; Jennings, G. S.; Heinrich, J.-C.; Herman, J.; Sprangers, B.; Louat, T.; Moreau, Y.; Schroeder, M. Structure-Based Drug Repositioning Explains Ibrutinib as VEGFR2 Inhibitor. *PLoS One* **2020,** *15* (5), e0233089.
50. Li, Y.; Xu, R.; Li, Z.; Mao, S. Global Dynamics of a Delayed HIV-1 Infection Model With CTL Immune Response. *Discrete Dyn. Nat. Soc.* **2011,** *2011*, 673843.
51. Lim, H. S.; Freemantle, N. Interpreting Comparisons Between Clinical Trials. *ASAIO J.* **2019,** *65* (7).
52. Chae, Y. K.; Pan, A. P.; Davis, A. A.; Patel, S. P.; Carneiro, B. A.; Kurzrock, R.; Giles, F. J. Path Toward Precision Oncology: Review of Targeted Therapy Studies and Tools to Aid in Defining "Actionability" of a Molecular Lesion and Patient Management Support. *Mol. Cancer Ther.* **2017,** *16* (12), 2645–2655.
53. Kimmelman, J.; Tannock, I. The Paradox of Precision Medicine. *Nat. Rev. Clin. Oncol.* **2018,** *15* (6), 341–342.

54. Calvo, E.; Walko, C.; Dees, E. C.; Valenzuela, B. Pharmacogenomics, Pharmacokinetics, and Pharmacodynamics in the Era of Targeted Therapies. *Am. Soc. Clin. Oncol. Educ. Book* **2016,** *35*, e175–e184.
55. Waldron, D. A Multi-Layer Omics Approach to Cancer. *Nat. Rev. Genet.* **2016,** *17* (8), 437–437.
56. Huang, H.; Zhang, P.; Qu, X. A.; Sanseau, P.; Yang, L. Systematic Prediction of Drug Combinations Based on Clinical Side-Effects. *Sci. Rep.* **2014,** *4* (1), 7160.
57. Kim, G.; McKee, A. E.; Ning, Y. M.; Hazarika, M.; Theoret, M.; Johnson, J. R.; Xu, Q. C.; Tang, S.; Sridhara, R.; Jiang, X.; He, K.; Roscoe, D.; McGuinn, W. D.; Helms, W. S.; Russell, A. M.; Miksinski, S. P.; Zirkelbach, J. F.; Earp, J.; Liu, Q.; Ibrahim, A.; Justice, R.; Pazdur, R. FDA Approval Summary: Vemurafenib for Treatment of Unresectable or Metastatic Melanoma With the BRAFV600E Mutation. *Clin. Cancer Res.* **2014,** *20* (19), 4994–5000.
58. Guo, L.; Zhang, H.; Shao, W.; Chen, B. Crizotinib as a Personalized Alternative for Targeted Anaplastic Lymphoma Kinase Rearrangement in Previously Treated Patients With Non-Small-Cell Lung Cancer. *Drug Des. Devel. Ther.* **2015,** *9*, 5491–5497.
59. Yao, J. C.; Shah, M. H.; Ito, T.; Bohas, C. L.; Wolin, E. M.; Van Cutsem, E.; Hobday, T. J.; Okusaka, T.; Capdevila, J.; de Vries, E. G. E.; Tomassetti, P.; Pavel, M. E.; Hoosen, S.; Haas, T.; Lincy, J.; Lebwohl, D.; Öberg, K., Rad001 in Advanced Neuroendocrine Tumors, T. T. S. G. Everolimus for Advanced Pancreatic Neuroendocrine Tumors. *N. Engl. J. Med.* **2011,** *364* (6), 514–523.
60. Zhang, Z.; Zhou, L.; Xie, N.; Nice, E. C.; Zhang, T.; Cui, Y.; Huang, C. Overcoming Cancer Therapeutic Bottleneck by Drug Repurposing. *Signal Transduct. Target. Ther.* **2020,** *5* (1), 113.
61. Hanahan, D.; Weinberg, R. A. The Hallmarks of Cancer. *Cell* **2000,** *100* (1), 57–70.
62. Hanahan, D.; Weinberg, R. A. Hallmarks of Cancer: The Next Generation. *Cell* **2011,** *144* (5), 646–674.
63. Weinstein, J. N.; Collisson, E. A.; Mills, G. B.; Shaw, K. R.; Ozenberger, B. A.; Ellrott, K.; Shmulevich, I.; Sander, C.; Stuart, J. M. The Cancer Genome Atlas Pan-Cancer Analysis Project. *Nat. Genet.* **2013,** *45* (10), 1113–1120.
64. Mottini, C.; Napolitano, F.; Li, Z.; Gao, X.; Cardone, L. Computer-Aided Drug Repurposing for Cancer Therapy: Approaches and Opportunities to Challenge Anticancer Targets. *Semin. Cancer Biol.* **2021,** *68*, 59–74.

## CHAPTER 6

# Drug Repurposing and Computational Drug Discovery for Inflammatory Diseases

VISHAL KUMAR SINGH, HIMANI CHAURASIA, JAYATI DWIVEDI, RICHA MISHRA, and RAMENDRA K SINGH

*Bioorganic Research Laboratory, Department of Chemistry, University of Allahabad, Prayagraj, Uttar Pradesh, India*

## ABSTRACT

Questing new molecular entities (NME) as drugs by traditional or *de novo* approach of drug discovery is a lengthy, arduous, and expensive venture. A powerful approach gaining momentum in pharmaceutical field regarding novel drug discovery that restricts the search unto existing drug candidates having authenticated and proven biological compatibility is the process of drug repurposing. This eventually eliminates the prolonged clinical trials and shortens the duration of drug availability under exigency conditions. It amplifies the therapeutic importance of a drug and subsequently intensifies the success rate. Thus, drug repositioning is an emphatic alternative tactic to traditional drug discovery process. Outcomes of several clinical analyses in therapeutics of inflammatory diseases alluded that the drugs acting via synergistic inhibition of multiple targets were likely to be more successful and promising. Keeping this hypothesis intact, this chapter dwells upon a representative set of currently used computational approaches to identify multi-targeted repositionable drugs for inflammatory diseases from a pool of drugs primarily approved for other microbial infections. Furthermore, a method to establish a successful relationship between computational

Drug Repurposing and Computational Drug Discovery: Strategies and Advances.
Mithun Rudrapal, PhD (Ed)

approaches and experimental studies is the integral part while focussing on a unified drug repurposing strategy for better pharmaceutical and biological results. The effective therapeutics thus developed may also act as promising agents in averting drug resistance.

## 6.1 INTRODUCTION

For more than a decade, a paradigm swing in drug development strategies unambiguously linked with a better mechanism of disease biology has been recorded, which permits better treatment of crucial diseases using targeted therapies.[1,2] Usually, a high attrition rate increases the duration of the development of new drugs and this factor becomes a major challenge for the pharmaceutical industries.[3] Acute or chronic inflammatory diseases pose a serious threat and an advanced level of scientific challenge that requires untiring efforts in developing anti-inflammatory drugs.[4,5]

Inflammation is one of the common events in the majority of acute as well as chronic debilitating diseases that represents a major cause of morbidity in the contemporary era of modern lifestyles.[6] It plays a very important role in the pathogenesis of various diseases such as allergies, atherosclerosis, rheumatoid arthritis, asthma, autoimmune diseases, coeliac disease, glomerulonephritis, hepatitis, inflammatory bowel disease, proper fusion injury, transplant rejection, and cancer. Inflammations involve immune cells, molecular mediators, and blood vessels as a protective response. It promotes the elimination of the initial cause of cell injury and also initiates tissue repair. Inflammatory mediators such as TLR-4, TLR-2, iNOS, and interleukins (ILs) drive the inflammation process.[7] Traditionally, drugs against inflammatory diseases were isolated from certain plants, and their extracts were used for relief from inflammations, fever, and pain. In the mid-19th century when salicylate, an anti-inflammatory agent, was discovered as the active form of *Willow* Spp., which triggered its synthesis, and then the acetylsalicylic acid or aspirin trademark was developed. The lack of anti-inflammatory drugs and vectors provokes the need for developing new molecules for the treatment of inflammatory disorders.[8] Current approaches to overcome inflammation include the use of non-steroidal anti-inflammatory drugs (NSAIDs), immune selective anti-inflammatory derivatives, selective glucocorticoid receptor agonists, resolvins/protectins, and TNF inhibitors.[9] These drugs are presently used in the treatment of diseases where cytokines and other non-prostaglandin components of chronic inflammatory and neurodegenerative diseases are manifested. Although drug treatment has been improved to some extent, it is still a challenge for pharmaceutical chemists to explore more effective, potent, and safe therapeutic regimens to treat inflammation and reduce the signs and symptoms of acute inflammation and chronic inflammatory diseases.

Developing novel inflammatory drugs with high efficacy may require a longer duration of research and development efforts. However, seeing the urgent need for drugs against inflammatory diseases, repurposing the existing drugs and focusing on the target may play an important role as it may lead to the development of a rapid and efficient method for combating fatal infections. This screening strategy of the existing drugs has various advantages over de novo drug development like it reduces the cost and associated risks as the pharmacokinetic data with toxicity profiles are already available.

Presently, many scientific groups are working on using the concept of both drug-associated and disease-associated gene sets to identify the novel uses of the existing drugs. The uses of amino acid sequences of target proteins, chemical structures, and chemical protein interaction networks can be utilized to find new molecular target proteins for the existing drugs. Several computational methods, such as molecular docking and dynamics simulations have proved to be very promising in identifying the novel target of the existing drugs in the current scenario. More than 10 online or licensed platforms are currently available for such types of simulations. Among them, Discovery Studio (DS) software for molecular docking studies and GROMACS software for molecular dynamics (MD) simulations are the most trending ones. These software packages are very useful in predicting the interactions between the existing drugs and target protein receptors and in studying the stability of the ligand-protein complexes.[10]

The present book chapter focuses on discussing the repurposed drugs used for the treatment of inflammatory diseases and possible in silico approaches for the identification of the existing drugs as anti-inflammatory agents, and the design and development of newer drugs against inflammatory diseases.

## 6.2 GLOBAL INCIDENCE OF INFLAMMATORY DISEASES

The advancement in the system of healthcare is helpful in the early diagnosis of various diseases. It also decreases the death rate and increases the life of the individual patient. For the past few centuries, clinical medicine complexity has been drastically increasing, leading to continuous refinement in the measurement of non-fatal health loss. As new diseases are rising, diagnostic categorization structure is expanding and the disability metrics are becoming better.[11]

The incidence of inflammatory disease burden is rising all over the world along with variations in disease trends in different regions in different countries. According to the data from the Global Burden of Diseases, Injuries, and Risk Factors Study (GBD) 2017, inflammatory disease cases reported in 2017 were 6.8 million. This rate has been increasing in age-standardized incidences from 79.5 (1990) to 84.3 (2017) per 100,000 population. However, the rate of death is decreasing from 0.61 (1990) to 0.51 (2017) per 100,000 population. Country-wise, the highest rate of age-standardized incidence has been recorded in the USA (464.5), and then the UK (449·6). The total year lived with a disability is almost double in inflammatory diseases from 0.56 million (1990) to 1.02 million (2017).[12]

Patients with inflammatory diseases are marked with the improved condition by introducing biological therapy using anti-TNF. Various antibodies of anti-TNF are approved in the USA and Europe for the clinical therapy of inflammatory diseases. However, lots of patients have to drop the treatment because of primary or secondary resistance caused within the year of the start of treatment. There are various factors such as pharmacological, clinical, patient-related, etc., identified as resistance to anti-TNF therapy (Fig. 6.1).

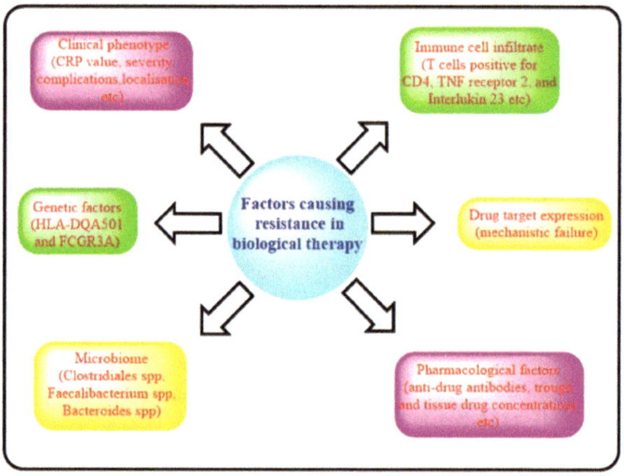


**FIGURE 6.1** Schematic presentation of factors responsible for resistance during inflammation therapy. CD4, cluster of differentiation 4; CRP, C-reactive protein; FCGRA, Fc fragment of IgG receptor IIIa; HLA-DQA, human leukocyte antigen-DQ alpha; TNF, tumor necrosis factor.

## 6.3 REPURPOSING APPROACHES FOR DRUG DISCOVERY

Considering the high attrition rates, hefty costs and slow pace of *de novo* drug discovery and development, repositioning of available drugs to treat both prevalent and rare diseases is becoming a flourishing proposition as it involves the usage of compounds with potentially shorter development timelines and lower development costs (Fig. 6.2).

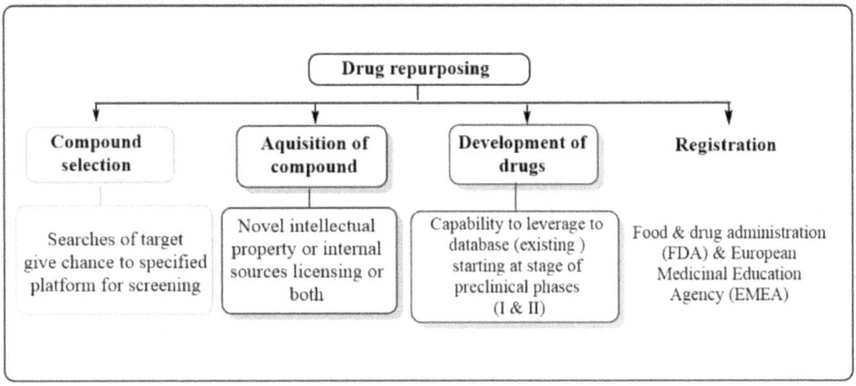


**FIGURE 6.2** Step-wise process of drug repurposing.

Generally, a drug repositioning approach consists of three steps—the first being the identification of a plausible candidate molecule (hypothesis generation), the second step is the mechanistic assessment of the drug in the preclinical model, and lastly, efficacy evaluation in phase II clinical trials (only if there is sufficient data regarding safety and toxicity from phase I). Of these three steps, the first step is where modern approaches for generating hypotheses could be most significantly used for the identification of the appropriate molecule for a particular interest. These systematic approaches can be classified into experimental approaches and computational approaches, both of which are elaborately used (Fig. 6.3). On the basis of clinical data, drug repurposing is encompassed into two broad categories. These approaches have led to the identification of a number of plausible drug candidates, some of which are already approved for disease and some are in advanced clinical stages.

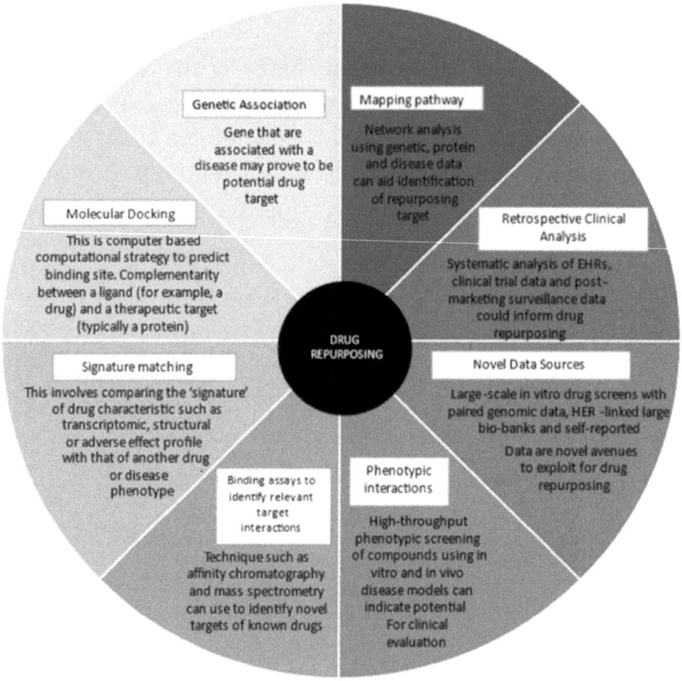


**FIGURE 6.3** Approaches used in drug repurposing.

An inflammatory response is mainly to protect the host from infection and injury and to maintain homeostasis of the body which is a reflex process.[13] Generally, ailment and fatality are primarily caused by inflammatory disorders such as allergies, asthma, autoimmune diseases, and sepsis.

By the use of zebrafish screening, 251 drugs have been identified with significant anti-inflammatory effects. This includes 22.4% of the drugs available in the library. Out of these, 43.9% are NSAIDs and 51.9% are corticosteroids.[14] Some repurposed drugs showing anti-inflammatory properties are listed in Table 6.1.

**TABLE 6.1** Repositioned Drugs for Targeting Various Diseases.

| S. No. | Medicine | Original indication | Repurposed for |
|---|---|---|---|
| 1 | *Artemisia apiacea* Hance | Malaria | Atopic dermatitis (AD) |
| 2 | Methylthiouracil (MTU) | Thyroid | Sepsis |
| 3 | Methotrexate | Cancer | Rheumatoid arthritis |

**TABLE 6.1** *(Continued)*

| S. No. | Medicine | Original indication | Repurposed for |
|---|---|---|---|
| 4 | Topiramate | Epilepsy | Inflammatory bowel disease |
| 5 | Niflumic acid | Analgesic | Osteoarthritis and rheumatoid arthritis |
| 6 | Heparin | Anti-coagulant/heart attack | Asthma |
| 7 | Mangiferin | Cancer | Sepsis |
| 8 | Rifampicin | Anti-biotic | AD |
| 9 | Simvastatin | Heart attack | Sepsis and asthma |
| 10 | Rapamycin | Anti-tumor | Asthma |

Anti-inflammatory drugs that have been repurposed for various types of diseases such as sepsis, asthma, AD, etc., are discussed in this section.

### 6.3.1 REPURPOSING DRUGS FOR SEPSIS

Sepsis is a systematic inflammatory response induced mainly by infection. Since 2001, FDA-approved, recombinant-activated protein C (APC), was the only available drug for sepsis and septic shock therapy. Later, in October 2011, as a result of side effects and lack of efficiency, APC was withdrawn,[15] hence search for novel therapeutics against sepsis is still a necessity. Drug repurposing can be utilized in severe situations where the currently prescribed drugs are not efficient.[16] Some common repurposed drugs for sepsis are discussed below.

#### 6.3.1.1 METHYLTHIOURACIL

MTU, an antithyroid drug was introduced as a thionamide for the treatment of hypothyroidism.[16–18] The anti-septic effect of MTU could be due to its ability to inhibit the release of high mobility group box 1 protein (HMGB1) and HMGB1-mediated inflammatory responses.[19] It has been repurposed for the treatment of sepsis involving multiple organ failure by CLP (cecum ligation and puncture) such as renal injury, liver injury, and overall tissue injury.[16]

#### 6.3.1.2 SIMVASTATIN

It is adopted for the treatment of hypercholesterolemia and hypertriglyceridemia. It lowers cholesterol synthesis by the inhibition of 3-hydroxy-3-methylglutaryl-CoA reductase and is significantly used in hyper-lipidermia to lower the risk of atherosclerotic complications.[20] Simvastatin has been repurposed as an anti-sepsis drug and has displayed improvement in survival rates of patients with multiple organ dysfunction syndrome.[21] Simvastatin has displayed signs of the prevention of loss of integrity of the blood–brain barrier (BBB), caused by polymicrobial sepsis. Additionally, it has anti-inflammatory and anti-oxidative properties.

#### 6.3.1.3 MANGIFERIN

Mangiferin possesses antioxidant, immunomodulatory, antitumor, and antiviral activities.[22] Additionally, it has been adopted for hypoglycemic activity. It has been repurposed for treating sepsis-induced acute kidney injury (AKI), which includes inflammatory reactions by systemic cytokine storm or the production of local cytokines.

### 6.3.2 REPURPOSING DRUGS FOR ASTHMA

Complex and multifactorial pathogenesis has been reported for asthma which has affected over 300 million people globally. Primary mediation of inflammatory response in asthma is by Th2-lymphocytes which are characterized by the production of Th2 cytokines, mucus hypersecretion, pulmonary eosinophilia, expression of inflammatory factors, and allergen-specific immunoglobulin E (IgE).[23,24] Th2 cytokines and IgE play a significant role in allergic asthma causing airway inflammation, airway hyperreactivity, etc. Some common repurposed drugs for asthma are discussed below:

#### 6.3.2.1 RAPAMYCIN

It is used to treat a rare lung disease called lymphangioleiomyomatosis and in preventing organ transplant rejection. Due to mTOR inhibition, this compound displays immunosuppressive functions and antiproliferative properties. It has been repurposed in the new clinical regimen for asthma.

Suppression of allergen-induced IL-13 and leukotriene levels has been reported for rapamycin. In addition to this, IL-13, and IgE are completely reduced by rapamycin.[25]

#### *6.3.2.2 HEPARIN*

Heparin is an anticoagulant and is primarily used in the treatment of arterial thromboembolism and to prevent deep vein thrombosis. Additionally, it has been used in the treatment of heart attacks and unstable angina. Anti-inflammatory properties, including asthma, have been reported in various low molecular weight heparin (LMWH).

### *6.3.3 REPURPOSING DRUGS FOR ATOPIC DERMATITIS*

AD is an inflammatory skin disorder, and it is accompanied by increased serum levels of IgE due to increased inflammatory infiltration.[26,27] Mast cells release histamine which is responsible for hypersensitivity and has the potential as a vasoactive agent. Some common repurposed drugs for AD are discussed in the following.

#### *6.3.3.1 RIFAMPICIN*

Rifampicin is used to treat various types of bacterial infections such as tuberculosis, leprosy, etc. It is reported to suppress inflammatory responses and play a key role in relieving neuropathic pain and helping in immune modulation. Rifampicin showed a decrease in the elevated serum levels of IgE and IL-4, which consequently led to anti-AD activity.[28]

#### *6.3.3.2 ARTEMISIA APIACEA HANCE*

It is a traditional medicine used to treat fever, eczema, and jaundice mostly in east Asian countries like China, Korea, and Japan. Artemisinin was isolated and developed as an active antimalarial drug. The repurposing of *A. apiacea* in the treatment of dermatitis was demonstrated.[29] Proinflammatory cytokines and chemokines expression was found to be regulated by ethanolic extracts of *A. apiacea* Hance (EAH) in allergic inflammation. EAH are reported to inhibit the formation of chemokines and pro-inflammatory cytokines.

## 6.4 COMPUTATIONAL APPROACHES FOR DRUG DISCOVERY

The computational approach of drug repurposing is also known as "in silico drug repurposing," which belongs to the field of computational pharmacology. This approach is classified into two parts – the first being the discovery of new indications for an existing drug (drug centric) and the second one is the identification of effective drugs for a disease (disease centric).

Desirable pharmacokinetic and toxicity profiles are of significant importance for the failure or success of drug development. Therefore, in silico techniques have been widely recognized and the drug absorption, distribution, metabolism, excretion, and toxicity (ADMET) properties are being considered at an early stage to reduce failure rates in the clinical phase of drug discovery. Some important tools and techniques used during in silico approaches are as follows:

- Molecular modeling: Molecular modeling focuses on predicting the strength of the interaction between a potent molecule and a transporter or metabolic enzyme.
- QSAR: It is used for the prediction of the pharmacokinetic properties relying mainly on traditional models or constructed data sets developed using the software.
- PBPK: It has been used to predict pharmacokinetics by using some software programs.
- ADMET: ADMET properties play a significant role and can be predicted using online software.
- Molecular docking: It focuses on the physical interactions between plausible drugs and their specific targets. Here, the chemical and physical binding of drugs to the protein of interest is studied.
- MD simulations: MD simulation represents a proper way to study atomic-level information about binding of ligands to targeted proteins. On the basis of MD trajectories, the root-mean-square deviation (RMSD), root-mean-square fluctuation (RMSF), the radius of gyration (Rg), number of hydrogen bonds and binding free energy of the complexes are predicted to analyze and get insight into their structural stabilities, binding modes, and binding strengths of the designed drug candidates.[30,31]

## 6.5 ADVANTAGES OF REPURPOSING OF DRUGS

Many invaluable advantages such as reduction in drug development cost, easy availability of drug components, higher success rates, and acceleration of the drug development process have been observed by integration or adoption of drug repositioning. Repurposed drugs are already toxicologically assessed and consideration of all safety measurements reduces the chances of failure to a great extent. Known drugs with new targets or those with known mechanisms for new indications are considered in the repurposing or repositioning of drugs.[32]

Drug companies have lots of advantages with drug repositioning because new drug synthesis using conventional technology is a time-consuming process.[33] However, the high rate of drug failure of newly synthesized drugs causes a major reason for using the repositioning strategy. A new drug can be obtained by better treatment options with comparatively low cost as well as a low time consumption period.[34–37] Many scientists are working to find out the alternative uses of generic or approved drugs for human welfare worldwide. Table 6.2 represents the drug repurposing target in many inflammatory mediators.

**TABLE 6.2** Repositioned Drugs in Inflammatory Diseases Targeting Inflammatory Mediators.

| Sl. No. | Drug | Inflammatory mediator | Disease |
|---|---|---|---|
| 1 | Simvastatin | IL-1, IL-6, IL-8, IL-12, CD4 T-cell, Th2, ICAM-1, and VCAM-1 | Sepsis and asthma |
| 2 | Mangiferin | Nrf2 expression; IL-1 and IL-18; and NLRP3 | Sepsis |
| 3 | *A. apiacea* Hance | IκIBσ, NF-κB p65, p38, RANTES, IL-8, IL-6, and TARC | AD |
| 4 | Rapamycin | mTOR, IL-13, and IgE | Asthma |
| 5 | Heparin | iNOS, Il-4, ARG1, and ARG2 | Asthma |
| 6 | Rifampicin | Histamine, β-HEX, PGD2, proinflammatory cytokines, TNF-α, and COX-2 | AD |
| 7 | Methylthiouracil | TLR2, TLR4, RAGE, and p38, NF-κB | Sepsis |

β-HEX, β-N acetylhexosaminidase Hexosaminidase A; ARG1, arginase 1; ARG2, arginase 2; ICAM-1, intercellular adhesion molecule 1; IgE, immunoglobin E; iNOS, inducible nitric oxide synthase; IL, interleukin; mTOR, mammalian target of rapamycin; NF-κB, nuclear factor kappa-light-chain-enhancer of activated B cells; Nrf2, nuclear factor erythroid 2 (NFE2)-related factor 2; p38, protein kinase 38; PGD2, prostaglandin D2; RAGE, receptor for advanced glycation end products; TLR2, toll-like receptor 2; TLR4, toll-like receptor 4; TNF-α, tumor necrosis factor alpha; VCAM-1, vascular cell adhesion protein 1.

For characterization as well as defining the drug activity, zebrafish presents a valuable platform. Larvae and embryos of zebrafish are used in the bioassay of drug discovery.[38] Table 6.3 represents the considerable top 12 hits of zebrafish using anti-inflammatory screening.

**TABLE 6.3** Some Significant Drugs and Their Role as Anti-Inflammatory Agents.

| S. No. | Drug name | Generic name | Therapeutic group | Mechanism of action |
|---|---|---|---|---|
| 1 | Amodiaquine | Amdaaquin, amobin, amochin, basoquin, trimalact, camoquine, larimal | Antimalarial | Heme polymerase inhibitor |
| 2 | Etodolac | Etodol, etodolac, apeotex, ETOFACT, etogesic | Anti-inflammatory (NSAIDs) | COX inhibitor |
| 3 | Pinacidil | Pindac | Vasodilator | $K^{+}$channel $Ca^{2+}$activator |
| | Mafenide hydrochloride | Sulfamylon, abamide, emilene, homonal malfamin, marfanil | Antibacterial | Inhibitors of folic acid biosynthesis |
| 4 | Clonidine hydrochloride | Clonidine, mylan, clonidural, cloniprex clonistada, clonnirit, clophelin | Antihypertensive | Alpha 2 antagonist/ imidazoline agonist |
| 5 | Acetohexamide | Dymelor | Antidiabetic (type-II non-insulin dependent) | Blocks ATP-sensitive K plus channels/ stimulates insulin release |
| 6 | Fludrocortisone acetate | Flrinef, astoni-h, cortineff, florineff acetate, astonin, merk, lonikan | Mineralocorticoid | Causes kidneys to retain sodium |
| 7 | Niflumic acid | Donalgin, flogovital, forenal, niflactol, niflam, landdruma | Anti-inflammatory (NSAIDs) | COX inhibitor |
| 8 | Methyldopa | Aldomet, aldochlor, aldopren, aldotensin, alfametildopa, datleal, dopagrand | Antihypertensive | L-aromatic amino acid Decarboxylase inhibitor |

**TABLE 6.3** *(Continued)*

| S. No. | Drug name | Generic name | Therapeutic group | Mechanism of action |
|---|---|---|---|---|
| 9 | Nefopam hydrochloride | Acupainlex, acupain, acuten, anton, benoton, glosic, ketopen, licopam, neforex, paton, sezen, tonfupin, xripa | Analgesic | Unknown |
| 10 | Alfuzosin hydrochloride | Uroxatral, rilif, tevax, unibenestan, urion, uroXatral OD, weiping, xatger, zofu | Antihypertensive | Alpha 1-adrenergic antagonist |
| 11 | Chlorphenesin carbamate | Cloricool, kalsont, maolate, muslax rinlaxer, skenesin, steacol | Muscle relaxant | Blocks nerve impulses |
| 12 | Etodolac | Etodol, etodolac, apeotex, ETOFACT, etogesic | Anti-inflammatory (NSAIDs) | COX inhibitor |

NSAIDs: non-steroidal anti-inflammatory drugs.

## 6.6 CONCLUSION

Increasing interest in drug repositioning has been developed due to the sustained high failure rates and costs required to bring new drugs to market. In the case of anti-inflammatory drugs, the risk is amplified by the looming threat of drug resistance and the pressing need for flawless strategies to tackle the problem. Indeed, in all likelihood, it requires a widespread effort from public-private partnerships, non-profit groups, academic researchers, and companies to successfully investigate and approve drugs for other indications. Academic laboratories and small biotech companies have frequently discovered new activities for existing drugs, but the translation of these discoveries to the clinic requires additional sophistication available with large pharmaceutical companies. There was hope that led discovery based on drug repurposing could deliver the next generation of anti-inflammatory agents. Indeed, new momentum is emerging from government-led initiatives such as the NIH program, the National Center for Advancing Translational Science, and more recently, the Medical Research Council in the United Kingdom has invested in this area. Overall, these initiatives lend credence to the potential of drug repurposing and should encourage repurposing in academia and large pharmaceutical companies alike.

## KEYWORDS


- **drug repurposing**
- **drug resistance**
- **antibiotics**
- **inflammatory diseases**
- **synergistic**


## REFERENCES


1. Elliott, M. J. Treatment of Rheumatoid Arthritis With Chimeric Monoclonal Antibodies to Tumor Necrosis Factor Alpha. *Arthritis Rheumatol.* **1993,** *36*, 1681–1690.
2. Weinblatt, M. E. A Trial of Etanercept, A Recombinant Tumor Necrosis Factor Receptor: Fc Fusion Protein, in Patients With Rheumatoid Arthritis Receiving Methotrexate. *N. Engl. J. Med.* **1999,** *340*, 253–259.
3. Kalunian, K. C. A Phase II Study of the Efficacy and Safety of Rontalizumab (rhuMAb Interferon-α) in Patients With Systemic Lupus Erythematosus (ROSE). *Ann. Rheum. Dis.* **2016,** *75*, 196–202.
4. Furie, R. A. Phase III, Randomized, Placebocontrolled Study of Belimumab, A Monoclonal Antibody That Inhibits B Lymphocyte Stimulator, in Patients With Systemic Lupus Erythematosus. *Arthritis Rheumatol.* **2011,** *63*, 3918–3930.
5. Mease, P. J. Etanercept Treatment of Psoriatic Arthritis: Safety, Efficacy, and Effect on Disease Progression. *Arthritis Rheumatol.* **2004,** *50*, 2264–2272.
6. Riphagen, S. Hyperinflammatory Shock in Children During COVID-19 Pandemic. *Lancet* **2020,** *395*, 1607–1608.
7. Chaban, T. I. Development of Effective Anti-Inflammatory Drug Candidates Among Novel Thiazolopyridines. *Ukr Biochem J.* **2020,** *92*, 132–139.
8. Nunes, C. D. R. Plants as Sources of Anti-Inflammatory Agents. *Molecules* **2020,** *25 (16),* 3726.
9. Roan, F.; Obata-Ninomiya, K.; Steven, F. Z. Epithelial Cell–Derived Cytokines: More Than Just Signaling the Alarm. *J. Clin. Investig.* **2019,** *129 (4)*, 1441–1451.
10. Singh, V. K.; Chaurasia, H.; Mishra, R.; Srivastava, R.; Naaz, F.; Kumar, P.; Singh, R. K. Docking, ADMET Prediction, DFT Analysis, Synthesis, Cytotoxicity, Antibacterial Screening and QSAR Analysis of Diarylpyrimidine Derivatives. *J. Mol. Struct.* **2022,** *1247*, 131400.
11. James, S. L.; Abate, D.; Abate, K. H.; Abay, S. M.; Abbafati, C.; Abbasi, N.; Abbastabar, H.; Abd-Allah, F.; Abdela, J.; Abdelalim, A.; Abdollahpour, I. Global, Regional and National Incidence, Prevalence, and Years Lived With Disability for 354 Diseases and

Injuries for 195 Countries and Territories, 1990–2017: A Systematic Analysis for the Global Burden of Disease Study 2017. *Lancet* **2018,** *392*, 1789–1858.
12. Atreya, R.; Neurath, M. F. Mechanisms of Molecular Resistance and Predictors of Response to Biological Therapy in Inflammatory Bowel Disease. *Lancet Gastroenterol. Hepatol.* **2018,** *3*, 790–802.
13. Medzhitov, R. Origin and Physiological Roles of Inflammation. *Nature* **2008,** *454*, 428–443.
14. Hall, C. J.; Wicker, S. M.; Chien, A. T.; Tromp, A.; Lawrence, L. M.; Sun, X.; Krissansen, G. W.; Crosier, K. E.; Crosier, P. S. Repositioning Drugs for Inflammatory Disease - Fishing for New Antiinflammatory Agents. *Dis. Model. Mech.* **2014,** *7*, 1069–1081.
15. Bernard, G. R.; Vincent, J. L.; Laterre, P. F.; LaRosa S. P.; Dhainau, J. F.; Lopez-Rodriguez, A.; Steingrub, J. S.; Garber, G. E.; Helterbrand, J. D.; Ely, E. W.; Fisher, C. J.; Recombinant Human Protein C Worldwide Evaluation in Severe Sepsis (PROWESS) Study Group. Efficacy and Safety of Recombinant Human Activated Protein C for Severe Sepsis. *N. Engl. J. Med.* **2001,** *344*, 699–709.
16. Kwak, S.; Ku, S. K.; Kang, H.; Baek. M. C.; Bae, J. S.; Methylthiouracil, A New Treatment Option for Sepsis. *Vascul. Pharmacol.* **2015,** *88*, 1–10.
17. Glock, G. E. Methyl-thiouracil and Thiouracil as Antithyroid Drugs. *Pharmacol. Chemother.* **1946,** *1*, 127–134.
18. Reddy, A. R.; Kaul, A.; Effect of Methyl Thiouracil on Radioiodine Thyroidal Retention in Rats. *Radiat. Environ. Biophys.* **1949,** *16*, 347–354.
19. Prakash, A. V.; Park, J. W.; Seong, J. W.; Kang, T. J. Repositioned Drugs for Inflammatory Diseases Such as Sepsis, Asthma, and Atopic Dermatitis. *Biomol. Ther.* **2020,** *28* (3), 222.
20. Robinson, J. G. Simvastatin: Present and Future Perspectives. *Expert Opin. Pharmacother.* **2007,** *8*, 2159–2127.
21. Schmidt, H.; Hennen, R.; Keller, A.; Russ, M.; Müller-Werdan, U.; Werdan, K.; Buerke, M. Association of Statin Therapy and Increased Survival in Patients With Multiple Organ Dysfunction Syndrome. *Intensive Care Med.* **2006,** *32*, 1248–1251.
22. Guha, S.; Ghosal, S.; Chattopadhyay, U. Antitumor, Immunomodulatory and Anti-HIV Effect of Mangiferin, a Naturally Occurring Glucosylxanthone. *Chemotherapy* **1996,** *42*, 443–445.
23. Barnes, P. J.; Cellular and Molecular Mechanisms of Asthma and COPD. *Clin. Sci. (Lond.)* **2017,** *131*, 1541–1558.
24. Barnes, P. J. Therapeutic Approaches to Asthma-Chronic Obstructive Pulmonary Disease Overlap Syndromes. *J. Allergy Clin. Immunol.* **2015,** *136*, 531–545.
25. Mushaben, E. M.; Kramer, E. L.; Brandt, E. B.; Khurana Hershey, G. K.; Le Cras, T. D. Rapamycin Attenuates Airway Hyperreactivity, Goblet Cells, and IgE in Experimental Allergic Asthma. *J. Immunol.* **2011,** *187*, 5756–5763.
26. Bergmann, R. L.; Edenharter, G.; Bergmann, K. E.; Forster, J.; Bauer, C. P.; Wahn, V.; Zepp, F.; Wahn, U. Atopic Dermatitis in Early Infancy Predicts Allergic Airway Disease at 5 Years. *Clin. Exp. Allergy* **1998,** *28*, 965–970.
27. Chang T. W.; Shiung, Y. Y. Anti-IgE as a Mast Cell-Stabilizing Therapeutic Agent. *J. Allergy Clin. Immunol.* **2006,** *117*, 1203–1212.
28. Kim, S. H.; Lee, K. M.; Lee, G. S.; Seong, J. W.; Kang, T. J.; Rifampicin Alleviates Atopic Dermatitis-Like Response In Vivo and In Vitro. *Biomol. Ther. (Seoul)* **2017,** *25*, 634–640.

29. Ryu, J. C.; Park, S. M.; Hwangbo, M.; Byun, S. H.; Ku, S. K.; Kim, Y. W.; Kim, S. C.; Jee, S. Y.; Cho, I. J. Methanol Extract of *Artemisia apiacea* Hance Attenuates the Expression of Inflammatory Mediators via NF-kappaB Inactivation. *Evid. Based Complement. Altern. Med.* **2013,** 494681.
30. Singh, V. K.; Chaurasia, H.; Kumari, P.; Som, A.; Mishra, R.; Srivastava, R.; Naaz, F.; Singh, A.; Singh, R. K. Design, Synthesis, and Molecular Dynamics Simulation Studies of Quinoline Derivatives as Protease Inhibitors Against SARS-CoV-2. *J. Biomol. Struct. Dyn.* **2021,** 1–24.
31. Singh, V. K.; Srivastava, R.; Gupta, P. S. S.; Naaz, F.; Chaurasia, H.; Mishra, R.; Rana, M. K.; Singh, R. K. Anti-HIV Potential of Diarylpyrimidine Derivatives as Non-nucleoside Reverse Transcriptase Inhibitors: Design, Synthesis, Docking, TOPKAT Analysis and Molecular Dynamics Simulations. *J. Biomol. Struct. Dyn.* **2021,** *39* (7), 2430–2446.
32. Ashburn, T. T.; Thor, K. B. Drug Repositioning: Identifying and Developing New Uses for Existing Drugs. *Nat. Rev. Drug Discov.* **2004,** *3* (8), 673–683.
33. Chong, C. R.; Sullivan, D. J. New Uses for Old Drugs. *Nature* **2008,** *448* (7154), 645–646.
34. Tobinick, E. L. The Value of Drug Repositioning in the Current Pharmaceutical Market. *Drug News Perspect.* **2007,** *22* (2), 119–125.
35. Sleigh, S. H.; Barton, C. L. Repurposing Strategies for Therapeutics. *Pharmaceut. Med.* **2010,** *24* (3), 151–159.
36. Phelps, K. Repositioning Drugs to Enhance a Product's Lifecycle. *Drug Discov. Today Ther. Strateg,* **2011,** *8* (3–4), 97–101.
37. Sardana, D.; Zhu, C.; Zhang, M.; Gudivada, R. C.; Yang, L.; Jegga, A. G. Drug Repositioning for Orphan Diseases. *Brief. Bioinform.* **2011,** *12* (4), 346–356.
38. Rihel, J.; Prober, D. A.; Arvanites, A.; Lam, K.; Zimmerman, S.; Jang, S.; Haggarty, S. J.; Kokel, D.; Rubin, L. L.; Peterson, R. T.; Schier, A. F. Zebrafish Behavioral Profiling Links Drugs to Biological Targets and Rest/Wake Regulation. *Science* **2010,** *327* (5963), 348–351.

## CHAPTER 7

# Drug Repurposing and Computational Drug Discovery for Cardiovascular Disorders

JOHRA KHAN[1,2] and MITHUN RUDRAPAL[3]

[1]*Department of Medical Laboratory Sciences, College of Applied Medical Sciences, Majmaah University, Al Majmaah, Saudi Arabia*

[2]*Health and Basic Sciences Research Center, Majmaah University, Al Majmaah, Saudi Arabia*

[3]*Department of Pharmaceutical Sciences, School of Biotechnology and Pharmaceutical Sciences, Vignan's Foundation for Science, Technology & Research (Deemed to be University), Guntur, Karnataka, India*

## ABSTRACT

The recent reports on developments in pharmaceutical research shows a decline in success of new pharmaceutical compounds in market last ten years due to which the time of new drug development also increased from 9 to 14 years. Drug repurposing is a method to use old drugs by repositioning them, which can reduce the new drug development time and cost. Drug repurposing also helps in predicting new therapeutic potentials of old drugs approved by FDA. Some of the commonly repurposed drugs are: metformin designed for diabetes type 2, now repurposed for cancer therapeutic and under clinical trial phase III, sildenafil and thalidomide designed for sickness and angina now repurposed for leprosy and erectile dysfunction. Due to limited data availability on cardiovascular profile of anti-inflammatory drugs the use of

Drug Repurposing and Computational Drug Discovery: Strategies and Advances.
Mithun Rudrapal, PhD (Ed)

these drugs are limited. Drug repurposing is a great tool to identify safety of different drugs in CVD condition. The data of different studies available till now shows possibility of monoclonal antibodies as target therapy for CVDs in coming years. Data of anti-inflammatory drugs like colchicine that have been repurposed for CVDs shows promising results in patients of STEMI. Similarly Metformin, an anti-diabetic drug also shows effective inflammatory process control and cardioprotective properties. The future studies need more evidences from clinical trials to better repurpose these drugs and their approval to be used in CVDs.

## 7.1 INTRODUCTION

The recent reports on developments in pharmaceutical research show a decline in the success of new pharmaceutical compounds in market in the last ten years due to which the time of new drug development also increased from 9 to 14 years.[1] Drug repurposing is a method to use old drugs by repositioning them, which can reduce the new drug development time and cost.[2] Drug repurposing also helps in predicting new therapeutic potentials of old drugs approved by FDA. Some of the commonly repurposed drugs are: Metformin designed for diabetes type 2, now repurposed for cancer therapeutic and under clinical trial phase III, Sildenafil and thalidomide were designed for sickness and angina now repurposed for leprosy and erectile dysfunction.[3]

### *7.1.1 CARDIOVASCULAR DISORDERS AND ITS CLASSIFICATION*

Cardiovascular disorders (CVDs) are a cluster of diseases including injuries that affect heart and blood vessels or cardiovascular system.[4] The risk of cardiovascular disease increases with age and most cardiologists believe that after 35 years of age if any risk related to CVDs present confirms the beginning of CVDs already begins.[4–5] CVDs are one of the leading causes of deaths around the world. The global burden of cardiovascular disease is on rise, especially in high-income countries at an alarming rate during the last decade.[6] The prevalence of CVDs increased from 257 million to 285 million in the last five years and the total death increased from 12.1 million to 19.7 million in the last decade.[7]

CVDs are classified on the basis of cardiovascular system affected. Some of the types are congenital heart disease, peripheral arterial disease, coronary heart disease, aortic aneurysm, angina, stroke, rheumatic heart disease, congenital heart disease, deep vein thrombosis, and many others less known forms.[8]

Congenital heart disease (CHD), also referred to as atherosclerotic heart and coronary artery disease, occurs due to deposition of atheromatous plaque in arteries supplying blood to myocardium.[9] The symptoms of CHD occur in the last state of disease and in most cases it remains silent and sudden heart attack only reveals its presence and ruptured plaque causes blockage in blood flow resulting in sudden death.[10] Peripheral arterial disease is caused by blockage due to fatty material built up in peripheral arterial supplying blood to legs. These blockages further narrow the arteries supplying blood to heart causing angina or sudden heart attack.[11] If the peripheral arteries to the neck also get affected, it affects the blood flow to brain causing stroke.[12] Congenital heart disease represents a group of abnormalities including structural and functional abnormalities in heart present before birth due to developmental errors or genetic disorders.[13] Coarctation of aorta is also congenital heart diseases that remain silent without causing any complications in the form of a small ventricular septal defect.[14] Some of the congenital heart diseases can be treated with medicine while for some it needs surgeries.[9] Due to development in medical technologies and recent surgery techniques the death rate due to congenital heart disease reduced to 5% in comparison with the 1970s data recorded which was 30%.[15]

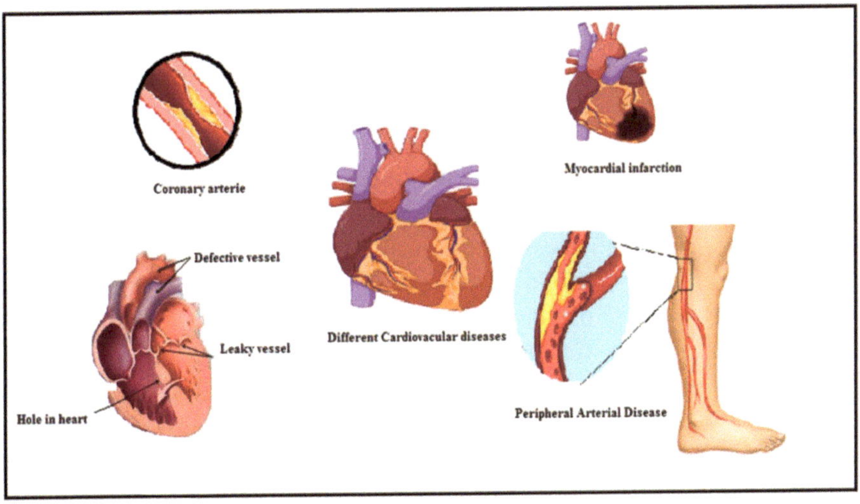


**FIGURE 7.1** Different types of cardiovascular diseases.

### *7.1.2 COMPUTATIONAL DRUG DISCOVERY REPURPOSING*

Computational drug repurposing involves data mining, target analysis, machine learning, and network analysis.[16] Computational repurposing investigates the relation between various biomedical factors such as diseases, genes, drug types, drug targets, and adverse drug reactions.[17] The target-based computational method tries to understand different binding site with compound–protein interfaces, and the disease-based approach tries to discover new indications for drug repurposing on the basis of differences and similarities of different diseases.[17c] Some of the common approaches used in computational repurposing are to study chemical and structural similarities of target-based approaches and binding sites to locate a new target.[18] Researchers such as Schroeder et al. and Moriaud et al. reviewed different tools like high-throughput screening and protein structure binding site analysis for better drug repurposing.[19–19b] MED-SuMo is a new approach some researchers used to scan whole protein surface without considering the prior data of protein docking sites.[20] Rare Disease Repurposing Database (RDBD) is a database provided by FDA with novel resources.[21] RDBD consists of a list of two hundred known products with a potential of repurposed for many rare disease conditions.[22] Jegga et al. discussed different intrinsic difficulties of using the computational method and its dependence on data related to old datasets, literature mining ontology modeling, and structure types.[23]

## 7.2 COMPUTATIONAL MODEL OF CARDIOVASCULAR DISEASE

Developments in treatment techniques made cardiac patient care and patient-specific modeling is entering into a new era of cardiac modeling.[24] Computational model based on patient-specific requirements helps to offer a framework that can address all the challenges of anatomy and pathophysiology of individual patient.[25] Computational model is advent due to their capacity of a combined effect on hemodynamic of different cardiovascular properties.[26] A lot of studies were carried out on an animal model and on a theoretical framework and on a mock circulatory system to better understand clinical data-based modeling.[27] Many recent studies expressed some features of cardiac functioning that can be measured by integrating some data available in a clinical setting in a cardiovascular model.[28]

### 7.2.1 COMPUTATIONAL HEART DISEASE MODELS

The first computational heart disease model was cellular model which delivered a physiological and physical constrained framework for quantitative combined measurement.[29] The cellular model was comprised of $Ca^{2+}$, $Na^+$, and $K^+$ channels with other physiological processes including pH, β-adrenergic stimulation, and $Ca^{2+}$ homeostasis in rabbit and human cardiac myocytes to forecast the effect of doxorubicin on $Ca^{2+}$ transient and action potential.[30] Another known computational model is based on fluid dynamics for analyzing different computer-based simulations such as heat transfer and fluid flow. Initially, computational fluid dynamics was limited only to high-technology engineering areas, but with modifications in technology it became a very powerful tool in many complex human anatomy and fluid behavior understanding.[31]

In many recent studies researchers used computational simulation tool to predict behavior of blood circulation and flow in the human body.[31] It also provides very specific information that cannot be obtained experimentally.[32] Computational fluid dynamics is mostly used to study the fluid movement-related phenomenon in the vascular system and to predict blood flow in abnormal or defected artery. Computational simulations of a circulatory system provide many benefits by lowering the chances of surgical complications, postoperative complications, in delivering better understanding of different biological processes, and in providing more efficient medical equipment like blood pumps.[33] Atherosclerosis development and its predisposing risk factors affect some regions of circulatory system.[34] Computational fluid dynamics is used to obtain information regarding spatial distribution related to intraluminal hemodynamics of coronary vascular tree.[35] The absence of right ventricle and its function in unique hemodynamics is known as Fontan circulation on the name of Fontan and Baudet who first described it.[36] The correction methods for Fontan circulation are by the separation of systemic and pulmonary venous with the establishment of passive, direct, and unobstructed assembly with systemic venous and pulmonary artery for the treatment of ventricle physiology.[37] Many studies tried to solve this problem using computational fluid dynamics along with medical information to establish artificial modeling of Fontan circulation. The modified Windkessel model is another computational model to calculate the work of heart (WHO) using the pressure volume curve.[38] The Windkessel model was used with the blood viscosity model to form a mathematical model to measure WHO by using pulse waves between 2 points of vessels.[39]

### *7.2.2 COMPUTATIONAL HERBAL DATABASE FOR CARDIOVASCULAR DRUG TARGETS*

A lot of traditional herbal medicines containing many biological compounds are in use against various cardiovascular diseases. These biological compounds are complex and their mechanism is also not fully understood.[40] With developments in biomedical techniques and increasing network of pharmacology, a need for the herbal medicine database was identified by many researchers for cardiovascular diseases for repurposing of drugs.[41] A cardiovascular disease herbal database (CVDHD) is based on natural product target protein interfaces of multi-level data for promoting drug discovery using herbal products.[42–42b]

The cardiovascular disease herbal database consists of six units of data including natural products, docking results, medicinal herbs, target proteins, disease, and clinical biomarkers.[43] The data consists of chemical name, CAS registration number, molecular formula, information and reference of author, and molecular weight of different compounds.[44] Open Babel was used to generate an absolute configuration of all molecules and all duplicate data was removed using InChIKey.[45] To study the molecular property like AlogP, hydrogen bond donor and acceptors of these compounds were measured using Discovery Studio.[46] Each human protein NMR ligand–protein complex and X-ray protein structures were collected from the RCSB protein data bank.[47] After downloading these structures were treated by molecular docking using Autodock. The binding sites of these proteins were defined on the basis of occupied space (40 × 40 × 40 Å) of original ligand with 0.375 Å between grid points.[48]

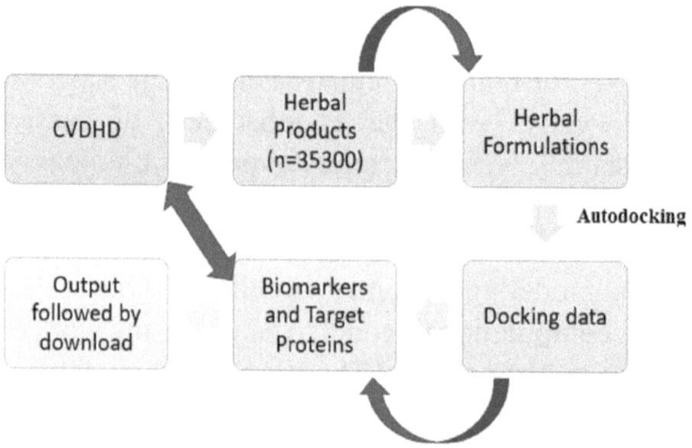


**FIGURE 7.2** Representation of steps of search flow chart in CVDHD.

The cardiovascular disease-related information can be collected from KEGG, TTD, and using other database by manual search.[49] CVDHD is accessible using Internet and API for Cytoscape was stored for future demand fulfilling.[50] Cytoscape and CentiBin are the two network analysis software that can be used for pharmacological analysis of CVDHD.[51]

## 7.3 REPURPOSING OF DIFFERENT DRUGS FOR CARDIOVASCULAR DISEASE

Drug repurposing approaches can be classified on the basis of drug-based, profile-based, and disease-based categories. For better results in repurposing pharmacological data like side effects of drugs, chemical structure helps in determining drug similarity.[50,52] In repurposing drug-based approaches help in predicting new drugs on the basis of disease–disease likeness, genetic and genomic disease data, and phenotypic data of disease.[53] Some researchers focus on profile-based repurposing in which gene expression data of the disease and changes in gene expression on specific drug exposure are studied.[54] This method is found to be very effective in many disease conditions like lung cancer and inflammatory bowel disease.[55]

### *7.3.1 COLCHICINE: ANTI-GOUT DRUG AS REPURPOSING FOR CARDIOVASCULAR DISEASES*

Colchicine is used as an anti-gout drug from long back as reported in Tralles "Therapeutica" around 550 AD.[56] Colchicine was first derived from crocus plant bulbs and used to treat various inflammations. The synthetic colchicine is used as a generic medication to treat gout. A low dose colchicine (LoDoCo) clinical trial reported that 0.5-mg colchicine one tie daily is safe and effective to prevent cardiovascular diseases especially in coronary artery disease and many related pathways.[57] Hartung reported excess use of colchicine to cause gastrointestinal defects and toxicity.[58] The modern use of colchicine was determined by Garrod in 1859 and Reverend Sydney Smith in 1838.[58–59] Colchicine is a potential drug as anti-inflammatory in acute gout flares. In the 19th century colchicine use for cardiovascular disease was limited to moderate function in pericarditis. In 1980, its use became general but inconsistent and its use was recommended on an evidence basis as reported in a randomized trial in the year 1990.[60]

Nidof and Thompson evaluated the use of colchicine in atherosclerosis and its different inflammatory constituents.[61] LoDoCo random trials two on more than 5000 patients with chronic coronary disease were treated with colchicine for a period of 30 days at 0.5 mg daily once.[62] The data was recorded on lipid lowering and antithrombotic endpoints.[63] After a 30-month follow-up of the study was analyzed on different endpoints, a steady trend was found in myocardial infraction and ischemia-related coronary revascularization and it was significantly low in colchicine-administered groups.[61,64] Out of the total number of patients 90% were reported tolerant to open label type colchicine and the rest intolerant patients reported GI tract infection and related symptoms.[65] The 5-year follow-up of the study reported LoDoCo have no side effect or risk that can let the patient's hospitalization and death.[66] Many studies confirmed that LoDoCo can be easily tolerated by patients and in the long run no side effects are reported.[67] The effect of the treatment becomes visible just after starting the therapy.[68] Similar results were also reported in the CANTOS and COLCOT trials. Tardif et al. (2019) performed the COLCOT trial to analyze the effectiveness of colchicine in patients with myocardial infraction.[63] During the trial it has been found that for the treatment is very important as to whom low dose colchicine is provided within 3 days of getting myocardial infraction reported to have reduction in risk associated to major cardiovascular events by 48% and the cost of treatment was reduced up to 47% during trial and the cost after recovery was reduced by 69% and diarrhea as side effect differ from group to group.[69] Many researchers verified reduction in the production of cytokines like IL-1β, IL-18, and IL-6 in acute colchicine on treatment with LoDoCo.[70]

Deftereos et al. revealed that LoDoCo reduce different inflammatory biomarkers like C-reactive protein and interlukin-6 circulating in blood and significantly affect the left ventricular restoration, but this study failed to show any significant development in function or risk reduction in patients with heart failure condition.[71] All these evidences and data from different studies show that colchicine is a suitable candidate for repurposing against ischemic disease condition.[72]

### *7.3.2 ANTI-CYTOKINE DRUGS*

Atherosclerosis is considered a disease condition in which lipoprotein gets deposited in arteries but with modern research it is confirmed that it is a type of chronic inflammation with a mixture of different pro-inflammatory cytokines, bioactive lipids, chemokines, and adhesion molecules.[73] Some studies consider Cysteinylleukotrienes (CysLts) to play a vital role in the progression and pathogenesis of this disease.[74] Montelukast is a receptor antagonist of CysLt1, which is under trial in animal models to be repurposed for cardiovascular diseases.[75] Data obtained from eight large cohort animal studies strongly recommend montelukast as pro-antherogenic and anti-antherogenic during different experimental conditions.[76] The review of this study also suggests that many immune-mediated inflammatory diseases like rheumatoid arthritis and lupus erythematous can increase risk of CVD and blocking of cytokines can help patients with atherosclerosis.[77]

### *7.3.3 ANTIDIABETIC DRUGS*

Modern lifestyle and increase in obesity increased prevalence of type 2 diabetes and cardiovascular disease. With diabetes risk of cardiovascular disease increases tremendously due to which life expectancy decreases.[78] The clinical studies in the last 10 years suggest strong co-relation between diabetes and heart failure as the rate of death due to cardiovascular diseases in a diabetic patient increased by 30–40%.[79] The use of metformin came in the market in 1995, but due to several side effects of many compounds in it specifically causing lactic acidosis they were retracted from the market.[80] The evidence from many studies provide evidence for metformin as a best T2D therapy due to its effect in weight reduction, tolerance, and low-degree hypoglycemia and acidosis is limited to some patients.[81] The UKPDS34 trial in 1998 studied the efficiency of metformin in overweight patients with their BMI more than 25 kg/m$^2$ having prediabetic and new T2D.[82] In this study patients with myocardial infraction and heart failure are not included. The result of the study showed a decrease in myocardial infraction up to 39% and 36% reduction in the rate of mortality due to T2D.[83] Some researchers also reported use of metformin in reducing the risk related to diabetes like sudden death due to hyper or hypoglycemia, stroke, kidney failure, myocardial infraction, vitreous hemorrhage, and heart failure.[84] Holman et al. stated that the effect of metformin remains on all risk factors related to T2D even after 10 years of therapy in comparison with HbA1c with no difference in the metformin and non-metformin groups.[85]

CAMERA is another study conducted on 173 non-diabetic cardiovascular disease patients to analyze the effect of metformin on atherosclerosis for a period of 18 months with an average BMI of 30 kg/m$^2$.[86] In this study progression of atherosclerosis were measured using cIMT, surrogate markers, and carotid plaque score and T2D.[87] The results of the study reported reduction of many factors related to obesity like body fat, body circumference, body weight, lowering of HbA1c, insulin level, and tissue plasminogen activators.[88] There are many other factors which remain unchanged including; HDL, C-reactive proteins, triglycerides, carotid score, and fasting glucose. A similar study by Katakami et al. (2004) reported the changes in cardiovascular biomarkers are independent of glucose reducing property of metformin and differs from group to group.[89] GIPS-III is another trial that evaluated 380 patients with systolic elevated myocardial infraction with no records of diabetes for a period of 4 months and at 1,000 mg daily doses.[90] Results did not show any effect on liver functioning even after 2 year follow up.[91]

#### *7.3.3.1 GLP1 (GLUCAGON LINKED PEPTIDE-1) CARDIO PROTECTION*

Glucagon linked peptide 1: A hormone secreted by our body when we eat food due to its incretin-like function to help in insulin secretion and inhibiting glucagon production was studies also in last 5 years on its cardioprotective function in T2D patients which provided good evidences in support of these activities.[92] GLP1-RA (receptor agonists) is also found helpful in limiting the risk of hyperglycaemia and it induces weight loss by reducing food consumption in overweight patients.[93] It was approved by FDA for treatment of obesity in 2014 and by European Medicines Agency in 2015.[94] 3PMACE study results confirmed the cardioprotective effect of GLP1 as it reduces the CVD mortality and the damages in myocardial infraction to various degrees.[95] Table 7.1 provides a summary of different diabetic drugs effective in CVD on the basis of trials conducted and effect of these drugs on CVD and related conditions.

**TABLE 7.1** Summary of Different T2D Drug Related Studies and Their Effect of CVD Types.

| Drug type | Study (trials) | Number of patients | Effectt2 on CVD | References |
|---|---|---|---|---|
| Metformin | UKPDS34 | 758 with 10 years follow up | Reduce T2D and effective in MI reduction by 0.61 HR | 97, 96a |
| | SAVOR TIMI 53 | 12,156 with 2 years follow up | Reduction in CVD death by 0.68 and MI by 0.79 HR | 96b, 97 |

**TABLE 7.1** *(Continued)*

| Drug type | Study (trials) | Number of patients | Effectt2 on CVD | References |
|---|---|---|---|---|
| GLP-1 RA | LEADER | 9,340 with 3 years follow up | Reduction in CVD death by 0.78, MI by 086 and HF by 0.78 HR | 98 |
| | SUSTAIN-6 | 3,297 with 2 years follow up | Reduce CVD by 0.98, MI by 0.81, and HF by 1.11 HR. | 99 |
| | PIONEER 6 | 3,183 with 1 years follow up | Reduction in CVD by 0.49, MI by 1.18, and HF by 0.71 HR | 100, 101 |
| | Harmony outcomes | 9,463 with 1.5 years follow up | Reduction in CVD by 0.93, MI by 0.75, and HF by 0.71 HR | 102 |
| | REWIND | 9,901 with 5 years follow up | Reduction in CVD death by 0.91, MI by 0.96, and HF by 0.93 HR. | 102, 103 |
| | EXSCEL | 14,752 with 3 years follow up | Reduction in CVD death by 0.88, MI by 0.97, and HF by 0.94 HR. | 85, 104 |
| DPP4-i | Carmelina | 6,979 with 2 years follow up | Reduction in CVD death by 0.96, MI by 1.12, and HF by 0.90 HR. | 105 |
| | Tecos | 14,671 with 3 years follow up | Reduction in CVD death by 1.03, MI by 0.95, and HF by 1.00 HR. | 105, 106 |
| | Savortimi 53 | 16,492 with 2 years follow up | Reduction in CVD death by 1.03, MI by 0.95, and HF by 1.27 HR. | 107 |
| | Examine | 5,380 with 1 year follow up | Reduction in CVD death by 0.85, MI by 1.10, and HF by 1.19 HR. | 108 |
| SGLT2-i | Empareg-outcome | 7,020 with 3 years follow up | Reduction in CVD death by 0.62, MI by 0.87, and HF by 0.65 HR. | 109 |
| | Canvas | 10,142 with 3.6 years follow up | Reduction in CVD death by 0.87, MI by 0.89, and HF by 0.67 HR. | 110 |
| | Declare-timi 58 | 17,160 with 4.2 years follow up | Reduction in CVD death by 0.98, MI by 0.89, and HF by 0.73 HR. | 111 |
| | Credence | 4,401 with 2.6 years follow up | Reduction in CVD death by 0.78, and HF by 0.61 HR. | 112 |

### 7.3.4 *ANTI-INTERLEUKIN DRUGS*

Different pharmacological drugs and agents have been approved by FDA to be used against many auto-inflammatory diseases including rheumatoid arthritis.[107a,113] Anakinra is a human recombinant interleukin-1 (IL-1) competitive receptor and inhibitor for IL-1α and IL-1β, whereas canakinumab is a monoclonal antibody which can target IL-1 receptors.[114] The use of these in cardiovascular disease treatment was first identified by Ikonomidis et al. (2008) in a studytrial on 23 patients at a concentration of 150 mg subcutaneously.[115] Treated group showed improvement in different CVD conditions like myocardial contractions and relaxations, improved endothelial functioning, and coronary flow reserve.[116] A study by CANTOS group on 556 clinical patients with T2DM with CVD risk using canakinumab at different concentrations as 5, 15, 50, and 150 mg per month.[117] In phase III trial of this study a total of 10,061 patients were enrolled having history of MI and high C-reactive protein.[118] After 1 month the results obtained showed significant reduction in inflammation and around 50% decrease in C-reactive protein level.[119] This study also demonstrated the use of IL-1 target therapy can significantly reduce the number of patient admission with HF and related mortality.[120]

Tocilizumab is another human recombinant antibody used against receptors of interleukin-6 (IL-6), approved by FDA for rheumatoid arthritis and giant cell arteritis.[121] A study named NSTEMI on 117 patients with percutaneous coronary intervention showed significant decrease in peri-procedural MI and expressed in terms of reduced troponin-T from 234 to 159 ng/L/h and C-reactive protein from 4.2 ng/L/h with no side effects reported.[122] Presently there are many other studies estimating the effect of tocilizumab on CVD, MI, and on giant cell arteritis. These studies are also considering the effect on cardiac injuries, different types of inflammations, CVD patient's cardiac arrest.[123]

## 7.4 CONCLUSION

Inflammation is considered to play central role in onset of many cardiovascular disease including atherosclerosis, coronary artery disease, and MI. Due to limited data availability on cardiovascular profile of anti-inflammatory drugs the use of these drugs are limited. Drug repurposing occurred as a great tool to identify safety of different drugs in CVD condition. The data of different studies available till now shows possibility of monoclonal antibodies as target therapy for CVD in coming years. Data of anti-inflammatory drugs like colchicine that have been repurposed for CVD shows promising results in patients of STEMI. Similarly Metformin, an anti-diabetic drug also shows effective inflammatory process control and cardioprotective properties. The future studies need more evidences from clinical trials to better repurpose these drugs and their approval to be used in CVD.

## KEYWORDS

- **cardiovascular diseases**
- **drug repurposing**
- **interleukins**
- **metformin**

## REFERENCES

1. DiMasi, J. A.; Grabowski, H. G.; Hansen, R. W. Innovation in the Pharmaceutical Industry: New Estimates of R&D Costs. *J. Heal. Eco.* **2016,** *47*, 20–33.
2. Ashburn, T. T.; Thor, K. B. Drug Repositioning: Identifying and Developing New Uses for Existing Drugs. *Nat. Rev. Drug Discov.* **2004,** *3*(8), 673–683.
3. Shim, J. S.; Liu, J. O. Recent Advances in Drug Repositioning for the Discovery of New Anticancer Drugs. *Int. J. Biol. Sci.* **2014,** *10*(7), 654.
4. Rudrapal, M.; Khairnar, J.; Jadhav, G., Drug Repurposing (DR): An Emerging Approach in Drug Discovery. *Drug Rep. Hypo. Mol. Asp. Ther. Appl.* **2020**.
5. Allarakhia, M. Open-Source Approaches for the Repurposing of Existing or Failed Candidate Drugs: Learning from and Applying the Lessons Across Diseases. *Drug Des. Devel. Ther.* **2013,** *7*, 753.
6. Gupta, S.; Rohatgi, A.; Ayers, C. R.; Willis, B. L.; Haskell, W. L.; Khera, A.; Drazner, M. H.; de Lemos, J. A.; Berry, J. D. Cardiorespiratory Fitness and Classification of Risk of Cardiovascular Disease Mortality. *Circul.* **2011,** *123*(13), 1377–1383.
7. Subanya, B.; Rajalaxmi, R. In *Feature Selection Using Artificial Bee Colony for Cardiovascular Disease Classification*, International Conference on Electronics and Communication Systems (ICECS), IEEE: 2014; 1–6.
8. Hosni, M.; Carrillo de Gea, J. M.; Idri, A.; El Bajta, M.; Fernandez Aleman, J. L.; García-Mateos, G.; Abnane, I. A Systematic Mapping Study for Ensemble Classification Methods in Cardiovascular Disease. *Artif. Intell. Rev.* **2021,** *54*(4), 2827–2861.
9. Van Der Bom, T.; Zomer, A. C.; Zwinderman, A. H.; Meijboom, F. J.; Bouma, B. J.; Mulder, B. J. The Changing Epidemiology of Congenital Heart Disease. *Nat. Rev. Cardiol.* **2011**, *8*(1), 50–60.

10. A Richards, A.; Garg, V., Genetics of Congenital Heart Disease. *Curr. Cardiol. Rev.* **2010,** *6*(2), 91–97.
11. Verheugt, C. L.; Uiterwaal, C. S.; van der Velde, E. T.; Meijboom, F. J.; Pieper, P. G.; van Dijk, A. P.; Vliegen, H. W.; Grobbee, D. E.; Mulder, B. J., Mortality in Adult Congenital Heart Disease. *Eur. Heart J.* **2010,** *31*(10), 1220–1229.
12. Gatzoulis, M. A.; Webb, G. D.; Daubeney, P. E., Diagnosis and Management of Adult Congenital Heart Disease E-Book. *Elsevier Sci.* 2010.
13. Hoffman, J. I.; Kaplan, S.; Liberthson, R. R., Prevalence of Congenital Heart Disease. *Am. Heart J.* **2004,** *147*(3), 425–439.
14. (a) Zaidi, S.; Brueckner, M., Genetics and Genomics of Congenital Heart Disease. *Cir. Res.* **2017,** *120*(6), 923–940; (b) McQuillen, P. S.; Miller, S. P. Congenital Heart Disease and Brain Development. *Ann. N. Y. Acad. Sci.* **2010,** *1184*(1), 68–86.
15. Bernier, P.-L.; Stefanescu, A.; Samoukovic, G.; Tchervenkov, C. I. In the Challenge of Congenital Heart Disease Worldwide: Epidemiologic and Demographic Facts, Seminars in Thoracic and Cardiovascular Surgery: Pediatric Cardiac Surgery Annual, *Els.* 2010; 26–34.
16. Park, K. A Review of Computational Drug Repurposing. *Transl. Clin. Pharmaco.* **2019,** *27*(2), 59–63.
17. Sanseau, P.; Koehler, J. *Computational Methods for Drug Repurposing.* Oxford University Press: 2011; (b) Saberian, N.; Peyvandipour, A.; Donato, M.; Ansari, S.; Draghici, S. A New Computational Drug Repurposing Method Using Established Disease-drug Pair Knowledge. *Bioinfor.* **2019,** *35*(19), 3672–3678; (c) Paranjpe, M. D.; Taubes, A.; Sirota, M., Insights into Computational Drug Repurposing for Neurodegenerative Disease. *Tre. Pharmaco. Sci.* **2019,** *40*(8), 565–576.
18. Peyvandipour, A.; Saberian, N.; Shafi, A.; Donato, M.; Draghici, S. A Novel Computational Approach for Drug Repurposing Using Systems Biology. *Bioinfor.* **2018,** *34*(16), 2817–2825.
19. (a) Haupt, V. J.; Daminelli, S.; Schroeder, M. Drug Promiscuity in PDB: Protein Binding Site Similarity is Key. *PL. One* **2013,** *8*(6), e65894; (b) Moriaud, F.; Richard, S. B.; Adcock, S. A.; Chanas-Martin, L.; Surgand, J.-S.; Ben Jelloul, M.; Delfaud, F. Identify Drug Repurposing Candidates by Mining the Protein Data Bank. *Brief. Bioinfor.* **2011,** *12*(4), 336–340.
20. Totrov, M., Ligand Binding Site Superposition and Comparison Based on Atomic Property Fields: Identification of Distant Homologues, Convergent Evolution and PDB-wide clustering of binding sites. *BMC Bioinfor.* **2011,** *12*(1), 1–9.
21. Xu, K.; Coté, T. R., Database Identifies FDA-approved Drugs with Potential to be Repurposed for Treatment of Orphan Diseases. *Brief Bioinform.* **2011,** *12*(4), 341–345.
22. Power, A.; Berger, A. C.; Ginsburg, G. S., Genomics-enabled Drug Repositioning and Repurposing: Insights from an IOM Roundtable Activity. *Jama* **2014,** *311*(20), 2063–2064.
23. Wu, C.; Gudivada, R. C.; Aronow, B. J.; Jegga, A. G., Computational Drug Repositioning Through Heterogeneous Network Clustering. *BMC Sys. Bio.* **2013,** *7*(5), 1–9.
24. Siddiqui, S. Y.; Athar, A.; Khan, M. A.; Abbas, S.; Saeed, Y.; Khan, M. F.; Hussain, M. Modelling, Simulation and Optimization of Diagnosis Cardiovascular Disease Using Computational Intelligence Approaches. *J. Med. Imag. Health Infor.* **2020,** *10*(5), 1005–1022.

25. MacLellan, W. R.; Wang, Y.; Lusis, A. J. Systems-based Approaches to Cardiovascular Disease. *Nat. Rev. Cardio.* **2012,** *9*(3), 172–184.
26. Taylor, C. A.; Draney, M. T. Experimental and Computational Methods in Cardiovascular Fluid Mechanics. *Annu. Rev. Fluid Mech.* **2004,** *36*, 197–231.
27. Rennels, G. D.; Shortliffe, E. H.; Stockdale, F. E.; Miller, P. L. Computational Model of Reasoning from the Clinical Literature. *Sel. Top. Med. Arti. Int.*, Springer: 1988, 125–140.
28. Kagiyama, N.; Shrestha, S.; Farjo, P. D.; Sengupta, P. P. Artificial Intelligence: Practical Primer for Clinical Research in Cardiovascular Disease. *J. Am. Heart Asso.* **2019,** *8*(17), e012788.
29. Biglino, G.; Capelli, C.; Bruse, J.; Bosi, G. M.; Taylor, A. M.; Schievano, S. Computational Modelling for Congenital Heart Disease: How Far are We from Clinical Translation. *Hear.* **2017,** *103*(2), 98–103.
30. Marsden, A. L.; Feinstein, J. A., Computational Modeling and Engineering in Pediatric and Congenital Heart Disease. *Cur. Opinion Ped.* **2015,** *27*(5), 587.
31. Vignon-Clementel, I. E.; Marsden, A. L.; Feinstein, J. A. A Primer on Computational Simulation in Congenital Heart Disease for the Clinician. *Pr. Pedia. Cardio.* **2010,** *30*(1–2), 3–13.
32. Ayon, S. I.; Islam, M. M.; Hossain, M. R. Coronary Artery Heart Disease Prediction: A Comparative Study of Computational Intelligence Techniques. *IETE J. Res.* **2020**, 1–20.
33. (a) Qian, Y.; Liu, J.; Itatani, K.; Miyaji, K.; Umezu, M. Computational Hemodynamic Analysis in Congenital Heart Disease: Simulation of the Norwood Procedure. *Ann. Biomed. Eng.* **2010,** *38*(7), 2302–2313; (b) Egbuna, C.; Awuchi, C. G.; Kushwaha, G.; Rudrapal, M.; Patrick-Iwuanyanwu, K. C.; Singh, O.; Odoh, U. E.; Khan, J.; Jeevanandam, J.; Kumarasamy, S. Bioactive Compounds Effective Against Type 2 Diabetes Mellitus: A Systematic Review. *Cur. Top. Med. Chem.* **2021,** *21*(12), 1067–1095.
34. Olechnowicz-Tietz, S.; Gluba, A.; Paradowska, A.; Banach, M.; Rysz, J. The Risk of Atherosclerosis in Patients with Chronic Kidney Disease. *Inter. Uro. Neph.* **2013,** *45*(6), 1605–1612.
35. (a) Gerrah, R.; Haller, S. J. Computational Fluid Dynamics: A Primer for Congenital Heart Disease Clinicians. *Asian Cardio. Thor. Ann.* **2020,** *28*(8), 520–532; (b) Egbuna, C.; Parmar, V. K.; Jeevanandam, J.; Ezzat, S. M.; Patrick-Iwuanyanwu, K. C.; Adetunji, C. O.; Khan, J.; Onyeike, E. N.; Uche, C. Z.; Akram, M. Toxicity of Nanoparticles in Biomedical Application: Nanotoxicology. *J. Tox.* **2021.**
36. Liang, F.; Senzaki, H.; Kurishima, C.; Sughimoto, K.; Inuzuka, R.; Liu, H. Hemodynamic Performance of the Fontan Circulation Compared with a Normal Biventricular Circulation: A Computational Model Study. *Am. J. Phy. Heart Cir. Phy.* **2014,** *307*(7), H1056–H1072.
37. Lin, W. P.; Doyle, M. G.; Roche, S. L.; Honjo, O.; Forbes, T. L.; Amon, C. H. Computational Fluid Dynamic Simulations of a Cavopulmonary Assist Device for Failing Fontan Circulation. *The J. Thoracic Cardio. Sur.* **2019,** *158*(5), 1424–1433. e5.
38. Westerhof, N.; Lankhaar, J.-W.; Westerhof, B. E. The Arterial Windkessel. *Med. Bio. Eng. Comput.* **2009,** *47*(2), 131–141.
39. Burkhoff, D.; Alexander Jr, J.; Schipke, J. Assessment of Windkessel as a Model of Aortic Input Impedance. *Am. J. Phy. Heart Cir. Phy.* **1988,** *255*(4), H742–H753.
40. Ru, J.; Li, P.; Wang, J.; Zhou, W.; Li, B.; Huang, C.; Li, P.; Guo, Z.; Tao, W.; Yang, Y. TCMSP: A Database of Systems Pharmacology for Drug Discovery from Herbal Medicines. *J. Cheminfor.* **2014,** *6*(1), 1–6.

41. Gu, J.; Gui, Y.; Chen, L.; Yuan, G.; Xu, X. CVDHD: A Cardiovascular Disease Herbal Database for Drug Discovery and Network Pharmacology. *J. Cheminfor.* **2013,** *5*(1), 1–6.
42. (a) Tnah, L.; Lee, S.; Tan, A.; Lee, C.; Ng, K.; Ng, C.; Farhanah, Z. N. DNA Barcode Database of Common Herbal Plants in the Tropics: A Resource for Herbal Product Authentication. *Food Cont.* **2019,** *95*, 318–326; (b) Chandran, U.; Mehendale, N.; Tillu, G.; Patwardhan, B. In Network Pharmacology: An Emerging Technique for Natural Product Drug Discovery and Scientific Research on Ayurveda. *Proc. Indian Natn. Sci. Acad.* 2015; 561–8.
43. Newman, D. J.; Cragg, G. M., Natural Products as Sources of New Drugs Over the 30 Years from 1981 to 2010. *J. Nat. Prod.* **2012,** *75*(3), 311–335.
44. (a) Chin, Y.-W.; Balunas, M. J.; Chai, H. B.; Kinghorn, A. D. Drug Discovery from Natural Sources. *The AAPS J.* **2006,** *8*(2), E239–E253; (b) Khan, J.; Deb, P. K.; Priya, S.; Medina, K. D.; Devi, R.; Walode, S. G.; Rudrapal, M. Dietary Flavonoids: Cardioprotective Potential with Antioxidant Effects and Their Pharmacokinetic, *Toxicolo. Ther. Con.* **2021**.
45. Gu, J.; Gui, Y.; Chen, L.; Yuan, G.; Lu, H.-Z.; Xu, X. Use of Natural Products as Chemical Library for Drug Discovery and Network Pharmacology. *Plo. One* **2013,** *8*(4), e62839.
46. Ekins, S.; Williams, A. J. Finding Promiscuous Old Drugs for New Uses. *Pharma. Res.* **2011,** *28*(8), 1785–1791.
47. Zhang, A.; Sun, H.; Yang, B.; Wang, X. Predicting New Molecular Targets for Rhein Using Network Pharmacology. *BMC Sys. Bio.* **2012,** *6*(1), 1–8.
48. Qiao, X.; Hou, T.; Zhang, W.; Guo, S.; Xu, X. A 3D Structure Database of Components from Chinese Traditional Medicinal Herbs. *J. Chem. Infor. Comp. Sci.* **2002,** *42*(3), 481–489.
49. Kanehisa, M.; Goto, S.; Sato, Y.; Furumichi, M.; Tanabe, M. KEGG for Integration and Interpretation of Large-scale Molecular Data Sets. *Nuc. Acids Res.* **2012,** *40*(D1), D109–D114.
50. Lipinski, C. A.; Lombardo, F.; Dominy, B. W.; Feeney, P. J. Experimental and Computational Approaches to Estimate Solubility and Permeability in Drug Discovery and Development Settings. *Adv. Drug Del. Rev.* **1997,** *23*(1–3), 3–25.
51. Jiang, X.; Kumar, K.; Hu, X.; Wallqvist, A.; Reifman, J. DOVIS 2.0: An Efficient and Easy to Use Parallel Virtual Screening Tool Based on AutoDock 4.0. *Chem. Central J.* **2008,** *2*(1), 1–7.
52. Mestres, J.; Gregori-Puigjané, E.; Valverde, S.; Solé, R. V. The Topology of Drug-target Interaction Networks: Implicit Dependence on Drug Properties and Target Families. *Mol. BioSys.* **2009,** *5*(9), 1051–1057.
53. Remez Vinogradov, N. *Drug Design at Biological Systems Level*. Universitat Pompeu Fabra, 2016.
54. Novick, P. A.; Ortiz, O. F.; Poelman, J.; Abdulhay, A. Y.; Pande, V. S. Sweetlead: An In Silico Database of Approved Drugs, Regulated Chemicals, and Herbal Isolates for Computer-aided Drug Discovery. *Plo. One* **2013,** *8*(11), e79568.
55. Campillos, M.; Kuhn, M.; Gavin, A.-C.; Jensen, L. J.; Bork, P. Drug Target Identification Using Side-effect Similarity. *Sci.* **2008,** *321*(5886), 263–266.

56. Sirota, M.; Dudley, J. T.; Kim, J.; Chiang, A. P.; Morgan, A. A.; Sweet-Cordero, A.; Sage, J.; Butte, A. J. Discovery and Preclinical Validation of Drug Indications using Compendia of Public Gene Expression Data. *Sci. Trans. Med.* **2011,** *3*(96), 96ra77–96ra77.
57. (a) Talevi, A.; Bellera, C. L. Challenges and Opportunities with Drug Repurposing: Finding Strategies to Find Alternative Uses of Therapeutics. *Exp. Opin. Drug Dis.* **2020,** *15*(4), 397–401; (b) Cha, Y.; Erez, T.; Reynolds, I.; Kumar, D.; Ross, J.; Koytiger, G.; Kusko, R.; Zeskind, B.; Risso, S.; Kagan, E. Drug Repurposing from the Perspective of Pharmaceutical Companies. *Br. J. Pharm.* **2018,** *175*(2), 168–180.
58. Hartung, E. F. History of the Use of Colchicum and Related Medicaments in Gout: With Suggestions for Further Research. *Ann. Rheum. Dis.* **1954,** *13*(3), 190.
59. Garrod, A. B. The Nature and Treatment of Gout, and Rheumatic gout. *Long. Comp.* **1876.**
60. Fiolet, A. T.; Nidorf, S. M.; Mosterd, A.; Cornel, J. H. Colchicine in Stable Coronary Artery Disease. *Clin. Therape.* **2019,** *41*(1), 30–40.
61. Nidorf, M.; Thompson, P. L. Effect of Colchicine (0.5 mg twice daily) on High-sensitivity C-reactive Protein Independent of Aspirin and Atorvastatin in Patients with Stable Coronary Artery Disease. *The Am. J. Cardio.* **2007,** *99*(6), 805–807.
62. Tardif, J.-C.; Kouz, S.; Waters, D. D.; Bertrand, O. F.; Diaz, R.; Maggioni, A. P.; Pinto, F. J.; Ibrahim, R.; Gamra, H.; Kiwan, G. S. Efficacy and Safety of Low-dose Colchicine After Myocardial Infarction. *New Eng. J. Med.* **2019,** *381*(26), 2497–2505.
63. Nidorf, S. M.; Fiolet, A. T.; Eikelboom, J. W.; Schut, A.; Opstal, T. S.; Bax, W. A.; Budgeon, C. A.; Tijssen, J. G.; Mosterd, A.; Cornel, J. H. The Effect of Low-dose Colchicine In Patients with Stable Coronary Artery Disease: The LoDoCo2 Trial Rationale, Design, and Baseline Characteristics. *Am. Heart J.* **2019,** *218*, 46–56.
64. Butt, A. K.; Cave, B.; Maturana, M.; Towers, W. F.; Khouzam, R. N. The Role of Colchicine in Coronary Artery Disease. *Cur. Pro. Car.* **2021,** *46*(3), 100690.
65. Xia, M.; Yang, X.; Qian, C. Meta-analysis Evaluating the Utility of Colchicine in Secondary Prevention of Coronary Artery Disease. *The Am. J. Cardi.* **2021,** *140*, 33–38.
66. Katsanos, A. H.; Palaiodimou, L.; Price, C.; Giannopoulos, S.; Lemmens, R.; Kosmidou, M.; Georgakis, M.; Weimar, C.; Kelly, P.; Tsivgoulis, G. Colchicine for Stroke Prevention in Patients with Coronary Artery Disease: A Systematic Review and Meta-analysis. *Eur. J. Neu.* **2020,** *27*(6), 1035–1038.
67. Nidorf, S. M.; Eikelboom, J. W.; Budgeon, C. A.; Thompson, P. L. Low–dose Colchicine for Secondary Prevention of Cardiovascular Disease. *J. Am. Coll. Cardio.* **2013,** *61*(4), 404–410.
68. Samuel, M.; Tardif, J.-C.; Khairy, P.; Roubille, F.; Waters, D. D.; Grégoire, J. C.; Pinto, F. J.; Maggioni, A. P.; Diaz, R.; Berry, C. Cost-effectiveness of Low-dose Colchicine After Myocardial Infarction in the Colchicine Cardiovascular Outcomes Trial (COLCOT). *Eur. Heart J. Qual. Care Clin. Out.* **2021,** *7*(5), 486–495.
69. Gaziano, J. M.; Brotons, C.; Coppolecchia, R.; Cricelli, C.; Darius, H.; Gorelick, P. B.; Howard, G.; Pearson, T. A.; Rothwell, P. M.; Ruilope, L. M. Use of Aspirin to Reduce Risk of Initial Vascular Events in Patients at Moderate Risk of Cardiovascular Disease (ARRIVE): A Randomised, Double-blind, Placebo-controlled Trial. *The Lan.* **2018,** *392*(10152), 1036–1046.
70. Opstal, T. S.; Hoogeveen, R. M.; Fiolet, A. T.; Silvis, M. J.; The, S. H.; Bax, W. A.; de Kleijn, D. P.; Mosterd, A.; Stroes, E. S.; Cornel, J. H., Colchicine Attenuates

Inflammation Beyond the Inflammasome in Chronic Coronary Artery Disease: A LoDoCo2 Proteomic Substudy. *Cir.* **2020,** *142*(20), 1996–1998.

71. (a) Liuzzo, G.; Patrono, C., Low-dose Colchicine: A New Tool in the Treatment of Chronic Coronary Disease, Comment on the Low-dose Colchicine (LoDoCo) 2 Trial. *Eur. Heart J.* **2020**; (b) Vrachatis, D. A.; Papathanasiou, K. A.; Giotaki, S. G.; Iliodromitis, K. E.; Papaioannou, T. G.; Stefanini, G. G.; Cleman, M.; Siasos, G.; Reimers, B.; Lansky, A., Repurposing Colchicine's Journey in View of Drug-to-drug Interactions. A Review. *Tox. Rep.* **2021,** *8*, 1389–1393.
72. Jia, X.; Al Rifai, M.; Hussain, A.; Martin, S.; Agarwala, A.; Virani, S. S. Highlights from Studies in Cardiovascular Disease Prevention Presented at the Digital 2020 European Society of Cardiology Congress: Prevention is Alive and Well. *Cur. Ather Rep.* **2020,** *22*(12), 1–9.
73. Kaptoge, S.; Seshasai, S. R. K.; Gao, P.; Freitag, D. F.; Butterworth, A. S.; Borglykke, A.; Di Angelantonio, E.; Gudnason, V.; Rumley, A.; Lowe, G. D. Inflammatory Cytokines and Risk of Coronary Heart Disease: New Prospective Study and Updated Meta-analysis. *Eur. Heart J.* **2014,** *35*(9), 578–589.
74. Lai, Y.; Dong, C. Therapeutic Antibodies that Target Inflammatory Cytokines in Autoimmune Diseases. *Inter. Immun.* **2016,** *28*(4), 181–188.
75. Barnabe, C.; Martin, B. J.; Ghali, W. A., Systematic Review and Meta-analysis: Anti-tumor Necrosis Factor α Therapy and Cardiovascular Events in Rheumatoid Arthritis. *Arth. Care Res.* **2011,** *63*(4), 522–529.
76. Di Minno, M. N. D.; Iervolino, S.; Peluso, R.; Scarpa, R.; Di Minno, G., Carotid Intima-media Thickness in Psoriatic Arthritis: Differences between Tumor Necrosis Factor-α Blockers and Traditional Disease-modifying Antirheumatic Drugs. *Arter. Thromb. Vas. Bio.* **2011,** *31*(3), 705–712.
77. Chung, E.; Packer, M.; Lo, K.; Fasanmade, A.; Willerson, J. Anti-TNF Therapy Against Congestive Heart Failure Investigators. Randomized, Double-blind, Placebo-controlled, Pilot Trial of Infliximab, a Chimeric Monoclonal Antibody to Tumor Necrosis Factor-alpha, in Patients with Moderate-to-severe Heart Failure: Results of the anti-TNF Therapy Against Congestive Heart Failure (ATTACH) trial. *Cir.* **2003,** *107*(25), 3133–3140.
78. El Messaoudi, S.; Nederlof, R.; Zuurbier, C. J.; van Swieten, H. A.; Pickkers, P.; Noyez, L.; Dieker, H.-J.; Coenen, M. J.; Donders, A. R. T.; Vos, A. Effect of Metformin Pretreatment on Myocardial Injury During Coronary Artery Bypass Surgery in Patients Without Diabetes (MetCAB): A Double-blind, Randomised Controlled Trial. *The Lan. Diab. Endo.* **2015,** *3*(8), 615–623.
79. Baartscheer, A.; Schumacher, C. A.; Wüst, R. C.; Fiolet, J. W.; Stienen, G. J.; Coronel, R.; Zuurbier, C. J. Empagliflozin Decreases Myocardial Cytoplasmic Na+ through Inhibition of the Cardiac Na+/H+ Exchanger in Rats and Rabbits. *Diab.* **2017,** *60*(3), 568–573.
80. Bergmark, B. A.; Bhatt, D. L.; McGuire, D. K.; Cahn, A.; Mosenzon, O.; Steg, P. G.; Im, K.; Kanevsky, E.; Gurmu, Y.; Raz, I. Metformin Use and Clinical Outcomes among Patients with Diabetes Mellitus with or Without Heart Failure or Kidney Dysfunction: Observations from the SAVOR-TIMI 53 Trial. *Cir.* **2019,** *140*(12), 1004–1014.
81. Bergmark, B. A.; Bhatt, D. L.; McGuire, D.; Cahn, A.; Mosenzon, O.; Steg, P. G.; Im, K.; Kanevsky, E.; Raz, I.; Braunwald, E. Metformin Use and Clinical Outcomes Among

Patients With Diabetes Mellitus and Heart Failure or Kidney Dysfunction–Observations From the SAVOR-TIMI 53 Trial. *Circ.* **2016,** *134*(Suppl_1), A16764–A16764.

82. Bonora, B.; Avogaro, A.; Fadini, G. Effects of Exenatide Long-acting Release on Cardiovascular Events and Mortality in Patients with Type 2 Diabetes: A Systematic Review and Meta-analysis of Randomized Controlled Trials. *Acta Diab.* **2019,** *56*(9), 1051–1060.
83. Foretz, M.; Guigas, B.; Viollet, B. Understanding the Glucoregulatory Mechanisms of Metformin in Type 2 Diabetes Mellitus. *N. Rev. Endo.* **2019,** *15*(10), 569–589.
84. Fu, X.; Pan, Y.; Cao, Q.; Li, B.; Wang, S.; Du, H.; Duan, N.; Li, X. Metformin Restores Electrophysiology of Small Conductance Calcium-activated Potassium Channels in the Atrium of GK Diabetic Rats. *BMC Cardio. Dis.* **2018,** *18*(1), 1–8.
85. Holman, R. R.; Bethel, M. A.; Mentz, R. J.; Thompson, V. P.; Lokhnygina, Y.; Buse, J. B.; Chan, J. C.; Choi, J.; Gustavson, S. M.; Iqbal, N. Effects of Once-weekly Exenatide on Cardiovascular Outcomes in Type 2 Diabetes. *New Eng. J. Med.* **2017,** *377*(13), 1228–1239.
86. Preiss, D.; Lloyd, S. M.; Ford, I.; McMurray, J. J.; Holman, R. R.; Welsh, P.; Fisher, M.; Packard, C. J.; Sattar, N. Metformin for Non-diabetic Patients with Coronary Heart Disease (the CAMERA study): A Randomised Controlled Trial. *The Lan. Diab. Endo.* **2014,** *2*(2), 116–124.
87. Nesti, L.; Natali, A. Metformin Effects on the Heart and the Cardiovascular System: A Review of Experimental and Clinical Data. *Nutr. Meta. Cardio. Dis.* **2017,** *27*(8), 657–669.
88. Zhou, J.; Massey, S.; Story, D.; Li, L. Metformin: An Old Drug with New Applications. *Int. J. Mol. Sci.* **2018,** *19*(10), 2863.
89. Katakami, N.; Yamasaki, Y.; Hayaishi-Okano, R.; Ohtoshi, K.; Kaneto, H.; Matsuhisa, M.; Kosugi, K.; Hori, M. Metformin or Gliclazide, Rather Than Glibenclamide, Attenuate Progression of Carotid Intima-media Thickness in Subjects with Type 2 Diabetes. *Diab.* **2004,** *47*(11), 1906–1913.
90. Bikbov, B.; Purcell, C. A.; Levey, A. S.; Smith, M.; Abdoli, A.; Abebe, M.; Adebayo, O. M.; Afarideh, M.; Agarwal, S. K.; Agudelo-Botero, M. Global, Regional, and National Burden of Chronic Kidney Disease, 1990–2017: A Systematic Analysis for the Global Burden of Disease Study 2017. *The Lan.* **2020,** *395*(10225), 709–733.
91. Kato, E. T.; Silverman, M. G.; Mosenzon, O.; Zelniker, T. A.; Cahn, A.; Furtado, R. H.; Kuder, J.; Murphy, S. A.; Bhatt, D. L.; Leiter, L. A. Effect of Dapagliflozin on Heart Failure and Mortality in Type 2 Diabetes Mellitus. *Circu.* **2019,** *139*(22), 2528–2536.
92. Donnelly, D. The Structure and Function of the Glucagon-like Peptide-1 Receptor and Its Ligands. *Bri. J. Pharma.* **2012,** *166*(1), 27–41.
93. Kristensen, S. L.; Rørth, R.; Jhund, P. S.; Docherty, K. F.; Sattar, N.; Preiss, D.; Køber, L.; Petrie, M. C.; McMurray, J. J. Cardiovascular, Mortality, and Kidney Outcomes with GLP-1 Receptor Agonists in Patients with Type 2 Diabetes: A Systematic Review and Meta-analysis of Cardiovascular Outcome Trials. *The Lan. Diab. Endo.* **2019,** *7*(10), 776–785.
94. Drucker, D. J. Mechanisms of Action and Therapeutic Application of Glucagon-like Peptide-1. *Cell Meta* **2018,** *27*(4), 740–756.
95. Sattar, N.; Lee, M. M.; Kristensen, S. L.; Branch, K. R.; Del Prato, S.; Khurmi, N. S.; Lam, C. S.; Lopes, R. D.; McMurray, J. J.; Pratley, R. E. Cardiovascular, Mortality, and Kidney Outcomes with GLP-1 Receptor Agonists in Patients with Type 2 Diabetes:

A Systematic Review and Meta-analysis of Randomised Trials. *The Lan. Diab. Endo.* **2021**.
96. (a) Group, U. P. D. S. Effect of Intensive Blood-glucose Control with Metformin on Complications in Overweight Patients with Type 2 Diabetes (UKPDS 34). *The Lan.* **1998,** *352*(9131), 854–865; (b) Holman, R. R.; Paul, S. K.; Bethel, M. A.; Matthews, D. R.; Neil, H. A. W. 10-year Follow-up of Intensive Glucose Control in Type 2 Diabetes. *New Eng. J. Med.* **2008,** *359*(15), 1577–1589.
97. Scirica, B. M.; Braunwald, E.; Raz, I.; Cavender, M. A.; Morrow, D. A.; Jarolim, P.; Udell, J. A.; Mosenzon, O.; Im, K.; Umez-Eronini, A. A. Heart Failure, Saxagliptin, and Diabetes Mellitus: Observations from the SAVOR-TIMI 53 Randomized Trial. *Circu.* **2014,** *130*(18), 1579–1588.
98. Marso, S. P.; Daniels, G. H.; Brown-Frandsen, K.; Kristensen, P.; Mann, J. F.; Nauck, M. A.; Nissen, S. E.; Pocock, S.; Poulter, N. R.; Ravn, L. S. Liraglutide and Cardiovascular Outcomes in Type 2 Diabetes. *New Eng. J. Med.* **2016,** *375*(4), 311–322.
99. Marso, S. P.; Bain, S. C.; Consoli, A.; Eliaschewitz, F. G.; Jódar, E.; Leiter, L. A.; Lingvay, I.; Rosenstock, J.; Seufert, J.; Warren, M. L. Semaglutide and Cardiovascular Outcomes in Patients with Type 2 Diabetes. *N. Engl. J. Med.* **2016,** *375*, 1834–1844.
100. Husain, M.; Birkenfeld, A. L.; Donsmark, M.; Dungan, K.; Eliaschewitz, F. G.; Franco, D. R.; Jeppesen, O. K.; Lingvay, I.; Mosenzon, O.; Pedersen, S. D. Oral Semaglutide and cardiovascular Outcomes in Patients with Type 2 Diabetes. *New Eng. J. Med.* **2019,** *381*(9), 841–851.
101. Ness, N.; Scearce, C.; Cantarano, S. Preliminary Results from the Pioneer 6 Magnetic Field Experiment. *J. Geophy. Res.* **1966,** *71*(13), 3305–3313.
102. Hernandez, A. F.; Green, J. B.; Janmohamed, S.; D'Agostino Sr, R. B.; Granger, C. B.; Jones, N. P.; Leiter, L. A.; Rosenberg, A. E.; Sigmon, K. N.; Somerville, M. C. Albiglutide and Cardiovascular Outcomes in Patients with Type 2 Diabetes and Cardiovascular Disease (Harmony Outcomes): A Double-blind, Randomised placebo-controlled Trial. *The Lan.* **2018,** *392*(10157), 1519–1529.
103. Gerstein, H. C.; Colhoun, H. M.; Dagenais, G. R.; Diaz, R.; Lakshmanan, M.; Pais, P.; Probstfield, J.; Riesmeyer, J. S.; Riddle, M. C.; Rydén, L. Dulaglutide and Cardiovascular Outcomes in Type 2 Diabetes (REWIND): A Double-blind, Randomised Placebo-controlled Trial. *The Lan.* **2019,** *394*(10193), 121–130.
104. (a) Holman, R. R.; Bethel, M. A.; George, J.; Sourij, H.; Doran, Z.; Keenan, J.; Khurmi, N. S.; Mentz, R. J.; Oulhaj, A.; Buse, J. B., Rationale and design of the EXenatide Study of Cardiovascular Event Lowering (EXSCEL) trial. *Ame. Heart J.* **2016,** *174*, 103–110; (b) Mentz, R. J.; Bethel, M. A.; Gustavson, S.; Thompson, V. P.; Pagidipati, N. J.; Buse, J. B.; Chan, J. C.; Iqbal, N.; Maggioni, A. P.; Marso, S. P. Baseline Characteristics of Patients Enrolled in the Exenatide Study of Cardiovascular Event Lowering (EXSCEL). *Am. Heart J.* **2017,** *187*, 1–9.
105. Rosenstock, J.; Perkovic, V.; Johansen, O. E.; Cooper, M. E.; Kahn, S. E.; Marx, N.; Alexander, J. H.; Pencina, M.; Toto, R. D.; Wanner, C. Effect of Linagliptin vs Placebo on Major Cardiovascular Events in Adults with Type 2 Diabetes and High Cardiovascular and Renal Risk: The CARMELINA Randomized Clinical Trial. *Jama* **2019,** *321*(1), 69–79.
106. Green, J. B.; Bethel, M. A.; Armstrong, P. W.; Buse, J. B.; Engel, S. S.; Garg, J.; Josse, R.; Kaufman, K. D.; Koglin, J.; Korn, S. Effect of Sitagliptin on Cardiovascular Outcomes in Type 2 Diabetes. *New Eng. J. Med.* **2015,** *373*(3), 232–242.

107. (a) Rosenstock, J.; Perkovic, V.; Alexander, J. H.; Cooper, M. E.; Marx, N.; Pencina, M. J.; Toto, R. D.; Wanner, C.; Zinman, B.; Baanstra, D. Rationale, Design, and Baseline Characteristics of the CArdiovascular safety and Renal Microvascular outcomE Study with LINAgliptin (CARMELINA®): A Randomized, Double-blind, Placebo-controlled Clinical Trial in Patients with Type 2 Diabetes and High Cardio-renal Risk. *Cardio. Diab.* **2018,** *17*(1), 1–15; (b) Scirica, B. M.; Bhatt, D. L.; Braunwald, E.; Steg, P. G.; Davidson, J.; Hirshberg, B.; Ohman, P.; Frederich, R.; Wiviott, S. D.; Hoffman, E. B. Saxagliptin and Cardiovascular Outcomes in Patients with Type 2 Diabetes Mellitus. *New Eng. J. Med.* **2013,** *369*(14), 1317–1326.
108. (a) White, W. B.; Cannon, C. P.; Heller, S. R.; Nissen, S. E.; Bergenstal, R. M.; Bakris, G. L.; Perez, A. T.; Fleck, P. R.; Mehta, C. R.; Kupfer, S. Alogliptin After Acute Coronary Syndrome in Patients with Type 2 Diabetes. *N. Engl. J. Med.* **2013,** *369*, 1327–1335; (b) Zannad, F.; Cannon, C. P.; Cushman, W. C.; Bakris, G. L.; Menon, V.; Perez, A. T.; Fleck, P. R.; Mehta, C. R.; Kupfer, S.; Wilson, C. Heart Failure and Mortality Outcomes in Patients with Type 2 Diabetes Taking Alogliptin versus Placebo in EXAMINE: A Multicentre, Randomised, Double-blind Trial. *The Lan.* **2015,** *385*(9982), 2067–2076.
109. Zinman, B.; Wanner, C.; Lachin, J. M.; Fitchett, D.; Bluhmki, E.; Hantel, S.; Mattheus, M.; Devins, T.; Johansen, O. E.; Woerle, H. J. Empagliflozin, Cardiovascular Outcomes, and Mortality in Type 2 Diabetes. *New Eng. J. Med.* **2015,** *373*(22), 2117–2128.
110. Neal, B.; Perkovic, V.; Mahaffey, K. W.; De Zeeuw, D.; Fulcher, G.; Erondu, N.; Shaw, W.; Law, G.; Desai, M.; Matthews, D. R. Canagliflozin and Cardiovascular and Renal Events in Type 2 Diabetes. *New Eng. J. Med.* **2017,** *377*(7), 644–657.
111. Wiviott, S. D.; Raz, I.; Bonaca, M. P.; Mosenzon, O.; Kato, E. T.; Cahn, A.; Silverman, M. G.; Zelniker, T. A.; Kuder, J. F.; Murphy, S. A. Dapagliflozin and Cardiovascular Outcomes in Type 2 Diabetes. *New Eng. J. Med.* **2019,** *380*(4), 347–357.
112. Perkovic, V.; Jardine, M. J.; Neal, B.; Bompoint, S.; Heerspink, H. J.; Charytan, D. M.; Edwards, R.; Agarwal, R.; Bakris, G.; Bull, S. Canagliflozin and Renal Outcomes in Type 2 Diabetes and Nephropathy. *New Eng. J. Med.* **2019,** *380*(24), 2295–2306.
113. Ehrt, C.; Brinkjost, T.; Koch, O. Impact of binding site comparisons on medicinal chemistry and rational molecular design. *J. Med. Chem.* **2016,** *59*(9), 4121–4151.
114. Rashmi, M.; Swati, D. In silico drug re-purposing against African sleeping sickness using GlcNAc-PI de-N-acetylase as an experimental target. *Comput. Bio. Chem.* **2015,** *59*, 87–94.
115. Van Tassell, B. W.; Raleigh, J. M. V.; Abbate, A. Targeting Interleukin-1 In Heart Failure and Inflammatory Heart Disease. *Cur. Heart Fail. Rep.* **2015,** *12*(1), 33–41.
116. Yamaji, N.; da Silva Lopes, K.; Shoda, T.; Ishitsuka, K.; Kobayashi, T.; Ota, E.; Mori, R. TNF-α Blockers for The treatment of Kawasaki Disease in Children. *Coch. Database System. Rev.* **2019** (8).
117. Ridker, P. M.; Howard, C. P.; Walter, V.; Everett, B.; Libby, P.; Hensen, J.; Thuren, T. Effects of Interleukin-1β Inhibition with Canakinumab on Hemoglobin A1c, Lipids, C-Reactive Protein, Interleukin-6, and Fibrinogen: A Phase IIb Randomized, Placebo-controlled Trial. *Circu.* **2012,** *126*(23), 2739–2748.
118. Everett, B. M.; Cornel, J. H.; Lainscak, M.; Anker, S. D.; Abbate, A.; Thuren, T.; Libby, P.; Glynn, R. J.; Ridker, P. M. Anti-inflammatory Therapy with Canakinumab for the Prevention of Hospitalization for Heart Failure. *Circu.* **2019,** *139*(10), 1289–1299.

119. Ridker, P. M.; Everett, B. M.; Thuren, T.; MacFadyen, J. G.; Chang, W. H.; Ballantyne, C.; Fonseca, F.; Nicolau, J.; Koenig, W.; Anker, S. D. Antiinflammatory Therapy with Canakinumab for Atherosclerotic Disease. *New Eng. J. Med.* **2017,** *377*(12), 1119–1131.
120. Zheng, Z. H.; Zeng, X.; Nie, X. Y.; Cheng, Y. J.; Liu, J.; Lin, X. X.; Yao, H.; Ji, C. C.; Chen, X. M.; Jun, F. Interleukin-1 Blockade Treatment Decreasing Cardiovascular Risk. *Clin. Cardio.* **2019,** *42*(10), 942.
121. Yano, T.; Osanami, A.; Shimizu, M.; Katano, S.; Nagano, N.; Kouzu, H.; Koyama, M.; Muranaka, A.; Harada, R.; Doi, H. Utility and Safety of Tocilizumab in Takayasu Arteritis with Severe Heart Failure and Muscle Wasting. *ESC Heart Fail.* **2019,** *6*(4), 894–897.
122. Kleveland, O.; Kunszt, G.; Bratlie, M.; Ueland, T.; Broch, K.; Holte, E.; Michelsen, A. E.; Bendz, B.; Amundsen, B. H.; Espevik, T. Effect of a Single Dose of the Interleukin-6 Receptor Antagonist Tocilizumab on Inflammation and Troponin T Release in Patients with Non-ST-elevation Myocardial Infarction: A Double-blind, Randomized, Placebo-Controlled Phase 2 Trial. *Eur. Heart J.* **2016,** *37*(30), 2406–2413.
123. Villiger, P. M.; Adler, S.; Kuchen, S.; Wermelinger, F.; Dan, D.; Fiege, V.; Bütikofer, L.; Seitz, M.; Reichenbach, S. Tocilizumab for Induction and Maintenance of Remission in Giant Cell Arteritis: A Phase 2, Randomised, Double-blind, Placebo-controlled Trial. *The Lan.* **2016,** *387*(10031), 1921–1927.

# CHAPTER 8

# Drug Repurposing and Computational Drug Discovery for Diabetes

MANISH KUMAR TRIPATHI[1], RAHUL KUMAR MAURYA[2], ALOK SHIOMURTI TRIPATHI[2], and MOHAMMAD YASIR[2]

[1]*Department of Pharmaceutical Engineering and Technology IIT (BHU) Varanasi, Uttar Pradesh, India*

[2]*Amity Institute of Pharmacy, Lucknow, Amity University Uttar Pradesh, Noida, Uttar Pradesh, India*

## ABSTRACT

Diabetes mellitus (DM) is a complicated metabolic condition that is defined by elevated blood glucose levels, as well as other symptoms. Type 2 diabetes is linked to sedentary lifestyles and obesity because of urbanization (T2D). A variety of methods have been used to control glucose metabolism and treat diabetes clinically. Controlling blood sugar levels with conventional anti-diabetic medicines is promising. Weight gain and decreased therapeutic effectiveness are also side effects of optimal usage of these medicines. Repurposing existing medications has been a major focus in pharmaceutics research and industry. It is possible that drugs designed to treat one condition will also work on others. It is possible to reuse, repurpose, reorganize, and redeploy your resources in a variety of ways. In the last 30 years, the practice of repurposing or repositioning medication has grown in popularity. Compared to previous methods, this one has lower research expenses and a more acceptable safety record. Certain drug regimens for certain illnesses often include side effects. In certain cases, antidiabetic drugs may have adverse effects that include porphyria. Those that suffer from porphyria are

Drug Repurposing and Computational Drug Discovery: Strategies and Advances.
Mithun Rudrapal, PhD (Ed)

quite rare. When an off phenotype is seen, it may imply that the medicine should be utilized more widely. A fortuitous discovery might be made because of biological tests or experimental screenings that target certain sickness models and drugs. When it comes to screening a broad range of drugs or illnesses, these approaches are more adaptable. By using computational biology methodologies and bioinformatic methods, *in-silico* drug screening can undertake virtual screenings of massive drug and chemical databases. The validation steps might be examined *in vitro* first, and subsequently *in vivo* models, after the medication has been selected. This chapter details about repurposing and computational drug discoveries in the field of diabetes.

## 8.1 INTRODUCTION

Diabetes mellitus (DM) is a complex metabolic syndrome characterized by increased blood glucose.[1] Prognosis of hyperglycemia is due to reduced insulin action or insulin secretion, loss of pancreatic β-cells causes decrease in the secretion of insulin lead to decline in its effect in type 1 diabetes (T1D) and insulin receptor sensitivity reduces which reduces utilization of glucose in adipose tissue due to impairment of signaling pathway in type 2 diabetes (T2D).[2] Prevalence of diabetes reveals that more than 380 million patients suffer from it throughout the globe, which will increase to approximately 592 million by 2035.[3] Moreover, T2D is the preliminary cause of this dramatic rise. Sedentary lifestyles and obesity due to an increase in urbanization contribute to the development of T2D.

Uncontrolled diabetes for a prolonged period contributes to the development of several complications associated with diabetes, such as microvascular complications (diabetic retinopathy, neuropathy, and nephropathy) and macrovascular complications (peripheral arterial disease, coronary artery disease, and stroke).[4] Chronic hyperglycemic conditions alter the cellular functions and thereby lead to damage to the tissues due to alteration in the status of reactive oxygen species, advanced glycation products, and aldose reductase. These changes in marker levels have a direct toxic effect on several body tissues. Moreover, metabolic products of glucose also alter the cell signaling pathways, such as the activation of protein kinase C, which causes damage to microvascular tissues.[5] These changes in microvasculature develop atherosclerosis, which causes several chronic complications.

## 8.2 PROBLEMS/COMPLICATIONS IN ANTIDIABETIC THERAPY

Management of diabetes clinically has been achieved by regulating glucose metabolism in a number of ways to achieve a low blood sugar level using several conventional drugs, which include insulin and oral hypoglycemic agents such as sulfonylureas, bigunides, thiazolidinediones, α glucosesidase inhibitors, DPP4 inhibitors, and amylin analogues.[6] These drugs, either by monotherapy or combination therapy, control glucose levels within the normal range. The systemic safety of antidiabetic agents gained attention due to a report by the United States Food and Drug Administration (FDA) on concomitant cardiovascular outcomes.[7]

Optimal use of these hypoglycemic agents causes weight gain and a reduction in therapeutic efficacy. The most common side effect is hypoglycemia.[8] Therapeutic efficacy of conventional drugs occurs due to reduced permeability, solubility, and lack of target specificity and an increase in drug metabolism.[9] However, conventionally available antidiabetic drugs show a promising role in controlling blood sugar levels, but there are still some major challenges for effective glycemic control by optimizing available therapies and reducing the complications associated with chronic antidiabetic drug use. There are several complications associated with each antidiabetic agent associated with chronic use of these drugs, which are given in Table 8.1.

### 8.2.1 *SULFONYLUREAS*

Sulfonylurea agents have also been called secretagogues agents, which stimulate the β-cells to secrete insulin.[10] Hypoglycemia is a major complication associated with the use of the sulfonylurea class of drugs in nephropathic, hepatic failure, malnourished, and elderly patients. Weight gain is another side effect commonly observed with these drugs. From the initiation of therapy to one year, weight gain is approximately 2 kg observed in patients taking sulfonylureas.[11] US-FDA warns the use of sulfonylureas as it enhances the risk of cardiovascular-associated death. Literature reveals that the use of the first generation of sulfonylurea tolbutamide enhances the risk of cardiovascular death more than placebo and insulin do.[12]

### 8.2.2 *MEGLITINIDES*

Repaglinide and nateglinide are meglitinide analogues used to treat T2D. They increase insulin secretion by antagonizing KATP channels.[13] Several clinical trials reveal that the use of meglitinide is reported to enhance weight gain by up to 2.1 g. Meglitinides are reported to have mild hypoglycemia as a common side effect, which also shows a lack of long-term cardiovascular outcomes.[14]

### 8.2.3 BIGUANIDES

Biguanides are an insulin sensitizer, that is, metformin approved for the management of T2D. Clearance of these drugs decreases in pathological conditions including dehydration, sepsis, congestive heart failure, renal and hepatic impairment, which enhances the risk of development of lactic acidosis. Metformin improves body weight even when combined with sulfonylurea. Congestive heart failure is not commonly observed with the long-term use of metformin, and it is one of the safest antidiabetic drugs.[15]

### 8.2.4 THIAZOLIDINEDIONES

Thiazolidinediones (TZD; pioglitazone and rosiglitazone) are insulin sensitizers that act as peroxisome proliferator–activated receptor (PPAR-) agonists. They are approved for the treatment of T2D. Thiazolidinediones are known to cause fluid retention and weight gain causes edema. Literature reveals that TZD in combination with insulin enhances weight gain by up to 5 g after 2 years of drug treatment. Strong evidence reveals that TZD appears to cause health failure as a side effect, and thus most drugs in this category are banned for clinical use.[16]

### 8.2.5 α-GLUCOSIDASE INHIBITOR

α-Glucosidase inhibitors (voglibose, acarbose, and meglitinide) inhibit the enzyme α-glucose hydroxylase and α-amylase enzyme, which contribute to the conversion of polysaccharide carbohydrates into monosaccharides. The most serious problem associated with the use of these drugs is diabetic ketoacidosis; however, this class of drug commonly metabolizes in the GI track by gastrointestinal enzymes and absorbs the least amount in the systemic circulation. Thus, the risk of weight gain and hypoglycemia was not observed with the use of α-glucosidase inhibitors. Moreover, there was a reduction in the risk of the development of congestive heart failure and hypertension of up to 49% observed with the use of these agents.[17]

### 8.2.6 GLUCAGON-LIKE PEPTIDE 1 (GLP-1) REGULATORS

Postprandial glucose homeostasis is regulated by GLP-1 in response to ingested nutrients. In systemic circulation, DPP-4 inhibits GLP-1 into an inactive metabolite. Thus, there are two different categories. GLP-1 receptor agonists and DPP-4 inhibitors were developed. DPP-4 inhibitors (Sitagliptin) are reported to act neutral on weight gain, and GLP-1 agonists reduce weight gain by more than 2 kg. GLP-1 receptor agonists have some role in pancreatitis, but an exact relationship has not been developed yet. It is also contraindicated in patients with thyroid cancer or a family history of it, as the GLP-1 agonist stimulates the proliferation of type C cells. DPP-4 expression has been found in T cells, suggesting that DPP-4 inhibitors may modulate T cells and have an immunomodulatory effect.[18]

### 8.2.7 SODIUM-GLUCOSE TRANSPORTER 2 INHIBITORS

Sodium-glucose transporter 2 (SGLT2) inhibitors modulate renal glucose handling and thereby lower the glucose level. SGLT2 inhibitors are contraindicated in patients with renal failure and hepatic insufficiency. Its consumption for more than 12 months reduces weight and fluid in the body and induces hypotension in the patient. Moreover, SGLT-2 also reported to modulate the level of lipoprotein, and thus its effect on cardiac complications needs to be studied in a detailed manner.[19]

**TABLE 8.1** Safety Consideration and Common Side Effects of Antidiabetic Agents.

| Antidiabetic agents | Hyperglycemia | Weight | Contraindication |
|---|---|---|---|
| Sulfonylurea drugs | Yes | Increases | – |
| Metformin/bigunides | No | Mildly decreases | Metabolic acidosis, hepatic failure, and renal failure |
| TZDs | No | Increases | Heart failure and hepatocellular diseases |
| α-Glucosidase inhibitors | No | Decreases | Renal failure, liver cirrhosis, and intestinal disease |
| GLP-1 receptor agonist | No | Decreases | Pancreatitis, renal failure, endocrine neoplasia |
| DPP-4 inhibitors | No | Neutral | Pancreatitis history |
| Amylin analog | Yes | Initially decreases | Gastroparesis |
| SGLT-2 inhibitors | No | Neutral | Renal failure |

## 8.3 NEED FOR DRUG DISCOVERY FOR DIABETES

Diabetes-associated morbidity and mortality occurs due to cardiovascular disease. Previously, there was no specific need for novel therapy for the management of T2D because existing therapies did not have any issues with cardiovascular safety.[20] However, a meta-analysis published in 2007 reveals the greater chances of stroke and myocardial infraction in patients treated with rosiglitazone than control.[21] Thereafter, this class of drug was completely withdrawn from the market for clinical use. The FDA changed the drug safety guidance related to therapies for T2D so that there is no cardiovascular risk associated with new drugs for the management of it. These regulations state that relative risk should be <1.8 and participants included in the study with diabetes (advanced age, renal impairment, and chronic diabetic) should be included in the study. The study also includes at least two years of safety data; stroke, myocardial infarction, and cardiovascular mortality should be included in the study. Meta-analysis will be performed for the unacceptable cardiovascular risk from the data of phase II and III clinical trials.[22] Presently, drugs available under development for T2D need to follow these FDA guidelines, which require patients with cardiovascular disorders to enroll in trials, and it requires longer and larger trials. This all contributes to greater expense and delays the time for the drug to come to the market.

T2D is a multifactorial disease in which several pathogenic pathways are involved in dysregulation of glycemic control. Oxidative stress, systemic and tissue-specific inflammation, and altered lipid metabolism[23] are all factors to consider. These pathogeneses are involved in the development of other diseases too, and drugs used for the management of those diseases that act on a given pathogenic pathway could be promising therapies for treating T2D.

## 8.4 DRUG REPURPOSING (DR) APPROACHES FOR ANTIDIABETIC DRUG DISCOVERY

Pharmaceutical research and industry have been paying close attention to the possibility of repurposing existing pharmaceuticals and producing novel drugs, as well as the efficiency of repurposing over conventional de novo drug development methods. Repositioning, reprofiling, redirection, retaking, and therapeutic switching are all examples of drug repurposing (Fig. 8.1).

Traditional studies on repurposing usually focused on identification of novel indications for drugs based on similarity of its mode of action. Conventional drugs target on new targets based on some characteristics like drug–disease relationship, drug side effects, and its chemical structures.

In general, there are two basic drug repositioning principles. First, because of the interdependence of various disorders, medications developed for one disease may also be effective for other diseases. Second, because medications are inherently confusing, they can be linked to a variety of targets and pathways.

In the next several decades, the number of individuals suffering from diabetes will climb substantially. As a chronic metabolic disease, T2D is characterized by insulin resistance and elevated blood sugar in most sufferers. Several underlying factors contribute to poor glycemic control in T2D, which is a complex condition. There has been a reconsideration of T2D therapy options that target comparable pathophysiological systems because of their adaptability.

When it comes to drug research, the creation of new medications and agents demands a significant investment of time and money. Repurposing or repositioning medicine has become more popular in the past 30 years due to a decrease in the number of FDA-approved pharmaceuticals. New drug development is a time-consuming and costly undertaking that is characterized by a significant degree of uncertainty about whether a medicine will succeed. Due to rising costs and time constraints, these gains have not been able to be translated into therapeutic improvements as quickly as planned. There are several methods to repurpose medications that have previously been approved by the FDA and/or are currently under research. It is a new technique that offers cheaper research costs and acceptable safety profiles since much of the trial data is already available in the public domain. Because the medicine has already undergone clinical testing, the chance of failure is almost nonexistent or very low. Due to extensive testing and safety evaluations previously completed, the time required to produce a new medicine is greatly reduced. Third, the medications that are currently on the market need relatively little investment. At this time, validation of targets and studies for on- and off-target drug repositioning have become natural allies in chemical biology. Repurposing drugs requires three primary phases before they can be tested in phase II clinical trials: Identification of the proper drug, rigorous study of the drug's action in clinical models, and appraisal of its utility.

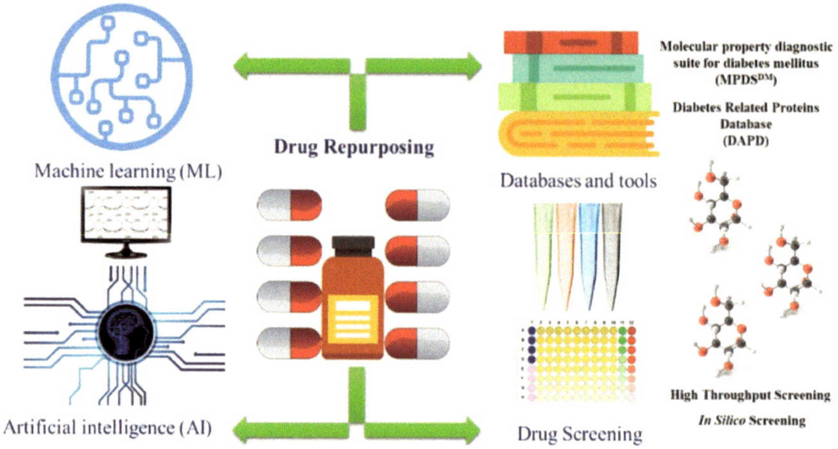


**FIGURE 8.1** Different approaches for drug repurposing.

Some of the present ways for applying drug-repurposing technology are targeted mechanism-based, pathway or network-based, signature-based, target-based, knowledge-based, and blind-based.[24] The above-mentioned medication repurposing strategies were successful in both clinical and preclinical research to find the drug's new role in treating DM.

### *8.4.1 PROTEINS/PATHWAYS TARGETED APPROACH*

Diabetic mellitus (DM) is the most prevalent metabolic disorder, and insulin insufficiency, inappropriate insulin usage, or both, may cause blood glucose levels to vary. An autoimmune attack on pancreatic islet cells causes type 1 diabetes; insulin signaling issues and environmental factors are to blame for T2D. The glucose metabolism of DM patients has recently been successfully controlled by glucagon-like peptide-1 (GLP-1) receptor agonists and DPP-4 inhibitors, in addition to insulin, alpha-glucosidase inhibitors, thiazolidinediones, meglitinides, sulfonylureas, and biguanides. Incretin hormones, including glucose-dependent insulinotropic peptide (GIP), are degraded by the DPP-4 enzyme, which is inhibited by DPP-4 inhibitors. Chemokines and growth factors are also destroyed by DPP-4, suggesting that DPP-4 inhibitors may help regulate inflammatory conditions. Incretin hormones increase insulin secretion and glucagon release, and they also lower beta cell mortality. It is not uncommon for patients on FDA-approved DPP-4 inhibitors like sitagliptin, vildagliptin, alogliptin, and saxgliptin to have pancreatitis, hepatitis, or renal damage. Over the course of therapy, complications such as diabetic retinopathy and painful diabetic nephropathy have been recorded, and they are more frequent in insulin-dependent diabetes patients.

Incretin effect regulates the blood glucose levels by enhancing secretion of insulin and controls the release of glucagon, which reduces the mortality of β-cells too. Patients on DPP-4 inhibitors are commonly associated with pancreatitis, hepatitis, or renal damage. Painful diabetic nephropathy and diabetic retinopathy are more frequent in insulin-dependent diabetes patient.

Diabetic nephropathy is a painful condition that may be treated with opioids, tapentadol, and lidocaine. Studies in both clinical and preclinical settings have shown that combination treatment for DM is beneficial. There has been evidence in clinical studies that NSAIDs may reduce the cyclo-oxygenase (COX) enzymes. However, several of these therapies have the potential to cause renal failure, and hence, NSAIDs are typically avoided in the treatment of diabetic mellitus. There is evidence that arthritis medications such as salsalate and ibuprofen, which are often prescribed to people with T2D, might reduce blood glucose levels by decreasing inflammatory mediators, suggesting that NSAIDs alone may be successful in doing so. In both clinical and preclinical trials, combination treatments for diabetes have been demonstrated to be successful. Salicylate, a pain-relieving medicine that already existed, was acetylated to make aspirin, which is the most popular and well-known drug in the world at the time. Salicylate improves glycemia and lowers inflammatory mediators in T2Dm.[25,26] Salicylate and the cellular senescence of IKK-b in the skeletal muscle restrict fat-induced insulin resistance, which means that fat-induced changes in insulin signaling activity could be a new class of T2D medicines.[27]

In a recent trial, HIV-1 and hepatitis B infections were originally treated with the anti-inflammatory drug known as nucleoside reverse transcriptase inhibitors (NRTIs). Preliminary studies in the C57BL/6J mouse model show that it enhances insulin sensitivity and slows the evolution of T2D.[28] Chen et al. found that the antidepressant medicine doxepin activates FAM3A signal transduction pathways in the hepatic and brown fatty tissue to ameliorate hyperglycemia and steatosis in diabetic mice. Doxepin may be preferentially suggested as an antidepressant in the treatment of people with diabetes who also have depression.[29] Telmisartan reduced obesity-induced insulin resistance, according to Liu et al., via suppressing ER stress through AMPK activation.[30] Similarly, multiple studies suggest that an existing class of drugs might be repurposed for the prevention of diabetes. Niclosamide ethanolamine (NEN) is an FDA-approved anthelmintic drug for the treatment of parasitic worms. A recent study by Tao et al. showed the potential of NEN as a therapeutic and management tool for people with T2D. Mice treated with NEN had better energy distribution and lipid metabolism.[31] It is a significant medical disorder known as diabetic kidney disease (DKD). There is a pressing need for new therapies to prevent or delay the progression of DKD. Many biological processes, such as cell viability and mortality, are regulated by potassium channel KCa3.1. The development and course of DKD are influenced by mitochondrial dysfunction. Inflammation, proliferation, and activation of cells are all linked to mitochondrial dysfunction and KCa3.1 dysregulation. Recently, it has been shown that KCa3.1 absence improves the mitochondrial dysfunction associated with DKD by regulating abnormal mitochondrial dynamics and rectifying deficient BNIP3-mediated mitophagy, which is attributed to the modification of mitochondrial quality control. DKD may benefit from pharmacological reduction of KCa3.1's involvement in fibrosis, inflammation, and mitochondrial dysfunction, which might lead to a new treatment option. Because of its high level of safety, repurposing senicapoc might have a positive impact on human health.[32]

### *8.4.2 TRADITIONAL MEDICINES-BASED APPROACH*

While great attention has been made on repurposing current pharmaceuticals to cure/manage illnesses, traditional medicines are an alternative and maybe neglected source of novel antidiabetic molecules. For hundreds of years, various pathological conditions have been treated with the use of these formulations and extracts. In the case of T2D, this prolonged history of use often provides strong empirical information for their advantages on glucoregulatory mechanisms as well as the absence of serious side effects. Furthermore, observations with an ancient herbal medicine led to the development of antidiabetic guanidine pharmaceuticals like metformin. It was discovered that guanidines in *Galega officinalis*, a plant that produces the chemical, might improve the symptoms of diabetes, including frequent urination.[33] Moving further with conventional medications requires identifying the active ingredients in these formulations. There are thousands of instances of traditional medicines being used to successfully find possible antidiabetic molecules. Berberine is a phytoconstituent that was first tested to treat viral diarrhea.[34] *Cortidisrhizoma* and *Cortex phellodendri* are richer in berberine content which possesses insulin-sensitizing and insulinotropic activities through the activation of AMPK.[35] Bitter melon (*Momordica charantia*) extracts have been used in traditional medicine to treat diabetes symptoms for many years, but the active ingredient has remained a mystery.

### 8.4.3 SIDE EFFECTS AND ADVERSE DRUG EVENT-BASED APPROACH

Documentation on the treatment of illnesses that arise in rodents is also utilized to repurpose drugs in humans because of the general parallels in targets. Side effects are common in drug regimens for specific disorders. Drug repurposing may also be triggered by clinical findings of unexpected adverse effects reported by patients or by clinicians' experiences of unexpected conditions cured by drugs. Although some side effects are considered infrequent, they are nevertheless implicated in the pathophysiology of a disease. Indeed, they may reveal a severe phenotype. Many associations include explicit repurposing theories, such as hypoglycemic medicines being possible diabetes candidates. Drugs that identify porphyria as a side effect, according to drug repositioning based on the side effect, may serve as antidiabetics. Porphyria is a genetic disease that affects only a few people.[36] Patients with this hereditary condition have a lower risk of developing porphyria when they become diabetic.[37] Antidiabetic medicines have been linked to an increased risk of developing porphyria. A subgroup of the population having a hereditary porphyria may conceivably be used as a "model" for assessing antidiabetic medications with porphyria as the goal. Consequently, a drug that produces porphyria in this subset with the genetic variation might be an effective therapy for diabetes in a wider population. As a result, a drug's off-phenotype in a subpopulation may indicate that it should be used in a larger population. Lithner et al. had reported in the study of 16 patients who developed diabetes in a study of 328 Swedish porphyria patients all had their porphyria symptoms alleviated.[38] Porphyria is a known side effect of the drugs valproic acid, pyrazinamide, naproxen, and estradiol; nevertheless, diabetic doctors do not suggest them because of the risk of developing the disease.[38] Valproic acid, an anticonvulsant, has been demonstrated to lower blood sugar levels in Wfs1 mutant mice in latest research.[39] Pyrazinamide is an anti-TB medicine, and T2D is a well-known risk factor for tuberculosis.[40] In prediabetic mice, naproxen has been demonstrated to delay or prevent the development of T2D from occurring.[39]

### 8.4.4 COMPUTATIONAL APPROACHES AND TECHNIQUES (FOR ANTIDIABETIC DRUG DISCOVERY)

Traditional techniques used for drug discovery have several limitations like time consuming and costly, and thus drug repurposing based on computer technologies is an option for pharmaceutical industry.

Computational intelligence with a broader application could be adapted to diabetes medicine repurposing. Network-based medication, molecular dynamics modeling, virtual screening, visualization tools, data mining, deep learning, artificial intelligence (AI), and machine learning are a few of the technologies that have influence on diabetes research. In the past few years, volume of publicly available health and biomedical-related electronic data such as high-performance computer, online health community, pharmaceutical databases, and microarrays gene expression signatures has grown exponentially, computational drug repurposing approaches have been developed.

The study of the link between diverse biomedical components is an important part of modern drug repurposing research. Drugs, diseases, genes, and adverse drug reactions (ADRs) are examples of biological entities. Distinguishing between a treatment's molecular targets and the hundreds to thousands of other gene products that respond indirectly to changes in the activity of the targets is a key challenge in drug repurposing. Unfortunately, due to the large number of genes, traditional statistical techniques are inadequate in finding the molecular targets of a medicine. Furthermore, traditional statistical methods rely on tiny datasets and biological networks derived from experiments conducted on a variety of platforms and conditions, which could lead to discrepancies in the results provided by some studies. Computational repurposing might be based on a target or a disease. Target-based approaches compare the characteristics and similarities of different diseases to infer similar binding sites by analyzing compound–protein interactions, whereas disease-based approaches link drugs to new indications by comparing the characteristics and similarities of different diseases.

### *8.4.5 DATABASES AND TOOLS*

Databases and tools that give chemical entity or expression of diabetes-associated genes are used in computational methodologies to access the drug molecules for medication repurposing. Because it contains approximately 1700 number of FDA approved and about 5000 experimental medications and is publicly available, DRUGSURV is one of the most essential tools in drug repurposing. Furthermore, two types of proteins are directly engaged in the turnover of human metabolites in vivo: Enzymes and transporters. By scanning the Human Metabolome Database (HMDB, http://www.hmdb.ca), we can uncover the enzymes and transporters linked to diabetes-associated chemicals discovered in prior metabolomics research. To visualize the interactions between diabetes metabolites and the enzymes or transporters that regulate them, the metabolites-proteins network was created using Cytoscape (www.cytoscape.org).

Researchers combined genes or proteins associated with diabetes metabolite proteins related to diabetes metabolites acquired from metabolomics data to construct a list of diabetic risk proteins retrieved from genomics and proteomics investigations. Furthermore, because the majority of drugs are antagonists or agonists, understanding the pathophysiology of target proteins is crucial for identifying whether a treatment will assist or aggravate an illness's phenotype. Researchers used the OMIM (http://www.omim.org) and perform a search of database for literature to learn about the pathophysiology of targets associated with diabetes (PubMed). Researchers looked at gain of function (GOF) and loss of function (LOF) functions in human and animal models to uncover antidiabetes protein candidates. Gene signatures are derived from illness omics data with the purpose of identifying potential markers. Data on genomics can be found in databases that are open to the public. The benefit of these approaches is that they may be used to investigate undiscovered pharmacological mechanisms of action. Signature-based methods, in contrast to knowledge-based methods, use computational methodologies to examine pharmacological processes at a molecular level, such as changes in gene expression.

### 8.4.6 *IN-SILICO APPROACHES*

Drug repurposing has considerably reduced the time and expense of drug research while also lowering the likelihood of failing because of the advent of bioinformatics/cheminformatics techniques and the access to large biological and structural databases. In-silico drug repurposing, on the other hand, uses computational biology and bioinformatics/cheminformatics techniques to perform virtual screenings of the databases of the libraries of the drug and chemical compounds. This method of identifying potential bioactive chemicals is based on the molecular interaction between therapeutic candidates and the target protein. To identify novel therapeutic targets, in-silico techniques that combine knowledge miming with molecular modeling technologies can be used. The validation steps might be evaluated in vitro first, then in vivo models once the drug has been chosen. Primaquine and simvastatin, for example, are two interesting drug possibilities that have been discovered because of virtual screening. The Broad Institute's Connectivity

Map (cMAP) covers gene expression patterns induced by drugs in different cell lines, allowing enabling drug repurposing predictions for antidiabetics. Genome-wide association studies (GWAS) have disclosed valuable insight into major genetic variants of human diseases, resulting in the identification of novel gene targets, some of which are shared by several diseases, resulting in drug repurposing. Zhiguo et al. have shown glucosidase to be a favorable target for the treatment of diabetes. Drug repurposing can help researchers find an active inhibitor faster. Using medication repurposing and in-silico techniques, the researchers set out to find a viable α-glucosidase inhibitor. The three α-glucosidase proteins also had critical amino acid residues identified. Furthermore, cross molecular docking investigations of three glucosidase proteins and drugs from the FDA database produced the hits with favorable binding affinities. An in vitro biological assay later revealed that raloxifene from the FDA database has exhibited a potent α-glucosidase inhibitor activity. Furthermore, in physiological conditions, mol. dynamics simulations of raloxifene and three α-glucosidase proteins revealed the stability of protein–hit interaction. The findings showed that raloxifene–protein interaction was stable, and that the amino acid residues of three proteins made stable connections with raloxifene.[40] Furthermore, Sitagliptin is a selective DPP-4 inhibitor that improves glycemic control and fasting effectiveness in people with type 2 diabetes. However, in patients treated with sitagliptin, substantial hypersensitivity reactions have been recorded. In this context, Crisan et al. show new medicines with improved characteristics that target DPP-4. Sitagliptin ((2R)-4-oxo-4-[3-(trifluoromethyl)-5,6-dihidro [1,2,4] triazolo[4,3-A]pirazin-7(8H)-yl]-1-(2,4,5-trifluorophenyl) butan-2-amine)) was utilized as a query in an approved DrugBank 3D-similarity search. Four of the DrugBank molecules, namely DB09195 (lorpiprazole), DB09089 (trimebutine), DB00298 (dapiprazole), and DB13858 (dimazole), were found effective antidiabetic agents for possible repurposing of marketed drugs after docking evaluations, based on the Tanimoto Combo parameter.[41]

### 8.4.7 NETWORK-BASED INTEGRATED APPROACHES

A pathway-based method can reveal genes that are useful for medication repurposing. To uncover novel targets, it entails creating drug or disease networks based on gene expression patterns, pathology, relationships, or GWAS data. The Human Genome Project (HGP) paved the way for a comprehensive understanding of biology at the systems and network levels.[42]

This must, however, be properly utilized to completely comprehend the pathophysiology of type 1 diabetes (T1D) and, as a result, to develop new treatments for the condition. Its comprehensive list of protein-coding genes ushered in a new era of noncoding genome research and opened the way for therapeutic discoveries. Importantly, the discoveries, wherein investigators explored the relationships between cellular building blocks, correlate also with the emergence of a processes understanding of biology, as opposed to the old single-gene approach.[43]

Martin et al. have explored that β-cell malfunction and death in T1D are driven by major gene regulatory networks. This could lead to the development of treatments that target the ultimate cause of T1D, rather than just treating hyperglycemia or attempting to delay disease by suppressing the immune system. It demonstrates the possibility of successful cell protective medicines that could change the course of diabetes' natural history. Many effective medications do not directly target illness genes; instead, they alter the outcomes of defective processes by targeting proteins.[44] Targeting the pathways that modulate β-cell responses to immunological reaction would either stimulate or suppress the immune response (for example, human leucosite antigen (HLA) class I and chemokine overexpression or neoantigen production for example, programmed death-ligand 1 (PDL1) and HLA-E expression).[45]

## 8.4.8 ARTIFICIAL INTELLIGENCE (AI) AND MACHINE LEARNING (ML) TECHNOLOGY

The adoption of in-silico approaches, as well as structure-based drug design (SBDD), artificial intelligence (AI) technology, and Machine learning (ML), has expedited the drug repurposing approach in recent years. ML is a computational technique that has contributed positively to various sections of the drug development pipeline. ML is the application of computing algorithms to learn from existing data to create new predictions and obtain new knowledge. By analyzing similar features, like clinical expression, toxicity, and analyzing the target for therapy, one can construct impartial algorithms to link seemingly disparate medications. AI and ML also help with the treatment of chronic diseases, such as diabetes.[46,47] In reality, ML and AI are already being used to forecast diabetes risk based on genetic data, to diagnose diabetes based on EHR data, to predict the risk of sequelae including nephropathy and retinopathy, and to diagnose diabetic retinopathy.[48] The Google AI research unit has already achieved significant progress in the field of automated diagnosis and grading of diabetic retinopathy based on fundus pictures in partnership with a few Indian ophthalmology institutes.[49,50] Diabetes AI now focuses on diabetes prediction, lifestyle counseling, insulin injection counseling, blood sugar monitoring, self-management, and complication monitoring. Elemento et al. present a novel medication repurposing strategy based on a platform called Creating A Translational Network for Indication Prediction (CATNIP). CATNIP is a computer model that can forecast new indications for individual medications or entire pharmacological classes. This method compares drugs based on a variety of biochemical, and therapeutic characteristics, providing a comprehensive insight of drug's mechanism and potential applications. Furthermore, CATNIP is a machine-learning method that uses 2576 distinct pharmaceuticals to learn how to anticipate whether two molecules have the same indication based solely on the biological and chemical features of the medicine. The application of CATNIP to chemical pairings develops a network with 4.6 million nodes that can be utilized to detect possible drug repurposing opportunities.[51] CATNIP provides a tool to discover several new drug classes that are expected to exhibit repurposing indications, such as Parkinson's disease and T2D.

### *8.4.9 MOLECULAR PROPERTY DIAGNOSTIC SUITE FOR DIABETES MELLITUS (MPDS$^{DM}$)*

Due to the increased demand for pharmaceutical products to produce safe and cost-effective drugs for existing as well as emerging diseases, the information generated from drug repurposing has been plummeting aggressively in recent years. Now, researchers and scientists must have easy access to the knowledge that has been developed. Comprehensive data comprehension and sharing is a growing concern since it reduces redundancy and allows for more in-depth research. In drug discovery methodologies, a platform for open research and innovation is critical. MPDS$^{DM}$ is a disease-specific open-source web portal for DM that includes a comprehensive database of T2D biomarkers, therapeutic targets, FDA-approved medications, and related genes.[52] The MPDS$^{DM}$ Galaxy tool is divided into three sections: Data library, data processing, and data analysis. It has the potential to be a valuable resource for researchers working on drug repurposing in general, as well as antidiabetes research. The galaxy-based web platform is a flexible, open-ended portal that accepts scripts and computational instructions written in any language. This portal is also designed to provide workflows, and connections between various technologies that are easily accessible. As a result, the MPDS$^{DM}$ website offers all-encompassing target information for a drug, a library of protein related to diabetes, a variety of information processing, molecular modeling, and drug discovery tools. Another digital platform, the Diabetes Related Proteins Database (DAPD), provides thorough information on diabetes-related target proteins and pathways.

## 8.5 CONCLUSIONS AND FUTURE PERSPECTIVES

Despite the rapid understanding of the disease's advanced molecular foundation, there is still a shortage of effective T2D treatments. As a result, alternative drug development tactics are being investigated, such as repurposing current medications to treat T2D. Computer-assisted technologies such as artificial intelligence (AI), big data analytics, data mining, visualization tools and methodologies, molecular dynamics, molecular modeling, and network-based integrated approaches have recently drawn interest in effective medication repurposing for the management of diabetes. Furthermore, sometimes, the provided evidence is not adequate and does not meet the need of regulatory bodies such as the FDA or EMA. According to regulatory guidelines fresh preclinical, clinical trials may be required for antidiabetic drugs evolved from repurposing. Another significant point is patent applications and intellectual property rights (IPR). According to IP and patent regulations, there is no IP protection for drug development using the repurposing technique. Some repurposed drugs are unable to enter the market due to intellectual property issues. Furthermore, some repurposing projects must be abandoned, resulting in a waste of time, money, and efforts.

## KEYWORDS

- **virtual sceening**
- **diabetes mellitus**
- **molecular docking**
- **drug repurposing**
- **bioinformatics**
- **antidiabetic**

## REFERENCES


1. Tripathi, A. S.; Mazumder, P. M.; Chandewar, A. V. Changes in the Pharmacokinetic of Sildenafil Citrate in Rats with Streptozotocin-Induced Diabetic Nephropathy. *J. Diabetes Metab. Disord.* **2014,** *13*, 8.
2. Tsalamandris, S.; Antonopoulos, A. S.; Oikonomou, E.; Papamikroulis, G. A.; Vogiatzi, G.; Papaioannou, S.; Deftereos, S.; Tousoulis, D. The Role of Inflammation in Diabetes: Current Concepts and Future Perspectives. *Eur. Cardiol. Rev.* **2019,** *14*, 50–59.
3. Forouhi, N. G.; Wareham, N. J. Epidemiology of Diabetes. *Medicine (United Kingdom)* **2014,** *42*, 698–702.
4. Cade, W. T. Diabetes-Related Microvascular and Macrovascular Diseases in the Physical Therapy Setting. *Phys. Ther.* **2008,** *88*, 1322–1335.
5. Giacco, F.; Brownlee, M. Oxidative Stress and Diabetic Complications. *Circ. Res.* **2010,** *107*, 1058–1070.
6. Krentz, A. J.; Bailey, C. J. Oral Antidiabetic Agents: Current Role in Type 2 Diabetes Mellitus. *Drugs* **2005,** *65*, 385–411.
7. Chaudhury, A.; Duvoor, C.; Reddy Dendi, V. S.; Kraleti, S.; Chada, A.; Ravilla, R.; Marco, A.; Shekhawat, N. S.; Montales, M. T.; Kuriakose, K.; Sasapu, A.; Beebe, A.; Patil, N.; Musham, C. K.; Lohani, G. P.; Mirza, W. Clinical Review of Antidiabetic Drugs: Implications for Type 2 Diabetes Mellitus Management. *Front. Endocrinol. (Lausanne).* **2017,** *8*, 6.
8. Lorenzati, B.; Zucco, C.; Miglietta, S.; Lamberti, F.; Bruno, G. Oral Hypoglycemic Drugs: Pathophysiological Basis of Their Mechanism of ActionOral Hypoglycemic Drugs: Pathophysiological Basis of Their Mechanism of Action. *Pharmacol.* **2010,** *3*, 3005–3020.
9. Hua, S.; de Matos, M. B. C.; Metselaar, J. M.; Storm, G. Current Trends and Challenges in the Clinical Translation of Nanoparticulate Nanomedicines: Pathways for Translational Development and Commercialization. *Front. Pharmacol.* **2018,** *9*.
10. Sola, D.; Rossi, L.; Schianca, G. P. C.; Maffioli, P.; Bigliocca, M.; Mella, R.; Corlianò, F.; Paolo Fra, G.; Bartoli, E.; Derosa, G. Sulfonylureas and Their Use in Clinical Practice. *Arch. Med. Sci.* **2015,** *11*, 840–848.
11. Marín-Peñalver, J. J.; Martín-Timón, I.; Sevillano-Collantes, C.; Cañizo-Gómez, F. J. del. Update on the Treatment of Type 2 Diabetes Mellitus. *World J. Diabetes* **2016,** *7*, 354.
12. Azoulay, L.; Suissa, S. Sulfonylureas and the Risks of Cardiovascular Events and Death: Amethodologicalmeta-Regression Analysis of the Observational Studies. *Diabetes Care* **2017,** *40*, 706–714.
13. Guardado-Mendoza, R.; Prioletta, A.; Jiménez-Ceja, L. M.; Sosale, A.; Folli, F. The Role of Nateglinide and Repaglinide, Derivatives of Meglitinide, in the Treatment of Type 2 Diabetes Mellitus. *Arch. Med. Sci.* **2013,** *9*, 936–943.
14. Xu, J.; Rajaratnam, R. Cardiovascular Safety of Non-Insulin Pharmacotherapy for Type 2 Diabetes. *Cardiovasc. Diabetol.* **2017,** *16*, 18.
15. Rao, C. V. Biguanides. In *Encyclopedia of Toxicology*, 3rd ed.; 2014; pp. 452–455.
16. Soccio, R. E.; Chen, E. R.; Lazar, M. A. Thiazolidinediones and the Promise of Insulin Sensitization in Type 2 Diabetes. *Cell Metab.* **2014,** *20*, 573–591.
17. Dabhi, A. S.; Bhatt, N. R.; Shah, M. J. Voglibose: An Alpha Glucosidase Inhibitor. *J. Clin. Diagnostic Res.* **2013,** *7*, 3023–3027.

18. Sharma, D.; Verma, S.; Vaidya, S.; Kalia, K.; Tiwari, V. Recent Updates on GLP-1 Agonists: Current Advancements & Challenges. *Biomed. Pharmacother.* **2018,** *108*, 952–962.
19. Tang, H.; Li, D.; Zhang, J.; Li, Y.; Wang, T.; Zhai, S.; Song, Y. Sodium-Glucose Co-Transporter-2 Inhibitors and Risk of Adverse Renal Outcomes among Patients with Type 2 Diabetes: A Network and Cumulative Meta-Analysis of Randomized Controlled Trials. *Diabetes, Obes. Metab.* **2017,** *19*, 1106–1115.
20. Cefalu, W. T.; Kaul, S.; Gerstein, H. C.; Holman, R. R.; Zinman, B.; Skyler, J. S.; Green, J. B.; Buse, J. B.; Inzucchi, S. E.; Leiter, L. A.; Raz, I.; Rosenstock, J.; Riddle, M. C. Cardiovascular Outcomes Trials in Type 2 Diabetes: Where Do We Go From Here? Ref Lections From a Diabetes Care Editors' Expert Forum. *Diabetes Care* **2018,** *41*, 14–31.
21. Nissen, S. E.; Wolski, K. Rosiglitazone Revisited: An Updated Meta-Analysis of Risk for Myocardial Infarction and Cardiovascular Mortality. *Arch. Int. Med.* **2010,** *170*, 1191–1201.
22. Sharma, A.; Pagidipati, N. J.; Califf, R. M.; McGuire, D. K.; Green, J. B.; Demets, D.; George, J. T.; Gerstein, H. C.; Hobbs, T.; Holman, R. R.; Lawson, F. C.; Leiter, L. A.; Pfeffer, M. A.; Reusch, J.; Riesmeyer, J. S.; Roe, M. T.; Rosenberg, Y.; Temple, R.; Wiviott, S.; McMurray, J.; Granger, C. Impact of Regulatory Guidance on Evaluating Cardiovascular Risk of New Glucose-Lowering Therapies to Treat Type 2 Diabetes Mellitus. *Circulation* **2020,** *141*, 843–862.
23. Galicia-Garcia, U.; Benito-Vicente, A.; Jebari, S.; Larrea-Sebal, A.; Siddiqi, H.; Uribe, K. B.; Ostolaza, H.; Martín, C. Pathophysiology of Type 2 Diabetes Mellitus. *Int. J. Mol. Sci.* **2020,** *21*, 1–34.
24. Jin, G.; Wong, S. T. C. Toward Better Drug Repositioning: Prioritizing and Integrating Existing Methods into Efficient Pipelines. *Drug Discov. Today* **2014,** *19*, 637–644.
25. Apovian, C. M.; Okemah, J.; O'Neil, P. M. Body Weight Considerations in the Management of Type 2 Diabetes. *Adv. Ther.* **2019,** *36*, 44–58.
26. Gentilella, R.; Pechtner, V.; Corcos, A.; Consoli, A. Glucagon-like Peptide-1 Receptor Agonists in Type 2 Diabetes Treatment: Are They All the Same? *Diabetes. Metab. Res. Rev.* **2019,** *35*, e3070.
27. Pereira, S.; Yu, W. Q.; Frigolet, M. E.; Beaudry, J. L.; Shpilberg, Y.; Park, E.; Dirlea, C.; Grégoire Nyomba, B. L.; Riddell, M. C.; George Fantus, I.; Giacca, A. Duration of Rise in Free Fatty Acids Determines Salicylate's Effect on Hepatic Insulin Sensitivity. *J. Endocrinol.* **2013,** *217*, 31–43.
28. Ambati, J.; Magagnoli, J.; Leung, H.; Wang, S.; Andrews, C. A.; Fu, D.; Pandey, A.; Sahu, S.; Narendran, S.; Hirahara, S.; Fukuda, S.; Sun, J.; Pandya, L.; Ambati, M.; Pereira, F.; Varshney, A.; Cummings, T.; Hardin, J. W.; Edun, B.; Bennett, C. L.; Ambati, K.; Fowler, B. J.; Kerur, N.; Röver, C.; Leitinger, N.; Werner, B. C.; Stein, J. D.; Sutton, S. S.; Gelfand, B. D. Repurposing Anti-Inflammasome NRTIs for Improving Insulin Sensitivity and Reducing Type 2 Diabetes Development. *Nat. Commun.* **2020,** *11*, 4737.
29. Chen, Z.; Liu, X.; Luo, Y.; Wang, J.; Meng, Y.; Sun, L.; Chang, Y.; Cui, Q.; Yang, J. Repurposing Doxepin to Ameliorate Steatosis and Hyperglycemia by Activating FAM3A Signaling Pathway. *Diabetes* **2020,** *69*, 1126–1139.
30. Liu, T.; Cui, L.; Xue, H.; Yang, X.; Liu, M.; Zhi, L.; Yang, H.; Liu, Z.; Zhang, M.; Guo, Q.; He, P.; Liu, Y.; Zhang, Y. Telmisartan Potentiates Insulin Secretion via Ion Channels, Independent of the AT1 Receptor and PPARγ. *Front. Pharmacol.* **2021,** *12*.

31. Tao, H.; Zhang, Y.; Zeng, X.; Shulman, G. I.; Jin, S. Niclosamide Ethanolamine–Induced Mild Mitochondrial Uncoupling Improves Diabetic Symptoms in Mice. *Nat. Med.* **2014,** *20*, 1263–1269.
32. Huang, C.; Chen, X.-M.; Pollock, C. A. KCa3.1 in Diabetic Kidney Disease. *Curr. Opin. Nephrol. Hypertens.* **2022,** *31*, 129–134.
33. Witters, L. A. The Blooming of the French Lilac. *J. Clin. Investig.* **2001,** *108*, 1105–1107.
34. Rabbani, G. H.; Butler, T.; Knight, J.; Sanyal, S. C.; Alam, K. Randomized Controlled Trial of Berberine Sulfate Therapy for Diarrhea Due to Enterotoxigenic Escherichia Coli and Vibrio Cholerae. *J. Infect. Dis.* **1987,** *155*, 979–984.
35. Lee, Y. S.; Kim, W. S.; Kim, K. H.; Yoon, M. J.; Cho, H. J.; Shen, Y.; Ye, J. M.; Lee, C. H.; Oh, W. K.; Kim, C. T.; Hohnen-Behrens, C.; Gosby, A.; Kraegen, E. W.; James, D. E.; Kim, J. B. Berberine, a Natural Plant Product, Activates AMP-Activated Protein Kinase with Beneficial Metabolic Effects in Diabetic and Insulin-Resistant States. *Diabetes* **2006,** *55*, 2256–2264.
36. Gross, U.; Hoffmann, G. F.; Doss, M. O. Erythropoietic and Hepatic Porphyrias. *J. Inherit. Metab. Dis.* **2000,** *23*, 641–661.
37. Yalouris, A. G.; Raptis, S. A. Effect of Diabetes on Porphyric Attacks. *Br. Med. J.* **1987,** *295*, 1237–1238.
38. Andersson, C.; Bylesjö, I.; Lithner, F. Effects of Diabetes Mellitus on Patients with Acute Intermittent Porphyria. *J. Intern. Med.* **1999,** *245*, 193–197.
39. Kendig, E. L.; Schneider, S. N.; Clegg, D. J.; Genter, M. B.; Shertzer, H. G. Over-the-Counter Analgesics Normalize Blood Glucose and Body Composition in Mice Fed a High Fat Diet. *Biochem. Pharmacol.* **2008,** *76*, 216–224.
40. Wu, J.; Hu, B.; Lu, S.; Duan, R.; Deng, H.; Li, L.; He, L.; Zhao, Y.; Wang, J.; Yu, Z. Identification of Raloxifene as a Novel α-Glucosidase Inhibitor Using a Systematic Drug Repurposing Approach in Combination with Cross Molecular Docking-Based Virtual Screening and Experimental Verification. *Carbohydr. Res.* **2022,** *511*, 108478.
41. Istrate, D.; Bora, A.; Crisan, L. A First Attempt to Identify Repurposable Drugs for Type 2 Diabetes: 3D-Similarity Search and Molecular Docking. *Chem. Proc.* **2020,** *3*, 7.
42. Gates, A. J.; Gysi, D. M.; Kellis, M.; Barabási, A. L. A Wealth of Discovery Built on the Human Genome Project — by the Numbers. *Nature* **2021,** *590*, 212–215.
43. Eizirik, D. L.; Szymczak, F.; Alvelos, M. I.; Martin, F. From Pancreatic β-Cell Gene Networks to Novel Therapies for Type 1 Diabetes. *Diabetes* **2021,** *70*, 1915–1925.
44. Yildirim, M. A.; Goh, K. Il; Cusick, M. E.; Barabási, A. L.; Vidal, M. Drug-Target Network. *Nat. Biotechnol.* **2007,** *25*, 1119–1126.
45. Piganelli, J. D.; Mamula, M. J.; James, E. A. The Role of β Cell Stress and Neo-Epitopes in the Immunopathology of Type 1 Diabetes. *Front. Endocrinol. (Lausanne).* **2021,** *11*.
46. Chiang, A. P.; Butte, A. J. Systematic Evaluation of Drug-Disease Relationships to Identify Leads for Novel Drug Uses. *Clin. Pharmacol. Ther.* **2009,** *86*, 507–510.
47. Gayvert, K. M.; Madhukar, N. S.; Elemento, O. A Data-Driven Approach to Predicting Successes and Failures of Clinical Trials. *Cell Chem. Biol.* **2016,** *23*, 1294–1301.
48. Singla, R.; Singla, A.; Gupta, Y.; Kalra, S. Artificial Intelligence/Machine Learning in Diabetes Care. *Indian J. Endocrinol. Metab.* **2019,** *23*, 495–497.
49. Voets, M.; Møllersen, K.; Bongo, L. A. Reproduction Study Using Public Data of: Development and Validation of a Deep Learning Algorithm for Detection of Diabetic Retinopathy in Retinal Fundus Photographs. *PLoS One* **2019,** *14*, e0217541.

50. Gulshan, V.; Peng, L.; Coram, M.; Stumpe, M. C.; Wu, D.; Narayanaswamy, A.; Venugopalan, S.; Widner, K.; Madams, T.; Cuadros, J.; Kim, R.; Raman, R.; Nelson, P. C.; Mega, J. L.; Webster, D. R. Development and Validation of a Deep Learning Algorithm for Detection of Diabetic Retinopathy in Retinal Fundus Photographs. *JAMA - J. Am. Med. Assoc.* **2016,** *316*, 2402–2410.
51. Gilvary, C.; Elkhader, J.; Madhukar, N.; Henchcliffe, C.; Goncalves, M. D.; Elemento, O. A Machine Learning and Network Framework to Discover New Indications for Small Molecules. *PLOS Comput. Biol.* **2020,** *16*, e1008098.
52. Gaur, A. S.; Nagamani, S.; Tanneeru, K.; Druzhilovskiy, D.; Rudik, A.; Poroikov, V.; Narahari Sastry, G. Molecular Property Diagnostic Suite for Diabetes Mellitus (MPDSDM): An Integrated Web Portal for Drug Discovery and Drug Repurposing. *J. Biomed. Inform.* **2018,** *85*, 114–125.

# CHAPTER 9

# Drug Repurposing and Computational Drug Discovery for Aging and Neurological Disorders

KUMAR NALLASIVAN PALANI, KARTHIKEYAN DEIVASIGAMANI, SIVASUBRAMANIAN PIRAMANAYAGAM, and VISHALI MURTHY

*Department of Pharmaceutical Sciences, SNS College of Pharmacy and Health Sciences, Coimbatore, India*

## ABSTRACT

The lack of therapeutic interventions in aging, neurological and neurodegenerative diseases has introduced drug repurposing in this domain. The rise of newer approaches and technologies in the field of proteomics, pharmacology, genetics, drug discovery, epigenetics, biochemistry, and molecular biology yielded a lot of greater insights on disease pathology, causal organisms, activation of genes, proteins, enzymes, up/downregulation of receptors, stimulation of signaling cascades, etc. Needless to mention, the advancement indicated in these domains pave the way for the genomic, cellular, and molecular solution of specific disease models along with the systematic scientific approach to gain upper hand against known/unknown causal organisms, and fatal diseases such as cancer, AIDS and neurodegenerative diseases, etc. Novel chemical entities derived utilizing several instrumentation approaches such as high-performance liquid chromatography (HPLC), nucleo magnetic resonance (NMR), and mass spectroscopy (MS) and its enhanced versions. However, the drug discovery paradigm towards neurological and neurodegenerative diseases is underway. In this case,

Drug Repurposing and Computational Drug Discovery: Strategies and Advances.
Mithun Rudrapal, PhD (Ed)

the inclusion of drug repurposing would shed light on documented small molecules already available in the market, thereby exploiting its use in neurological and neurodegenerative diseases. Moreover, the incorporation of omics, toxicogenomics, pharmacodynamics, and pharmacokinetics along with drug repurposing would be a viable approach in the drug discovery of neurological and neurodegenerative diseases.

## 9.1 INTRODUCTION

During ancient times, sick people were prescribed traditional medicines from Ayurveda, Unani, Siddha, etc. However, the diseases and their causal organisms evolved with time, and the same traditional medicines became ineffective. With the discovery of antibiotics and the smallpox vaccine, the use of medicines becomes widespread, yet, the cure for diseases was not certain. In the 19th century, the discovery of new compounds was largely based on serendipity and was often carried out by apothecaries and chemists. The advancement and research in the field of genetics, recombinant DNA (rDNA) technology, epigenetics, biochemistry, and molecular biology (gene sequencing) provided a deep understanding of disease pathology, causal organisms, activation of genes, proteins, enzymes, up/downregulation of receptors, stimulation of signaling cascades, etc. This scientific era provided insights into genomic, cellular, and molecular details of a human organism and provided a systematic and scientific approach to deal with diseases caused by known/unknown causal organisms, and fatal diseases, such as cancer, AIDS, etc. New bioactive compounds were isolated using modern sensitive techniques, such as high-performance liquid chromatography (HPLC), nucleo magnetic resonance (NMR), and mass spectroscopy. Still, at this stage, the screening of new compounds was a "hit" and "trial" process. The merger of physical, biophysical, chemical sciences (combinatorial chemistry), and computational resources (gene and protein databases) gave birth to computer-aided drug design (CADD) which includes molecular modeling, docking, molecular dynamics, high-throughput screening, virtual screening, etc. This revolutionized the field of drug discovery and development. The onset of omics, toxicogenomics, pharmacodynamics, and pharmacokinetics provided a new dimension to the field of pharmacology and the biopharmaceutical industry.[1–3]

## 9.2 AGING, NEUROLOGICAL, AND NEURODEGENERATIVE DISEASES

Aging is a physiological process in human life, where cellular machinery slows down—biochemical signaling takes time to respond, mitochondria become sluggish, the regenerative capacity of cells, and neurons decrease. In order to study and recognize these features distinctively, scientists have allotted nine hallmarks to aging, viz., genomic instability, epigenetic alterations, telomere attrition, loss of proteostasis, cellular senescence, stem cell exhaustion, mitochondrial dysfunction, deregulated nutrient sensing, and altered intercellular communication. Various factors drive toward these conditions and succumb old age people to suffer from frailties, disabilities, and severe health ailments. These factors are oxi-inflamm-aging, neuroinflammation, mitochondrial genome damage, genetic instability, telomere shortening, and genetic aging programs. In aged people, it is observed that the level of reactive oxygen species (ROS) increases than the normal level (mitochondria reduce the 2–3% of cellular oxygen into ROS). This causes DNA, cell membrane, mitochondria (cell membrane, proteins, and DNA), and cellular protein damage; further ROS activates inflammatory signaling pathway via activation of NF-κB. This triggers the transcriptional activation of several pro-inflammatory cytokines, such as IL-1, IL-6, and tumor necrosis factor-α (TNF-α). On the other hand, expression of IL-2 decreases. We know that immune cells and cytokines function in a coordinated manner. So, the imbalance between pro- and anti-inflammatory cytokine levels, and decreased IL-2, dismantles the normal immune homeostasis. This cellular condition in old age people is known as oxi-inflamm-aging. Excess NF-κB causes overactivation of immune cells, and this consequently affects the homeostasis of the nervous system and neuroendocrine system. These circulating immune cells cross the blood–brain barrier (BBB) and activate the immune cells of the central nervous system (CNS); this causes—neuroinflammation or neuro-inflamm-aging. ROS maintains a persistent low level of chronic stressful cellular condition—this becomes a primary cause of health deterioration in old age people.[4,5]

### *9.2.1 NEUROINFLAMMATION/NEURO-INFLAMM-AGING*

With increasing age, the activity of the adaptive immune system decreases, while the activity of innate immune cells remains the same. Aging is also characterized by "immunosenescence."[5] The central nervous system and immune system function in a coordinated and organized manner. In the brain, the immune cells perform various vital functions. They regulate neurogenesis, maintain homeostasis, retain proper integrity, and maneuver appropriate intellectual and mental levels.[6] In old age people, genomic instability, oxidative damage, gene expression alteration, activation of inflammatory cytokines and immune cells, disruption of the blood–brain barrier—activate microglia (resident immune cells): disorganizes the normal orchestration of CNS-immune system. The activation of microglia is a characteristic feature of neuroinflammation/neuro-inflamm-aging. In the inflammatory milieu, the functional ability and activity of microglia get affected. Either they fail to clear off the debris, or while phagocytosing the debris they release inflammatory cytokines such as TNF-α, ROS, and nitrogen-free radicals. This worsens the stressful condition in brain and expediates the degradation of neurons. Hence, old age people become more susceptible to neurological or neurodegenerative diseases, such as Alzheimer's disease, Parkinson's disease, Huntington's disease, multiple sclerosis (MS), and amyotrophic lateral sclerosis (ALS).[6,7]

### *9.2.2 PARKINSON'S DISEASE*

Parkinson's disease (PD) is caused by the accumulation of α-synuclein (α-Syn) in the brain cells. The cellular aggregates of α-Syn form Lewy bodies. PD can be caused by gene mutation, virus infection, aging, and environmental factors. Mutation in Parkinson's or PARK genes such as *SNCA*, α-synuclein (PARK1/PARK4); *DJ-1*, Parkinsonism-associated deglycase (PARK7); *LRRK2*, Leucine-rich repeat kinase 2 (PARK8); *PINK1*, PTEN-induced putative kinase 1 (PARK6), etc., is responsible for PD.[7] α-synuclein is a small protein present in the terminal region of a nerve cell or presynaptic vesicles. α-synuclein regulates the synthesis, storage, clearance, and efflux of dopamine. The dopamine-nigrostriatal system is intricately involved in voluntary and nonvoluntary movements. α-synuclein aggregation decreases dopamine neurotransmission and in severe cases, it has been observed that dopamine neurons that emanate from the substantia nigra pars compacta and traverse to the caudate-putamen in the striatum die. This leads to the loss of dopamine neurotransmission. Hence, PD sufferers suffer from a loss of motor activity, instability, tremor, bradykinesia, body rigidity; inability to move from one place to another, etc. As PD progresses, people also suffer from dementia, insomnia, depression, loss of smell (anosmia), and other psychological problems.[8] In elderly PD patients, neuroinflammation becomes a primary cause of neurodegeneration. Microglia, astrocytes, α-synuclein, and LRRK2 are the principal players of neuroinflammation in PD. In PD, change in the conformation of α-synuclein and intrinsic regulation of LRRK2 activates microglia and astrocytes. Activated microglia secrete pro-inflammatory cytokines, such as TNFα, IL-6, IL-1β, IFNγ; elevated levels of CXCR4 receptor and chemokine CXCL12 have been observed in the brain of patients with PD. Besides this, increased expression of anti-inflammatory cytokines, such as Tgfβ and IL-10 are also reported in the brain samples of PD patients. Due to the neuronal damage, microglia express toll-like receptors (TLRs) that prime the peripheral immune system. These primed immune cells along with activated microglia amplify neurodegeneration (nigrostriatal pathway injury). Further, it has been found that astrocytes of PD patients express high level of S100b (calcium-binding protein). S100b behaves like cytokine and augments the expression of inducible nitric oxide synthase (iNOS)—cyclooxygenase-2 (COX-2, a pro-inflammatory enzyme), nitric oxide, and superoxide radicals in microglia.[9]

### 9.2.3 ALZHEIMER'S DISEASE

The aggregation of amyloid β (Aβ) peptides and hyperphosphorylated tau proteins are peculiar features of Alzheimer's disease. The hyperphosphorylation of tau proteins forms paired helical filaments or neurofibrillary tangles (NFT). Their deposition in neurons and extracellular Aβ plaques are the principal cause of neurodegeneration. The formation of PHF destabilizes the microtubular organization of the CNS, ruptures nuclear membrane, and damages neurons. NFTs are distributed throughout the amygdale, para hippocampal gyrus, hippocampal formation, and temporal association cortex, they cause major impairment in the entorhinal cortex, CA1, and the subiculum of hippocampus, amygdala, and the temporal association cortex. The extracellular Aβ plaques are found in the association neocortex and striatum.[8] Aging, neuroinflammation, genetic (HLA-DRB5-DRB1, TREM2, CD33), epigenetic factors, etc., lead to the formation of Aβ plaques and phosphorylation of tau proteins. In old age people, the extracellular Aβ plaques, NFTs, neuronal/membrane damage, or damage-associated molecular patterns (DAMPs) stimulate microglia (DAMPs/pathogen-associated molecular patterns (PAMPs)/Aβ plaques/NFTs) and interact with toll-like receptors

(TLRs), RIG-like receptors (RLRs), AIM2-like receptors (ALRs), NOD-like receptors (NLRs), etc., present on the surface of microglia.[7] The activated microglia secrete inflammatory cytokines and express more toll-like receptors (as a response toward neuronal damage or tissue injuries), such as TLR 2, 4, 6, 9. Moreover, inefficient phagocytosis of these aggregates by the resident microglial cells also leads to the activation of pro-inflammatory cytokines, such as IL-1β, IL-6, tumor necrosis factor-α (TNF-α). The activated microglia express chemokine ligands: CCL-2, CCL-4, CCL-11, CCR3, CCR5, CXCL8, CCL3, etc., in order to recruit more microglia and astrocytes at the site of inflammation. Additionally, the inflammatory milieu polarizes microglia toward the M1 phenotype. The inability of M1 microglia to clear off amyloid-beta aggregates, further adds to neuroinflammation and neurodegeneration. In AD, there are reports that astrocytes express receptors for inflammatory cytokines (IL-1β, TNF-α) and chemokines. The C-terminal end (100 amino acids) of β-amyloid precursor protein (APP) stimulate astrocytes (astrogliosis) and enhances the death of neurons. In the later stages of Alzheimer's, the condition becomes worse as blood–brain barrier collapses, excessive infiltration of peripheral innate and adaptive immune cells occurs, it increases the neurotoxicity, and accelerates the neuronal damage.[8,9]

### *9.2.4 AMYOTROPHIC LATERAL SCLEROSIS (ALS)*

ALS is majorly considered to be an idiopathic neuroinflammatory disorder. In 90–95% of the cases, the cause of occurrence is unknown, this type of ALS is known as sporadic ALS (SALS), while in 5–10% of the cases ALS occurs due to genetic mutations-familial type ALS (FALS). ALS is highly prevalent in old age people. The exact mechanism of the disease remains unknown—categorized by FDA as an orphan disease. The low chronic inflammatory environment in elderly people might act as a trigger for ALS. The glial cells, peripherally activated immune cells are the major producers of ROS and reactive nitrogen species (RNS). The increase in intracellular oxidative stress causes mitochondrial disintegration (decrease in the competency of electron transport chain (complex I–IV) and reduced intracellular ATP production has been observed in ALS patients), and activates resident microglial cells and astrocytes. Activated microglia exacerbates neurotoxicity by secreting IL-1β, IL-12, IFNγ, TNF-α, insulin growth factor-1 (neurotrophic growth factor), monocyte chemoattractant protein (MCP)-1, macrophage colony-stimulating factor (M-CSF). Besides these cytokines and growth factors, elevated levels of IL-2, IL-15, IL-17, IL-6 vascular endothelial growth factor (VEGF), TGFβ, macrophage inflammatory protein-1α (MIP-1α) have also been observed in ALS patients. These cytokines are also responsible for generating reactive A1 astrocyte subtypes, and recruitment of reactive astrocytes and oligodendrocytes at the site of inflammation. A1 astrocytes act as a key player in the degeneration of motor neurons (astrogliosis). Further, they secrete cyclooxygenase-2 (COX-2), inducible nitric oxide synthase (i-NOS), neuronal NOS and prevent the repair of damaged axons. These events lead to the loss of upper and lower motor neurons in the brain and spinal cord, which causes muscle cramps, muscle twitching, and loss of muscular strength. In advanced cases, patients become paralyzed, suffer from respiratory problems, and may even die due to respiratory failure. As its prevalence remains unknown, hence, this dilapidating, progressive, and fatal disorder becomes a cause of great concern.[12] Multiple sclerosis is an autoimmune disorder where lymphocytes (majorly T cells) attack myelin sheath, axon, and nerve cells in the CNS. The loss in myelin sheath occurs due to the generation of self-reactive antibodies against antigens of oligodendrocytes present in the myelinated axons. The disruption of BBB and neuroinflammation are the primary cause of multiple sclerosis. These two factors are prominent in old age people; hence, they are prone to neurodegeneration and disabilities associated with multiple sclerosis. Multiple sclerosis is divided into two phases: (1) relapsing-remitting phase (RRMS) eventually develops into secondary progressive phase (SPMS), (2) primary progressive multiple sclerosis (PPMS). The older ones generally develop primary progressive multiple sclerosis. The principal cause for the development and progression of multiple sclerosis in old age people is low-grade chronic intracellular inflammatory milieu maintained by pro-inflammatory cytokines, ROS generation, mitochondrial degeneration (inflammaging), immunosenescence, decrease in the remyelination process, and excessive accumulation of iron in the human body. The different stages of multiple sclerosis are as follows[10,11]:

- Inflammatory milieu and T-cell activation – Tissue injury or inflammatory cellular condition activates peripheral innate immune cells. The activated dendritic cells and macrophages secrete pro-inflammatory cytokines: IL-β, IL-23, and IL-6. The activated dendritic cells and macrophages, and pro-inflammatory cytokines activate CD4 T cells, Th1, Th17, Treg, γδT cells (secrete GM-CSF), and B cells. The breach in BBB allows clear passage of these peripherally activated innate and adaptive immune cells to the CNS (brain and spinal cord). These immune cells activate microglia and astrocytes. They further increase the level of pro-inflammatory cytokines and advance the process of neurodegeneration.

- Gliosis and formation of demyelinating plaques – The loss of myelin sheath from multiple areas of axon form lesions or plaques. This causes neuroaxonal degeneration.
- Synaptopathy – The loss of neurons and myelin sheath minimizes the synaptic transmission.
- Impairment/loss of neuronal signaling—Synaptopathy ultimately leads to loss of neuronal signaling.
- Neurodegeneration – These aforementioned events finally lead to the loss of neurons.

## 9.3 DRUG REPURPOSING AND COMPUTATIONAL DRUG DISCOVERY

Drug discovery program may be divided into two stages – preclinical and clinical which involve the following steps:

- Selection of bioactive compounds (2–3 years)—To proceed with the discovery of a new compound, selection of bioactive compounds is done via the application of genomics, proteomics, databases (and other bioinformatics tools), in vitro, and in vivo assays.
- Screening (0.5–1 year)—Among these compounds, the screening is done using structural biology (molecular modeling/docking/simulation), combinatorial chemistry, high-throughput screening (HTS), or high content screening (HCS), in vitro, ex vivo, and in vivo methods.
- Lead optimization (1–3 years)—Through the screening process, we get the "lead compound." Now, the lead is chemically modified (traditional chemistry approach) or undergoes through rational drug design approach—quaternary structure–activity relationships (QSAR) that involve software and structural biology tools. These methods are required to bring about desired changes in the pharmacokinetic (ADMET and drug–drug interactions) properties of the chosen compound. This enhances the lead compound`s specificity, selectivity, and potency—to be categorized as a drug.

- ADMET (absorption, distribution, metabolism, excretion, and toxicity) (1–2 years)—Further, the lead compound's ADMET analysis is done through structural biology and experimental (in vitro and in vivo) approach. The ADMET analysis infers the bioavailability and efficacy and determines the toxicity of the lead compound.[4]

The in vivo experiments performed during the preclinical stage determine the following properties of the lead compound: (1) the safe drug range to be applied in humans during the clinical trial study, (2) the pharmacodynamics and pharmacokinetics of the drug. Pharmacodynamics is the branch of pharmacology that deals with the mechanism of action, and biochemical and physiological effects of drugs in humans. To test the efficacy of the lead compound in a human trial, the in vivo and other preclinical data need approval from the Investigational New Drug (IND). If IND gives approval, then the test compound moves to phase I, II, and III trials.[5]

Clinical trials (5–6 years)—The test compound must go through these phases.[4–6]

**TABLE 9.1** Different Phases of Clinical Trials.

| Phase I | Phase II | Phase III |
|---|---|---|
| The new lead compound is tested on the human that includes a small number (20–100) of healthy and/or unhealthy people. In this phase, the subject's response toward the drug is evaluated. Based on this response—the maximum tolerated dose (MTD) is calculated. | This phase is also known as the "therapeutic exploratory" trial. The number of diseased people involved in this phase is greater than phase I, that is, 100–500. They are tested for the safety of the test compound, pharmacodynamics, and pharmacokinetics.<br>Based on these results, the planning for the phase III trial is done regarding—the involvement of a number of volunteers (trial size), trial design, data collection, data analysis, trial outcomes, and safety of volunteers | This phase is also known as "therapeutic confirmatory," "pivotal trial," or "comparative efficacy." In this phase, generally, 1000–5000 volunteers are involved |

After the phase III trial, New Drug Application (NDA) is submitted to the Food and Drug Administration (FDA). Whether the test compound will enter phase IV clinical trial or not is decided by the intricate evaluation of the FDA and the external committee members. Sometimes, to ensure the efficacy of the compound with no/fewer side effects, more than one phase III trial is conducted by the FDA.

Phase IV—This phase is also known as post-marketing surveillance. In this phase, the efficacy of a safe dose of the test compound is monitored on the envisioned population for a long time. This phase is regulated by safety surveillance (pharmacovigilance) and a technical support committee. Here, the volunteers are exposed to the drug for a longer period than phase I–III trials and undergo extensive monitoring for the test compound's side effects or adverse effects. If adverse effects are reported, then depending upon the case reports either the use is restricted or withdrawn from the further drug discovery process. If the drug passes this phase, then to have acceptance across the United States, Europe, and Japan, the approval is processed through the International Conference on Harmonization of Technical Requirements of Pharmaceuticals for Human Use (ICH).[5]

- Registration (1–2 years)—The registration of the drug in the USA is done by the FDA, in Brazil by The National Health Surveillance Agency or Agencia Nacional de Vigilancia Sanitaria (ANVISA), in Europe by European Medicine Agency (EMEA), and Japan`s regulatory body is the Ministry of Health, Labour, and Welfare (MHLW).
- Marketing—After registration, the drug is marketed by the concerned regulatory bodies.

We have observed that drug discovery in total takes 10–17 years and its cost is too high. In the USA, the average drug discovery process surmounts to 2–3 billion $. Even at the stake of time and money, it is uncertain that for how long the drug will sustain itself in the market and be beneficial to mankind.[4,7,8] Owing to these facts, the scientists, researchers, and pharmacists have converged on drug repositioning. In accordance with the 10th version of the International Classification of Diseases (ICD-10), drug repositioning or repurposing is defined as the remodeling of existing, obsolete, or banned drugs for a particular medical condition. Drug repositioning can involve new drugs, but they must show the required pharmacological activity in the human system by hitting different targets. It is also termed drug re-tasking, reprofiling, redirection, reformulation, combination, or therapeutic switching. The therapeutic dosage, safety profile, adverse effects, pharmacodynamics, and pharmacokinetics properties of the drug are well characterized, hence, in comparison to the traditional drug discovery process, the drug repurposing saves time and money. The repositioned drug can escape the preclinical phase and can directly enter the clinical stage. Hence, generally within 3–12 years,[4] repositioned drug flourishes in the market. The reason behind selecting this approach is to increase the sustainability of the drug in the market and pay attention to rare/orphan or neglected diseases.[9] Neglected or rare diseases are unlucrative for pharmaceutical companies, but people may die due to the absence of suitable drugs, and with time, there are chances that rare diseases can become prominent in the human population. The World Health Organization (WHO) estimated that 1 billion people (approx.) and 420–560 million people are affected by neglected and rare diseases, respectively.[4]

In general, the drug repositioning process takes 3–12 years. The steps involved in DR are as follows:

- Search for test compound (1–2 years)—The test compound can be found by the high-throughput screening process, virtual screening, molecular docking, chemical structures, or serendipity.
- Acquisition of test compound (0–2 years)—The test compound is acquired either via licensing or novel intellectual property or via both.
- Drug development (1–6 years)—In drug repurposing, the drug development starts either from preclinical phase I or phase II trial.
- Registration (1–2 years)—In the USA, Europe, Brazil, and Japan, the registration of the drug is done by the FDA, European Medicine Agency (EMEA), The National Health Surveillance Agency or Agencia Nacional de Vigilancia Sanitaria (ANVISA), the Ministry of Health, Labour, and Welfare (MHLW), respectively.
- Marketing—After registration, the drugs are marketed by the respective companies.

To find one drug for a particular target, virtual screening, high-throughput screening or other screening methods are applied. It was found that the screened compounds showed unanticipated and unwanted results. During screening, the test compound is exposed to virtual conditions. But under physiological environment, there is a high probability of interaction of the test compound with other bioactive compounds (several proteins, enzymes, receptors) having a similar structure or located at the off-target site. This phenomenon is known as drug promiscuity (multi-targeting). The drug promiscuity signifies good or bad effects produced by the drug–target binding at the therapeutic or nontherapeutic site. For example, as the 3D structure of isoform IX and II of carbonic anhydrase (CA) are similar (31% sequence identity), so the cancer drug against CA IX binds with CA II also and causes adverse effects, such as blurred vision, irritation in the eye, eye-watering, constipation, and diarrhea.[10,11]

The principle of polypharmacology has been derived from the concept of drug promiscuity. Polypharmacology harnesses the ability of the drug to interact with multiple therapeutic targets (or off-targets).[10] Scientists believe that polypharmacology has a connection with protein evolution (simple proteins attained complex structures due to changing chemical and cellular environments) which makes them susceptible to chemical promiscuity.[11,12] However, with advancements in technology (computational biology and ligand-based approaches), scientists are keeping track of both on- and off-target effects of the compound. This knowledge can be used to target multi-factorial diseases with one benefit, that is, these types of drugs remain less responsive toward drug-resistance mutations.[11] The promising strategy of drug repositioning is based on polypharmacology.[7,10] In drug repositioning, the off-target action of the drug is repurposed to produce beneficial effects.

## 9.4 COMPLICATIONS IN CURRENT THERAPY FOR AGING, NEUROLOGICAL, AND NEURODEGENERATIVE DISEASES

For the treatment of Alzheimer's disease, the following drugs have been prescribed by the FDA: Donepezil (Aricept), Galantamine (Razadyne), Memantine (Namenda and Namenda XR), Rivastigmine (Exelon Patch), and Memantine plus donepezil (Namzaric). These drugs are either cholinesterase inhibitors or N-methyl-D-aspartate (NMDA) receptor antagonists. In AD patients, cholinesterase inhibitors elevate the levels of acetylcholine neurotransmitters. NMDA receptor modulates glutamatergic transmission. Generally, elderly patients suffer either by a single problem or a combination of problems, such as diabetes, hypertension, cardiovascular problems, hypercholesterolemia, and hyperhomocysteinemia. These factors not only accelerate AD development or progression but also pose challenges and complications in the treatment of AD. Due to these reasons, older patients remain in the disease phase for a longer time and finally they suffer from complete cognitive and memory loss.[13,14] Psychosocial problems, emotional distress, depression, insomnia, lack of care and affection, immunosenescence, comorbidities, and inability to heal faster make management of motor symptoms, such as tremors, bradykinesia, muscle rigidity, and fatigue difficult in older PD patients. Levodopa-dopa decarboxylase inhibitor (levodopa-carbidopa, levodopa-benserazide), monoamine oxidase-B (MAO-B) inhibitors (selegiline, safinamide, rasagiline), dopamine agonists (non-ergot derivatives: pramipexole immediate release, rotigotine, apomorphine, pramipexole controlled release; ergot derivatives: cabergoline, bromocriptine), NMDA-antagonist (amantadine), anticholinergics (benztropine, trihexyphenidyl), catechol-o-methyltransferase (COMT) inhibitors (entacapone) medications are used to treat PD. Generally, levodopa is prescribed for the elderly, however, depending on the comorbid problems and physical conditions, the range of therapy management options is determined. Older PD patients show reluctance against clinical trial registration, so there is a lack of broad range of therapeutic strategies for them. So, the frequency of dementia, disabilities, and frailties is higher in older people suffering from PD.[15] Cognitive deficits/ dysregulated motor activities or both – These factors culminate in the loss of cognitive and motor activities. The majority of old age men having multiple sclerosis suffer from lower urinary tract symptoms (LUTS). Comorbid problems, such as cardiovascular disease, diabetes, cancer (leukemia), psychiatric problems worsen the dilapidating multiple sclerosis disease. FDA-approved drugs for multiple sclerosis are interferon-beta products, glatiramer acetate, fingolimod, teriflunomide, dimethyl fumarate, ocrelizumab, alemtuzumab, natalizumab, and daclizumab.[16–18] Currently, riluzole is prescribed for ALS. Riluzole maintains normal glutamate levels by decreasing excess glutamate release from presynaptic terminals and by clearing off glutamate from synapses.[12] Old age people are vulnerable to developing this medical condition (mechanism explained above). Age-related problems, such as diabetes, muscle weakness, arthritis, and heart risks increase the chance of disease development. Further, these problems not only complicate the therapy management but also prolong patients' hospitalization time. This leads to mental stress and generates economic problems. This vicious circle breaks poor old age people, renders them helpless, and forces them to choose death over life.

## 9.5 DRUG REPURPOSING: APPROACHES, MECHANISMS, AND ITS APPLICATIONS IN DISEASE MANAGEMENT

There are several examples of the repositioned drugs such as

- Sildenafil: The relaxation of coronary arteries is controlled by phosphodiesterase-5 (PDE-5). In the 1980s, sildenafil (PDE-5 inhibitor)

was developed to cure chest pain by increasing the blood flow by inhibiting the PDE-5 activity. In the clinical trials, it was found that the drug was ineffective against angina, and patients were also reported to have the problem of prolonged erections. The sildenafil problem was turned into a solution for patients suffering from erectile dysfunction and became popular by the name "Viagra." The drug is also repurposed successfully for pulmonary hypertension patients.[9]

- Thalidomide: Thalidomide is structurally close to barbiturates and was first used in Europe (1957) as a sedative. The safety and toxicity of the drug were tested on rodents. Based on this result, thalidomide was declared safe and was prescribed to pregnant women having morning sickness. This turned into a disaster-as women having thalidomide gave birth to babies with skeletal birth defects, and over 15,000 babies had anatomical malformations. The drug was quickly withdrawn. Later, Sheskin showed another side of this drug by revealing that thalidomide acts as a pain killer and reduced soreness in patients suffering from erythema nodosum leprosum. Thalidomide is also used as an anticancer drug for multiple myeloma. It degrades Ikaros and Aiolos transcription factors, active in multiple myeloma. In multiple myeloma and other blood cancer, thalidomide analogs—lenalidomide and pomalidomide—collectively named immune-modulating drugs (IMiDs) are used extensively. Thalidomide and its analogs exert an anticancer effect by (1) reducing angiogenesis, (2) inducing oxidative stress, and (3) activating T cells (via increasing IL-2 level), and natural killer (NK) cells.[9,11]

Other examples of repositioned drugs have been presented in Table 9.2.[11,13–18] Apart from these drugs, extensive drug repositioning has been done for neurological disorders as well, which has been discussed in the later section.

**TABLE 9.2** The Available Repurposed Small Molecules in the Market.

| Drug | Original practice | Repurposed for |
|---|---|---|
| **Aspirin** | Thrombosis | Cardiovascular diseases, diabetes, central nervous system disorders, cancer, and cryptococcal infection |
| **Mebendazole** | Antihelminthic drug | Metastatic colon cancer effective against active kinase ABL and BRAF (v-raf murine sarcoma viral oncogene homolog B1) in colon cancer |

**TABLE 9.2** *(Continued)*

| Drug | Original practice | Repurposed for |
|---|---|---|
| **Lithium** | Gout | Controlling mood swing, used as an antimaniac drug |
| **Methotrexate** | Cancer | Psoriasis and rheumatoid arthritis |
| **Cyclosporine** | Antifungal drug | Lymphocytes (immune regulatory drug) |
| **Metformin** | Type II diabetes | Aging, cancer, obesity, and depression, PCOS |
| **Warfarin** | Pesticides | Anticoagulant |
| **β-blockers** | Antiepileptic | Migraine |
| **Topiramate** | Antiepileptic | Migraine |
| **Itraconazole** | Antifungal drug | Lung cancer (Phase II) |
| **Saracatinib** | Anticancer | Alzheimer`s disease (Phase I) |
| **Hydroxychloroquine** | Antimalarial | COVID-19 |
| **Remdesivir** | Antiviral (Ebola) | COVID-19 |
| **Tocilizumab** | Rheumatoid arthritis | Lung fibrosis in COVID-19 |

### 9.5.1 APPROACHES FOR DRUG REPOSITIONING AND ITS MECHANISMS

There are three approaches applied for drug repositioning, such as disease-centric, drug-centric, and target-centric approaches. When the drug repositioning database was analyzed, then it was found that more than 60%, almost 30%, and less than 10% of definition was allocated to disease-centric, target-centric, and drug-centric approaches, respectively. The disease-centric approach is defined as the repositioning of drugs by keeping the disease at the point of focus. This technique easily identifies an association between an old and a new indication. The target-centric approach connects an already known target and its well-known drug to a new indication, while the drug-centric approach repurposes a known drug with a new target and an indication related to it.[8]

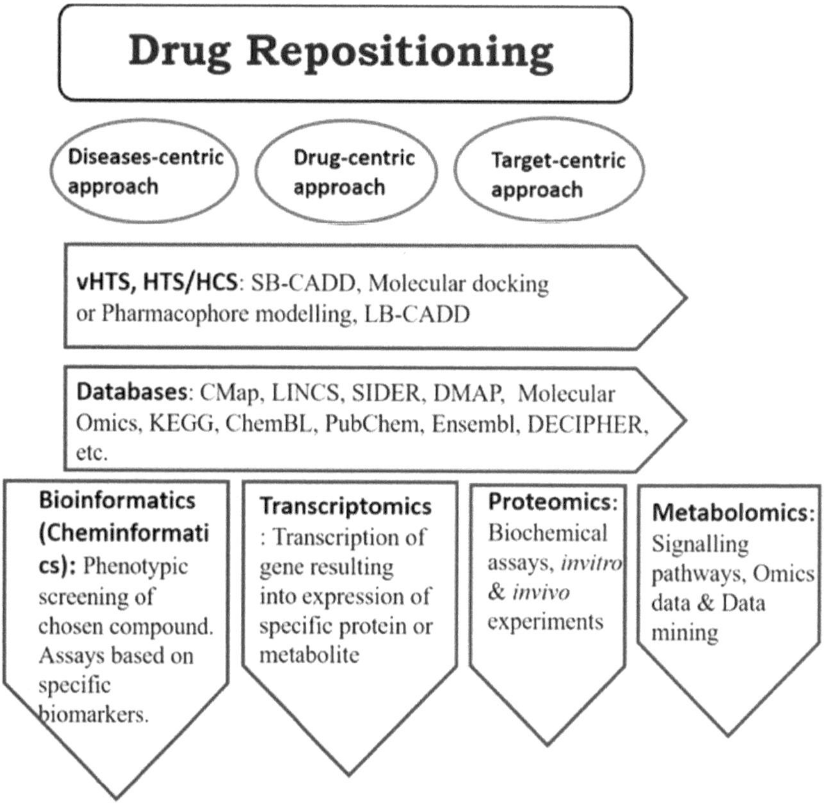


**FIGURE 9.1** Overview of drug repurposing.

## 9.6 KEY DRUG DISCOVERY TOOLS FOR DRUG REPURPOSING

### *9.6.1 VIRTUAL HIGH-THROUGHPUT SCREENING (HTS) WITH EXAMPLES*

This is the primary technique used in the drug discovery process. The biological activity of Several compounds ($10^3$–$10^6$) is screened at the molecular and cellular levels. HTS screens oligonucleotides, antibodies, extract of natural products, and chemical mixtures present in the library of compounds and gives the results in the form of "hits." HTS performs tests in microtiter plates (96/384/1536 well) at a single (traditional method) or multiple concentrations (quantitative HTS) to generate concentration–response curves.[17]

### 9.6.2 COMPUTER-AIDED DRUG DESIGN (CADD) AND ITS TECHNIQUES

It is the most frequently used technique in drug discovery and drug repositioning. Instead of searching for a suitable drug candidate for clinical trials, virtual HTS (vHTS) aims at finding a lead with diverse chemical structures against a particular target. Like traditional HTS, if the activity of the compound is near 10 mM, then that compound is considered a "hit," among several others present in the virtual compound library. The ranking obtained after vHTS is based on either percentage similarity between the virtual library compound and the test compound or based on the lowest energy score of the test ligand-target pose, in case of docking. QSAR modeling, pharmacophore mapping, docking, and similarity prediction between chemical structures using fingerprints or topology can also be used for carrying out vHTS. CADD contributes to the drug discovery process in the following ways: (1) by providing filtered active compounds from large and diverse compound libraries, (2) modulating affinity or optimizing ADMET (pharmacokinetics) and drug metabolism properties, (3) growing novel chemotypes either from one functional group or by reinforcing various fragments. Virtual HTS (vHTS) collects information on carbohydrates, enzymes, reactants, natural, and chemical compounds from the databases, such as LIGAND, Chem DB, PDBeChem, Drug Bank, Accelrys Available chemicals directory (ACD), WOMBAT (World of molecular bioactivity) database, 3D MIND, MDDR (MDL drug Data Report), and Zinc.[18]

**TABLE 9.3** The Comparison Between CADD and Traditional HTS.[18]

| CADD | Traditional HTS |
|---|---|
| CADD produces effective results even by screening a lesser number of compounds in lesser time. | HTS produces a low hit rate. Needs more time to produce concentration–response curves, as traditional HTS requires development and validation |
| Dorzolamide (carbonic anhydrase inhibitor), captopril (angiotensin-converting enzyme inhibitor), and saquinavir, ritonavir, and indinavir (HIV drugs) | Ritonavir and indinavir have also been verified and approved via HTS |
| While screening inhibitors of tyrosine phosphatase-1B, a hit rate of ~35% was produced | Parallel screening was done with traditional HTS, hit rate of 0.021% was obtained |
| Economic | Costly |

CADD is divided into two groups—structure-based and ligand-based CADD. Structure-based CADD includes molecular docking, molecular dynamics (MD), de novo design, pharmacophore modeling, and artificial intelligence (AI), and it requires the knowledge of target structure. Ligand-based CADD includes QSAR, pharmacophore modeling, and ligand-based virtual screening and the structure of the target enzyme is unknown.[18,19]

**TABLE 9.4** Difference Between Structure-Based and Ligand-Based CADD.

| Structure-based CADD (SB-CADD) | Ligand-based CADD (LB-CADD) |
|---|---|
| The direct approach to drug discovery: the structure of the target is known | An indirect approach to drug discovery: as the structure of the target is unknown |
| SB-CADD is applied when the structural information of the target protein is available | LB-CADD is applied when negligible or little structural information is available (e.g., membrane proteins) |
| This tool is inefficient if the structure of the target is unknown | This tool is very much effective when the structure of the target is not known |
| Based on interaction energies of all test compounds | Based upon QSAR models, and chemical similarities between known active and inactive molecules |
| SB-vHTS less effective than LB-vHTS | LB-vHTS- identifies legitimate active compounds |

### 9.6.2.1 STRUCTURE-BASED CADD

It is one of the relevant methods to study ligand–target interactions and has played a pivotal role in the drug discovery process since 1970s. In the 1980s (NIH-structure-based), SB-CADD was used to analyze the feasible ligand–receptor interactions for haem-myoglobin/metmyoglobin and thyroxine/prealbumin—that is determined by favorable geometrical configurations. This was the first study to give information related to the bonding and behavioral pattern of protein and ligand, such as "hydrogen bonding" and "hard sphere repulsions" and referred to the receptor as a "solid rigid body" whose binding site contains "pockets." Since then, developments in the field of genomics-proteomics, application of X-ray crystallography, and NMR spectroscopy have totally changed the face of SB-CADD. It is because of biophysical techniques, such as NMR spectroscopy, X-ray crystallography that the 3D structure of various proteins is well-known and easily available to us. The database banks, such as protein database (PDB), PDBBIND, and protein-ligand database banks incorporate 81,000, 5671, and 129 (2003) protein structures, respectively.[18–20] Several ways of SB-CADD are being used in the discovery of several drugs and drug repositioning:

- Comparative modeling: The comparative modeling is based on sequence comparison and follows the fact that "similar sequence proteins have similar structures." The template identification is a critical step in comparative modeling. Homology modeling is the special type of comparative modeling where query and known sequences are evolutionary connected. During template selection, the structural biologists should cater to the alignment length (or sequence similarity), sequence identity, template structure resolution, and constancy of secondary structures between query and known sequence. The template sequence is obtained from the PDB database (PDB-BLAST search). Then the alignment between a query sequence and a known sequence (templates) is done using ClustalW- multiple-sequence alignment tool. The scores obtained after multiple-sequence alignments build and improve the quality of comparative models. There are several software tools and web servers, such as PSIPRED and MODELER that fasten the comparative modeling process. Oliviera et al. have used homology modeling and molecular docking to reposition drugs against fungal infection Paracoccidioidomycosis (PCM) caused by a fungus belonging to the genus *Paracoccidioides*.[21]

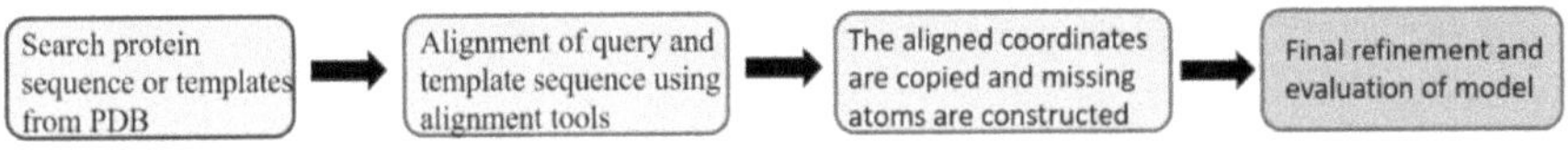


**FIGURE 9.2** Comparative modeling process.

- Molecular docking: Molecular docking refers to the binding of receptors that may be protein, deoxyribonucleic acid (DNA), ribonucleic acid (RNA) to a ligand viz., protein, peptides, or other small molecules, etc. Among SB-CADD, molecular docking is one of the widely used in silico techniques used in drug discovery and drug repurposing. In molecular docking, favorable binding poses of the ligand inside the binding cavity of the target protein are selected which mimic physiological conditions.

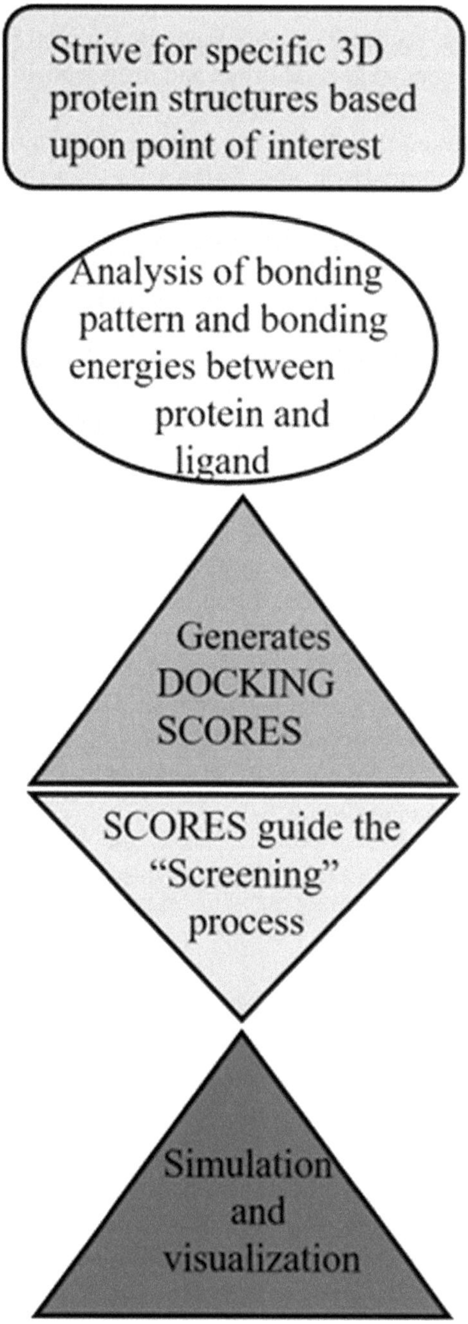


**FIGURE 9.3** Molecular docking method.

**TABLE 9.5** Various Docking Approaches.

| Docking programs | Met | Applications |
|---|---|---|
| Glide | • Belong to an exhaustive search algorithm<br>• Accurate docking module<br>• The module precomputes grid representation of protein and ligand<br>• Ligand conformations are generated in ligand torsion-angle space, that have low potential energies<br>• Glide scoring methods are based upon geometry and shape of various poses<br>• Feasible poses are shortlisted by using a scoring method and approximate positioning<br>• Then high-resolution search is done using molecular mechanics energy function<br>• This is followed by the Monte-Carlo process that examines torsional minima | Used to design inhibitors of cytochrome P450, Falcipain, and Aurora kinases |
| **GOLD**[18,23] (Genetic Optimization for Ligand Docking) | • It is a genetic algorithm<br>• GOLD Suite Software includes GOLD, Goldmine, and Hermes. Hermes is required to input data, while Goldmine is required for analysis and post-processing of results, during the application of GOLD docking module<br>• It is used for docking of flexible ligand and protein or target-bearing hydroxyl groups or hydrogen bonds<br>• The scoring function of GOLD is based upon Cambridge Structural database and on results (empirical) due to chemical interactions (weak)<br>• The GOLD results are evaluated on the basis of hydrogen bonds, steric energy (due to the interaction between protein and ligand), and ligand's internal energy (Lennard-Jones potential) | GOLD module was used to design Casein Kinase inhibitors and SKLB1002, a VEGFR2 inhibitor.<br>Insights into phase I metabolism reaction by cytochrome P450 enzymes |

**TABLE 9.5** *(Continued)*

| Docking programs | Met | Applications |
|---|---|---|
| SEED | • Belong to fragmentation algorithm.<br>• Based upon "anchor and grow method" | Module has led to the development of FK506 immunophilin and inhibitor of BCL6, a B-cell lymphoma oncogene |
| **DOCK** | • Belong to fragmentation algorithm<br>• In 1986, Desjarlais et al. did modifications in DOCK algorithm<br>• In the binding sites, the fragments of ligand were separately docked and then they were joined further—this was done to permit the ligand flexibility. | Inhibitors of anthrax toxin.<br>Optimization is going on for topoisomerase- I as an anticancer drug |
| **Surflex** | • Belong to fragmentation algorithm.<br>• Based upon "anchor and grow method."<br>• Accurate docking algorithm. | FLEXX has targeted malaria, anthrax, and bacterial infections by leading to the development of Plasmepsin II and IV inhibitors, Anthrax edema factor, and Pneumococcal peptidoglycan deacetylase inhibitors, respectively |
| **FLEXX** | • Belong to fragmentation algorithm<br>• Based upon "anchor and grow method"<br>• In this, a fragment of the ligand is docked in the protein binding site. Then the fragment is extended in all directions and all possible conformations. The extended fragments with no/minimal steric hindrances are selected | Inhibitors of α-glucosidase (screening INTERBIOSCREEN chemical database) and RNA Editing Ligase-1 of *Trypanosoma brucei* (causal organism of African trypanosomiasis) |
| **Auto DOCK**[18,23] | • It is also a genetic algorithm and uses a Lamarckian genetic algorithm<br>• Autodock functions in the following way: (1) with the use of AutoDock tools coordinate files are prepared, (2) atomic affinities are precalculated using AutoGrid, (3) ligands are docked via AutoDock module, (4) results are analyzed and interpreted using AutoDock tools | Insights into phase I metabolism reaction by cytochrome P450 enzymes |

- Molecular dynamics (MD) simulation: MD, based on Newtonian physics, was first developed in the 1970s. After molecular docking or comparative/homology modeling or computerized models of molecular system-generated after NMR/X-ray crystallography, the protein–ligand pair (or any other pair in question) is subjected to MD simulations. MD is performed in the presence of a solvent to mimic the physiological condition and leverage the pair to interact in every possible conformation in the allowed virtual conformational space. Here, the molecular forces acting on each atom of ligand and proteins are calculated. The force exerted by atoms is quantified following Newton's second law of motion. Newton's second law of motion states that the acceleration of an object is dependent upon two variables, the net force acting upon the object and the mass of the object. The acceleration is always in the same direction as the net force.[18,20]

$$\text{acceleration} = \frac{\text{Force}_{\text{net}}}{\text{mass}}$$

$$\text{or } F_{\text{net}} = m * a$$

In this system, Newton's laws of motion also govern the atom's position change. This process is repeated several times, and every time, the simulation time is increased by only 1 or 2 quadrillionths of a second. The system runs through time, and as time lapses, the potential energy of the atom changes with reference to time. The differentiation of potential energy of every atom with a change in position is calculated using a defined set of equations and parameters, which collectively generate a force field.[18,24] In other words, the force fields are determinants of different forces acting between different atoms during the molecular dynamics simulations. The force fields are generated via association between bonded and nonbonded atoms. The chemical bonds, atomic angles, and dihedral angles are responsible for bonded force. The simple virtual springs and sinusoidal function are used to model chemical bonds and atomic angles, and dihedral angles, respectively.[25] On the other hand, Vander Waals, electrostatic interactions contribute to nonbonded forces. Lennard-Jones 6-12 potential and Coulomb's law are used to model Vander Waals and electrostatic interactions, respectively. To compute the results of MD, many mathematical calculations are required, hence, to perform this study, clusters of computers or supercomputers (having several parallel processors) are used. In MD, different types of force fields, such as AMBER, NAMD (Nanoscale Molecular Dynamics), CHARMM, and GROMOS are frequently used. The Message Passing Interface (MPI) facilitates easy and simultaneous implementation of these force fields on multiple processors. There are different types of MD viz., metadynamics, umbrella sampling, and replica-exchange MD. They give better results and help in clear visualization of docking results. Further, an amalgamation of molecular docking with molecular dynamic simulation and binding free energies, for example, BEAR (Binding Estimation After Refinement) is a promising approach in the molecular docking technique. BEAR, a post-docking tool, provides clear and refined MD results and predicts the stability of ligand–protein complex via binding free energy methods—MM-PBSA and MM-GBSA. Other advanced free-energy predictors are thermodynamic integration (TI), funnel metadynamics, and free energy perturbation (FEP). They can be applied to interpret the post-docking result, but they are too costly.[19] Docking results are also analyzed by root-mean-square deviation (RMSD), B-factor analysis, and template modeling score (TMS).[26–28] Maestro is a Schrodinger software visualization tool. The results obtained after docking and simulation are analyzed, interpreted, and visualized using a powerful and versatile Maestro computational tool.[23]

In Alzheimer's disease (AD) β-amyloid aggregates are responsible for neurodegeneration in brain. Docking and simulation studies were applied to identify β-amyloid inhibitors. Several compounds were identified among them Aβ42, and beta-site amyloid precursor protein (APP) cleaving enzyme (BACE) inhibitors (significant role in AD pathogenesis) were obtained.[20–22]

#### *9.6.2.2 LIGAND-BASED CADD*

In 1990, Johnson et al. proposed a ligand-based approach, based on the "Similarity Property Principle." According to this principle, structurally similar molecules are expected to have similar properties. In this method, the structure of the target remains unknown. With the help of a set of reference ligands, it is determined whether test compounds can interact with the target or not. The test compounds that interact with the target are screened, and their 2D and 3D structures are analyzed. LB-CADD functions in the following way: (1) similarity measures—the chemical composition of the active site is known, so the compounds that are chemically similar w.r.t target's active site are chosen, (2) quantitative structure–activity relationship (QSAR) model—These models are constructed by correlating the chemical structures with their experimentally determined biological activity, so that the resultant QSAR model can predict the biological activity of unsynthesized chemical structures. In the QSAR approach, biological activity predominates the entire hit to lead, lead optimization, and finally DMPK/ADMET optimization process, while in the similarity measure approach, chemical structures are given preference.[18]

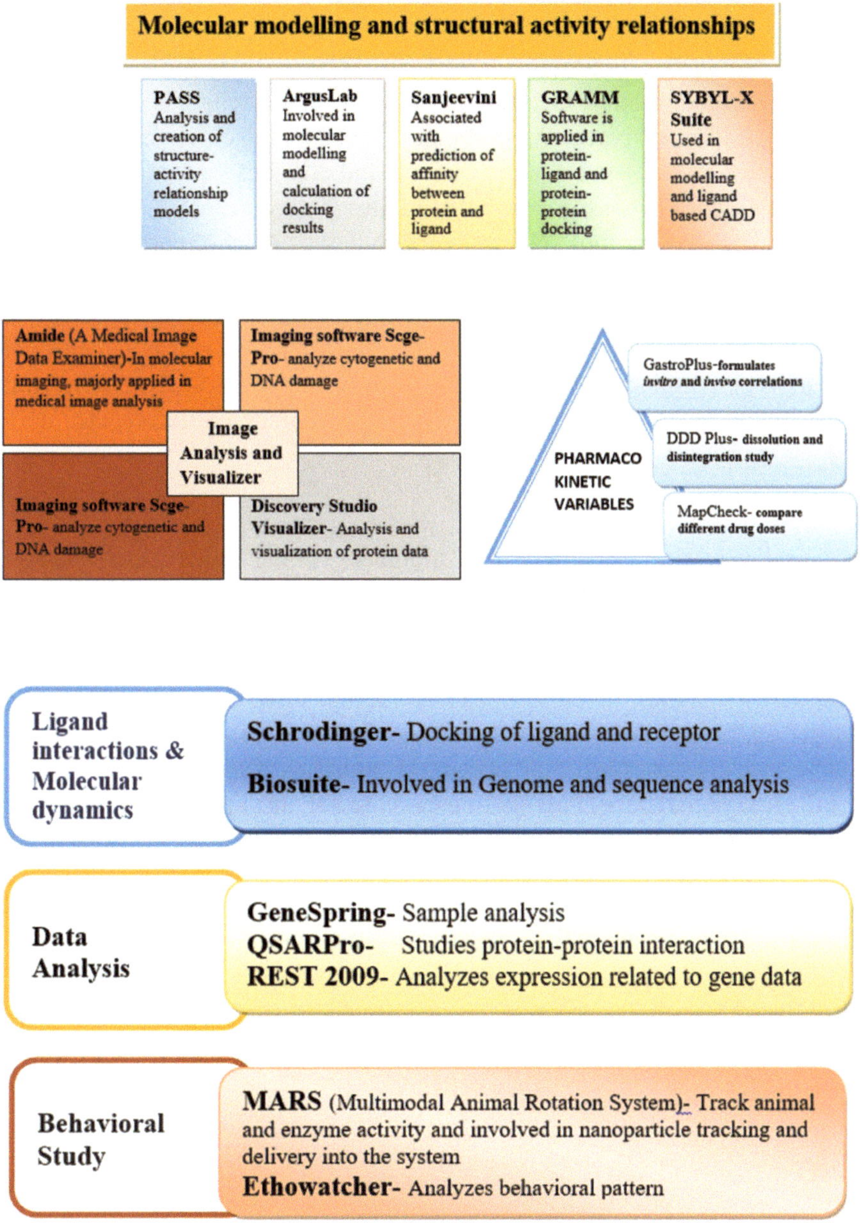


**FIGURE 9.4** Software and computer-associated drug discovery and development tools and techniques.[23]

#### 9.6.2.3 DETERMINANTS OF LB-CADD

There are various determinants of LB-CADD, such as molecular descriptors (functional groups, electronegativity and partial charges, polarizability, and octanol/water partition coefficient), binary molecular fingerprints, 2D descriptors of molecular constitution, and 3D descriptors of molecular configuration and conformation that guides the co-ordinated functioning of this drug discovery tool. The knowledge-based, graph-theoretical methods , molecular mechanical, and quantum-mechanical tools determine the following chemical determinants: functional group, geometry, ring content, volume, molecular weight, surface areas, interatomic and bond distances, atom types, molecular walk counts, electronegativities, symmetry, atom distribution, topological charge indices, and aromaticity indices, etc.[18]

#### 9.6.2.4 SIMILARITY MEASURES

In similarity searches, fingerprint methods, similarity networks, off-target predictions (polypharmacology approach), and fingerprint extensions (MTree) are used. In 2006, Bologa et al. identified new agonists of GPR30 (estradiol receptor family) by applying 2D fingerprint and 3D shape-similarity searches. In 2009, Keiser et al. improvised the chemical similarity approach by introducing statistical models into this method. This approach is known as the similarity ensemble approach (SEA). In this method, statistical models are used to foresee whether the ligand will bind to the target or not. The comparison between the sets of interacting ligands is done based on Tanimoto coefficients, which are based on standard 2D Daylight fingerprints (Daylight Chemical Information Systems, 2013). The evaluation of competent ligands is done via Raw Similarity Scores, calculated by summing up all Tanimoto coefficients (for the sets having a value greater than 0.57). Another method MTree, a topological model, is an amalgamation of polypharmacology and similarity measures, which predicts and identifies new chemotypes and molecular scaffolds.[18]

#### 9.6.2.5 QUANTITATIVE STRUCTURE–ACTIVITY RELATIONSHIP (QSAR)

Mueller et al. applied qQuantitative structure–activity relationship (QSAR) to decipher modulators of mGlu5/mGluR5 signaling. In the central nervous system (CNS) and peripheral nervous system (PNS), glutamate is the principal neurotransmitter. The glutamate transmits neuronal signals via ionotropic (N-methyl-D-aspartate (NMDA), kainate, and amino-3-hydroxy-5-methyl-4-isoxazolepropionic acid (AMPA) receptors), and metabotropic receptors (mGluRs).[29] Metabotropic receptors are G-protein-coupled receptors (GPCRs), which are responsible for neuronal cell migration, proliferation, and inflammation in microglia. In CNS, mGluRs critically regulate the presynaptic transmission (release of neurotransmitters) depending upon the activity of human beings. Based on the signaling pathways and mode of action of drugs, mGluRs are divided into three groups: group I (mGluR1, mGluR5), group II (mGluR2, mGluR3), and group III (mGluR4, mGluR6, mGluR7, and mGluR8). Among them, mGluR5 is prominently expressed in the astrocytes, helps in the modulation of neuronal cell proliferation and is also associated with schizophrenia. The group I mGluRs evoke neuroinflammatory responses and neurodegeneration (by secreting excess glutamate) in Alzheimer's, Parkinson's, Huntington's motor neuron disease (MND), and amyotrophic lateral sclerosis (ALS). Therefore, the treatment of neurological problems needs the use of either positive allosteric modulators (PAM) or negative allosteric modulators (NAM). The treatment modality is to decrease the expression of group I receptors and increase the expression of group II and III receptors, except in schizophrenia where PAM is used. Figures 5 and 6 depict the efficiency of ANN QSAR models in finding the agonists (PAM) and antagonists (NAM) of mGluR5 signaling in comparison to traditional HTS. Autotaxin (autocrine motility factor) is linked with AD, cancer, diabetes, and other chronic diseases. QSAR has been utilized to synthesize autotaxin inhibitors. CoMFA and CoMSIA 3D-QSAR methods have been applied to develop anticancer, anti-hepatitis B-virus (HBV) drugs.[24–28]

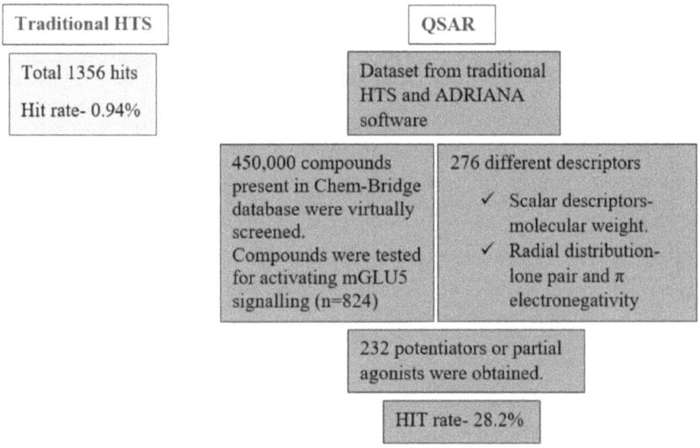


**FIGURE 9.5** The traditional and QSAR method to deduce mGluR5 agonists.[28]

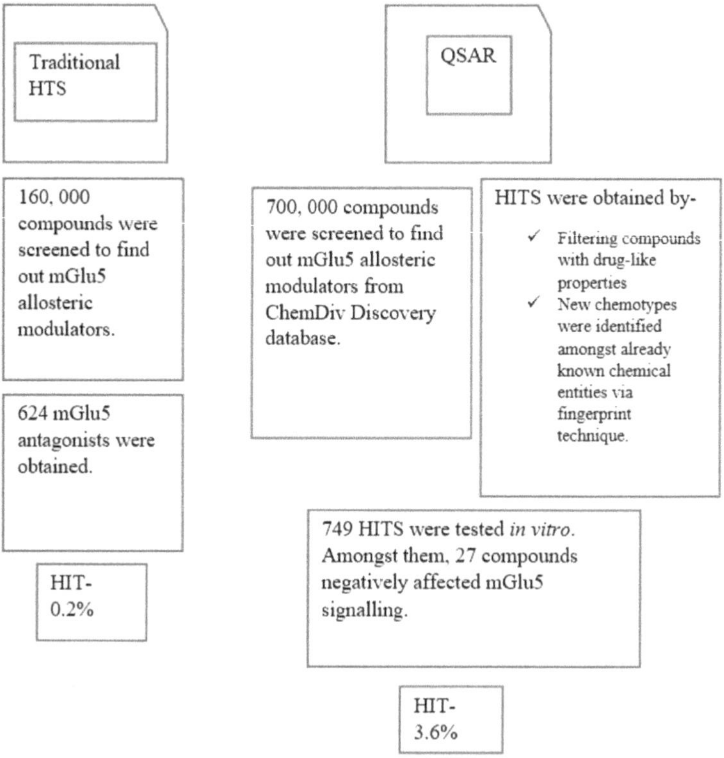


**FIGURE 9.6** The traditional and QSAR method to deduce mGluR5 antagonists.[18]

### 9.6.2.6 *COMPUTATIONAL ANALYSIS OF NOVEL DRUG OPPORTUNITIES (CANDO)*

CANDO (http://protinfo.org/cando) is an open system analysis software tool that generates leads with evolutionary connection and has multiple modes of action—potent enough to be categorized as a "drug." This tool efficiently characterizes the "on" and "off" target effects, which pave the way for drug repurposing. CANDO is based on Moore's Law of pharmacotherapeutics.[30] The Computational Analysis of Novel Drug Opportunities determines the lead compounds in the following way[31]:

- Evolutionary and bioinformatic analysis—A rough prediction is drawn about whether the ligand will interact with the protein (target) or not, based on the evolutionary connections and databases.

- Molecular docking and simulations—The rough poses are refined and finalized using fragment-based docking and molecular dynamic simulation (all-atom knowledge-based force fields) studies. During docking and simulation studies, the binding is minutely assessed via digitized interface interaction with each biomolecule. Interaction with different biomolecules improves the hits of query compounds.
- Calculation of binding free energy—The stable interaction between the query and target structure is analyzed by calculating the binding free energy.

CANDO is a way ahead in the drug discovery approach as compared with virtual screening or traditional high-throughput screening (HTS).[30]

Features of CANDO:

- Drug discovery based on evolutionary lineage—CANDO platform rests its belief in the fact that human beings, plants, other organisms, and microorganisms living in the same environment exert beneficiary as well adverse effects on each other. This affirms that the natural products (query compounds) and potent drug candidates function simultaneously via interaction with several biomolecules and activation of multiple mechanisms of action in the human system. The evaluation of "signatures of interaction" is done numerically or digitally. This indicates the bonding between compound–target protein (protein library) and envisages the multiscale relationship a compound can have with other biomolecules. Further "on-target" (positive) and "off-" or "anti-target" (adverse/negative) effects are analyzed through similarity and dissimilarity of query compound-–proteome interaction signatures, respectively. This sets the ground for lead discovery and drug repurposing.[30]
- Predictive bioanalytics—The evolutionary basis of lead discovery directs the predictive bioanalytics, which converts a query compound into a drug (targeting a particular disease or showing a definite indication) by combining a homology-driven method (functional at the atomic scale) and heterogeneous biodatabases (recognizes signature interactions at various levels). It utilizes data mining methods to predict the behavioral pattern of biomolecules and search for NCEs (new chemical entities). CANDO platforms generate a number of leads/computing cycles via statistical multiplier effects. Its efficiency is on par with whole-genome shotgun sequencing. This has led to the

success of CANDO first version (v1), as it has predicted >1 billion interactions in humans (3733 approved compounds and 48,278 protein structures), and drug predictions have been done for 2030 indications.

- In-house interaction scoring protocol[32]—It facilitates faster assessment of query compound–protein interaction score and their interaction similarities.
- Benchmark accuracy[30–32]—The benchmarking validates the accuracy of the CANDO platform and opens the gateway for shotgun drug repurposing and novel discovery. It applies different types of benchmarking protocols that recognize the relationship between known drugs with respect to a particular disease or an indication for which they have received approval. In total, CANDO v1 has done benchmarking for 1439 indications (≥2 approved compounds). After bench validation, the compounds move toward in vitro and in vivo screening assays. When the query compound passes this phase, then it is forwarded to the clinical trials. Besides v1 pipeline, ligand and structure-based pipelines are also applied for drug repurposing and calculating the benchmarking performance of putative drug candidates.[31]

CANDO has been effective in predicting apernyl, prednisolone, prednisone, and cloquinate for autoimmune disorder systemic lupus erythematosus (SLE). The indications for Alzheimer's disease and diabetes mellitus 2 have been obtained.[30] CANDO platform has also been explored to repurpose drugs for Ebola virus disease (EVD).[33]

**TABLE 9.6** Difference Between CANDO and Virtual Screening Techniques.

| Virtual screening/traditional HTS | CANDO |
|---|---|
| Closed system | Open system |
| In virtual screening, the interaction of the ligand with the active site of protein is given preference. | In CANDO, the interaction of the ligand with the active site of protein is a first step toward determining the lead compound |
| The binding interaction is possible with a limited set of biomolecules, cells, and tissues. Hence, the total performance of the test compound remains incomplete. So, the chances of failure are high during clinical trials or even after | The binding interaction is possible with all macro and micro biomolecules such as nucleic acid, RNA, proteins (enzymes, receptors), lipids, carbohydrates, etc. This provides a complete framework about the performance of the test compounds; hence, the chances of failure are low/minimal during a clinical trial or even after |

**TABLE 9.6** *(Continued)*

| Virtual screening/traditional HTS | CANDO |
|---|---|
| Not based on polypharmacology | Based on polypharmacology |
| Off- and anti-target effects of the test compound remain unknown. So, the chances of drug failure increases | Off- and anti-target effects are known as wider space is provided for interaction. This surfaces the pleiotropic effect of the compound, as biologists and scientists become aware of other mechanisms of action as well. This not only kindles the beneficial effects but also safeguards from the adverse effects of the compound |
| Focusses on single disease aetiologies | Open system, big databases, docking/simulations, and benchmarking study leverages CANDO to focus on multiple diseases aetiologies |
| Molecular docking and simulation study play a prominent role in drug discovery via virtual screening | Molecular docking and simulations are important computer-aided tools among several other software tools |
| Virtual screening is one of the computational tools that is applied in drug repurposing | CANDO provides a complete platform for drug repurposing |

### *9.6.2.7 REVERSE SCREENING*

Reverse screening, also known as in silico/computational target fishing, tries to find out the appropriate target structure from a query ligand structure. It comprises of shape screening, pharmacophore screening, and reverse docking. The first two methods are used by overall comparison of shape or pharmacophore when crystal structure of protein is not available. In reverse docking, the target structure is searched based on known ligand (crystal structure of protein available). Despite the development of Big Data, evolving computational tools and techniques, reverse screening has garnered attention because with every new drug discovery, new drug indications (20% of new drugs launched in 2013) pops-up. These new indications (majorly multi-target) need a potent target to be categorized as a drug. Here, reverse screening is the suitable choice, where these new indications will be repurposed as a novel drug. Besides this, after passing clinical trials, several compounds fail due to their off-target or anti-target effects. So, a new target can be assigned to those failed compounds and studying their proper mechanism of action will give a new life to failed or withdrawn drugs. This will save a lot of time, and money. Hence, reverse screening has a significant contribution in the drug repurposing/drug repositioning.[34]

#### 9.6.2.8 SHAPE SCREENING

The principle of shape screening is based upon two viewpoints- (1) 2D: the molecules that have a similar structure, target similar proteins, and show similar biological activity, (2) 3D: molecules having same volume can fit in the same space or volume of the active site present in the target protein. The screening involves two steps; first is to map the ligands present in the database with the query target molecules and then second is the validation step, where mapped ligands undergo further mapping with their annotated targets (proteins whose information is already present in the database). The selection of potent targets is based on similarity scores. The software or programs used in the shape screening are as follows: **2D**: FingerPrint 2D (FP2) - extended connectivity fingerprints (ECFPs), and Molecular ACCess System (MACCS), the MDL structural key, ChemProt 3.0. **3D**: gWEGA, WEGA, SHAFTS (encoded in ChemMapper). ROCS, Target Hunter, CSNAP3D, SEA, and Swiss Target Prediction are programs and algorithms used in the shape screening method.[34–37]

#### 9.6.2.9 PHARMACOPHORE SCREENING

The presence of chemical entities, chemical structures, or bonds such as hydrogen bond acceptor (HBA) or donor (HBD) vector, hydrophobic center (H), positively (P) or negatively (N) charged centers, create a space where ligands and target can interact effectively—space is known as pharmacophore. The pharmacophore screening follows a basic principle—pharmacophore directs the ligand and target interaction. The first step in the pharmacophore screening is to map pharmacophore models of a query (target) and ligands present in the database, and the second is to map the first step results with their annotated targets. Pharmacophore modeling – (1) ligand-based, (2) structure-based, (3) complex-based builds pharmacophore database. Ligand-based modeling incorporates QSAR; structure-based models is built upon by Pocket v.2, and Catalyst SBP in Discovery Studio (DS) (BIOVIA, 2017); complex-based models are constructed by PharmTargetDB (PharmMapper), and PharmaDB in Discovery Studio.[34]

#### 9.6.2.10 REVERSE DOCKING

In reverse docking, generally, the active site of the target is known or derived from a co-crystal ligand (small-molecule). The steps involved in reverse docking are the following[34,20,35]:

- For target fishing: Potential Drug Target Database (PDTD) sc-PDB, Protein Data Bank (PDB) (Ligand Expo), Therapeutic Target Database (TTD), Pocketome, ZINC, DrugBank, ChEMBL, PubChem & PDSPKi (BindingDB), UniProt, and ChEBI are used.
- Reverse docking (RD) is performed using the following programs: INVDOCK, DOCK, AutoDOCK, AutoDOCK Vina, ACTP, TarFisDOCK, idTarget, SELNERGY, GlamDock, GOLD, MDock (uses PDTD database and computes via ITScore), and MEDock are applied. The ligand is docked with the grid database of the protein targets.
- Docking score (GLIDE) and binding/docking energy (binding strength/interaction energy) between small molecule ligand and query target are calculated.
- Docking energy helps in ranking the potent target molecules.

Machine learning methods, such as protein atom score contributions derived interaction fingerprint (PADIF), support vector machine (SVM), and neural networks (NN) are also being explored in target fishing to enhance the sensitivity and efficacy of reverse docking method.[20]

### 9.6.2.11 APPLICATIONS OF REVERSE SCREENING

Reverse screening has been successfully applied to identify target receptors for marine or terrestrial natural compounds, off- and anti-target effects (drug toxicity/adverse actions), drug repositioning, and in the investigation of molecular mechanisms against chronic infectious diseases and cancer.[34]

### 9.6.2.12 APPLICATIONS OF REVERSE DOCKING

- SELNERGY, a reverse docking tool was used to repurpose tofisopam, an antianxiety drug.[36]
- Off-target effects of torcetrapib-cholesteryl ester transfer protein (CETP) inhibitor were identified using CDOCKER (Accelrys Software Inc., Discovery Studio Modeling Environment, CA, USA).[34]
- Target fishing was done by using the PDTD database and TarFisDock against *Helicobacter pylori*.[34]
- Identification of anticancer targets.[20]

**TABLE 9.7** Applications of Shape and Pharmacophore Screening.[34]

| Screening method | Applications |
|---|---|
| Shape screening | Drug repurposing: identified receptors (target) for the following drugs—Prozac, Validex, and Rescriptor. |
| | Identified targets for the following small molecules - Plumbagin, Obacunone, 5-aza-dc, Sini decoction (aconitine, liquiritin, 6-gingerol), Salvinorin A, Lignan, and Wuweizi |
| Pharmacophore screening | Identified targets for the following small molecules: BBR, Phytoestrogens (genistein, daidzein, secoisolariciresinol), UA, NCI 748494/1, HSYA, Arctigenin, CT, and 5,7, -dihydroxy 4 ′- methoxy -8-prenylflavanone |

### *9.6.2.13 SIGNATURE-BASED APPROACH*

Based on structure or ligand-based programs or algorithms, the compounds are sorted on the grounds of genomic, transcriptomic, or proteomic signatures. There are several sensitive and efficient data repositories, databases, and software designed to accomplish the task of drug discovery and drug repositioning. These tools help us analyze how these compounds will operate in the human system by envisioning test compound–protein, test compound–protein–gene/ test compound–gene interaction, and activation of signaling pathways; hence, elucidating on-, off-, and anti-target effects of the test compound. There are public repositories, such as National Center for Biotechnology Information (NCBI), DNA Data Bank of Japan (DDBJ), European Bioinformatics Institute (EBI), etc., that provide information about diseases related to changes produced in the transcriptome; yet, there is a need for the development of a transcriptomic-based data repository. The reasons are as follows:[37]

- Preclinical: To determine the mechanism of action (MoA) and other effects of perturbagens, in vitro, and in vivo studies are performed. Clinical: Before administering the test compounds, a range of effective doses are required which must be determined in preclinical stage so, preclinical requirements are costly and time-consuming, whereas clinical one is tedious and requires perfection. Hence, to minimize the preclinical issues and solve the clinical issues, transcriptomic-based data repository is required.
- Research related to lead discovery is carried out simultaneously in several laboratories, and the results of each lab may vary. So, a constant tool is required which is devoid of human errors, minimizes

discrepancy and disputes, and provides a platform for benchmarking and validation of results.

### *9.6.2.14 CONNECTIVITY MAP (CMAP)*

The connectivity map establishes the connections between diseases, genes, and perturbagens (drugs, small molecules, shRNA, cDNA, or other biologics). The aim is to perturb the existing condition using perturbagens, modify the disease-associated gene expression, and study the downstream signaling pathways inside the cell. This first-generation data repository was constructed by treating different cancer cell lines—human prostate cancer (PC3), breast cancer (MCF-7), leukemia (HL-60), and melanoma (SKMEL5) with perturbagens. The cell lines were treated with varying doses of perturbagens at varying time points.[37–40]

CMap approach has been extensively utilized in the field of cancer; its applications are listed below.

Lead discovery of the following drugs (in vitro validation):

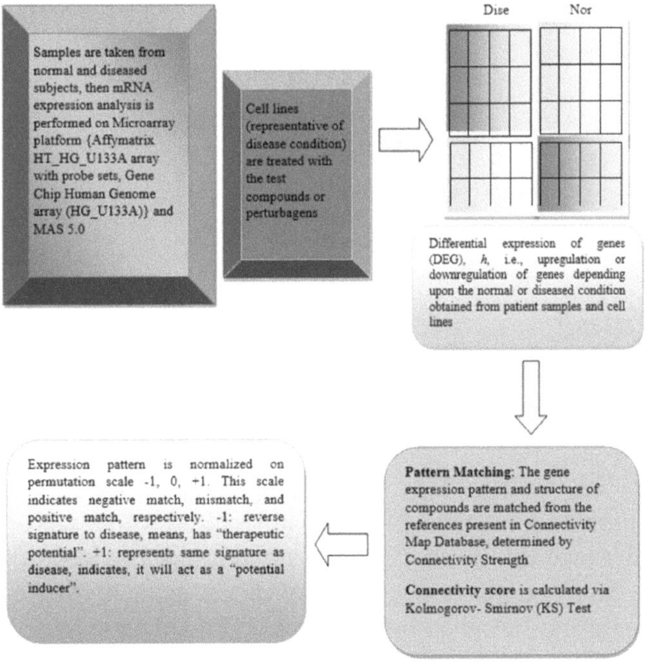


**FIGURE 9.7** Connectivity map in drug discovery or drug repositioning.[39,40]

Explored mechanism of action (MoA) of the following drugs: VXL50 (ovarian cancer), b-AP15 (AML), thioridazine (ovarian cancer), epoxy anthraquinone derivatives (EAD) (neuroblastoma), celastrol and gedunin (prostate cancer), cisplatin (chemotherapy resistance), etc. Apart from cancer, it has also been explored in Down's syndrome and obesity.

**TABLE 9.8** Role of CMap in Drug Repositioning.

| Drug | Initial therapy | Repurposed for | Preclinical validation |
|---|---|---|---|
| **Piperazine** | Antipsychotic | CNS injury | In vitro |
| **Fluphenazine** | Antipsychotic | Hair growth | In vivo (rodent) |
| **Topiramate** | Anticonvulsant | Inflammatory bowel disease (IBD) | In vivo (rodent) |
| **Phenothiazines** | Antipsychotic and antihistamine | Breast cancer | In vitro |
| **Cimetidine** | Antiulcer | Lung cancer | In vitro and in vivo (rodent) |
| **Chlorpromazine and trifluoperazine** | Schizophrenia/ psychotic disorders | Hepatocellular carcinoma (HCC) | In vitro and in vivo (rodent) |
| **Ursolic acid** | Indications for anti-inflammatory, antioxidant, antiapoptotic, and anticarcinogenic role[41] | Muscle atrophy | In vitro and in vivo (rodent) |
| **Phenoxybenzamine** | Antihypertensive | Osteoarthritic pain | In vivo (rodent) |
| **Vorinostat** | Cutaneous T-cell lymphoma | Gastric cancer | In vitro |

Library Integrated Network-based Cellular Signatures (LINCS): For CMap, Lamb et al. tested 164 small molecules on four cancer cell lines. Further, to make this system diverse, more molecules and cell lines were included. The sensitivity and efficacy of the system were enhanced by using fluorescent-colored microspheres and the flow cytometry detection technique. This advanced technique was named L1000 (or large-scale connectivity map) containing 1058 probes. These probes were confirmed by targeting landmark transcripts; 955 shRNAs were developed for 978 landmark transcripts and 80 control transcripts. CMap-L1000v1 generated a 1000-fold large dataset in comparison to CMap. Total 1319138 L1000 profiles and 473647 gene signatures were created by testing 42,080 perturbagens (19,811 small molecule test compounds, 314 biologics, ~18,500 shRNAs, and 3462 cDNAs against 5075 genes in >100 cell lines) (experiments were performed in triplicate, treatment timeline was in between 6 and 24 h). L-1000 is now termed as Library of Integrated Network-Based Cellular Signatures (LINCS), which is also known as high-throughput reduced representation of transcriptome profiling method and its efficiency is at par with RNA sequencing.[41–45] In the LINCS database primarily breast adenocarcinoma (MCF-7), pancreatic carcinoma (YAPC), colorectal adenocarcinoma (HT29), malignant lung melanoma (A375), prostate cancer cell lines – prostate adenocarcinoma (PC3), and metastatic prostate (VCAP), lung cancer cell lines – non-small cell carcinoma (NSCLC) (A549) and non-small cell adenocarcinoma (HCC515), and hepatocellular cancer cell line (HEPG2) are present, but NeuroLINCS has opened the gateway for the central nervous system (CNS) disorders as well. NeuroLINCS has been maintained and established by culturing organoids and induced pluripotent stem cells (iPSCs) from the neuronal cells of patients suffering from CNS disorders, such as Alzheimer's diseases, spinal muscular atrophy (SMA), and amyotrophic lateral sclerosis (ALS). NeuroLINCS encapsulates datasets for all "omics," viz, genomic (epigenomics), transcriptomic, proteomic, and imaging.[37] NeuroLINCS and CMap have explored drug candidates that can be repurposed for Alzheimer's disease.[43] The tools and software designed to browse, visualize, and analyze the LINC database are Enrichr, LINCS Data Portal, Slicr (LINCS L1000 Slicer GSE70138), LIFE, L1000CDS, LINCS Canvas Browser, and iLINCS.[39,62]

Genome-Wide Associated Studies (GWAS): Due to diverse alleles, the genomic variation is prominent in neuropsychiatric disorders (include Alzheimer's disease and dementia, Parkinson's disease, schizophrenia, Huntington's disease, depression, and bipolar disorders). These alleles and single-nucleotide polymorphisms (SNPs) increase the risk of developing neuropsychiatric disorders. GWAS approach is based on SNPs and has a significant contribution to drug discovery and drug repurposing for neuropsychiatric disorders. In the meta-analysis study (796 studies), GWAS identified small-molecule targetable genes, and biopharmable genes (generate peptides/transmembrane domains modified into pharmaceutical agents or drugs).[46–50] GWAS has been intricately explored in drug discovery and repurposing for neurological disorders, discussed in detail below.

Artificial Intelligence and Machine Learning: Artificial intelligence has given a new dimension and horizon to the computational drug designing. This platform is based on the mathematical approach and provides new indications with precision and accuracy. Artificial intelligence and machine learning (ML) is a combination of applied mathematics and software/databases. In the computer science, ML is a sub-field of artificial intelligence (AI). Machine learning performs its operations or tasks in three ways.[70–75]

- Supervised learning algorithms: Supervised learning algorithms, such as SVM or (Deep) neural networks have been applied in the biological field.
- Unsupervised learning: Unsupervised learning determines the related relationships or patterns present in the unlabeled data. It is performed by dimension reduction methods, such as PCA (principal component analysis), collaborative filtering, clustering/grouping data, and density estimation.
- Sequential learning: In this task, the input for data generation is obtained from the previous interacting environment. The goal-oriented entity that interacts with its surrounding environment, makes the data selection choice based on input. In sequence learning, from the stream of data, only one data is processed at one time by algorithms. Further, the decision made by algorithms is a trial-and-error process. Multi-Armed Bandit (MAB) algorithms belong to the sequence learning family and are associated with recommender systems. Recommenders have a significant role in the prediction of drug–target interactions (DTI) and drug repurposing. In the recommender system, the goal agent recommends the test compound (or the compounds that can be chosen as a new drug candidate). The selection criteria are based either on disease or the chemical structure/chemical composition of the test compound.[50–53]

The baseline regularization algorithm of ML repurposes drugs from electronic health record (EHR) data.[45,77]

Artificial intelligence and ML has several implications in Alzheimer's disease,[46–48] Parkinson's disease,[49,50] and multiple sclerosis.[51,52]

## 9.7 CASE STUDIES OF DRUG REPURPOSING-BASED SMALL MOLECULES FORAGING, IN NEUROLOGICAL, AND NEURODEGENERATIVE DISORDERS

**TABLE 9.9** List of Drug Repurposed Small Molecules in Varieties of Neurological and Neurodegenerative Diseases.

| Diseases | Repurposed drugs, identification of transcription factors/targets/MoA | Initial intervention | Approach |
|---|---|---|---|
| **Alzheimer's disease** | • Drugs proposed for repurposing: Vorinostat, Trimethadione, Cyproterone, Metrizamide, Cefuroxime, and Dydrogesterone[53]<br>• Identification of master regulators of AD-ATF2 and PARK2[53] | | Connectivity map |
| | Valsartan[10] | Hypertension | |
| | Galantamine[10] | Paralysis, Polio | |
| | Amyloid precursor protein (n=14) | | Proteomics[54] |
| | Dopamine D1 receptor (n=13) | | Metabolomics[54] |
| | Acetylcholinesterase (n=10) | | Metabolomics[54] |
| | Plasminogen (n=7) | | Proteomics[54] |
| | CGMP-specific 3′, 5′- cyclic phosphodiesterase (n=6) | | Metabolomics[54] |
| | Myeloid cell surface antigen CD33 (n=6) | | GWAS[54] |
| | Proposed for anti-AD[54]:<br>**Bupivacaine** (target *SCN10A*). *SCN10A* associated with tau protein (Microtubule Associated Protein Tau, *MAPT*) {hallmark of AD}<br>**Topiramate** (target *SCN1*). SCN1 is linked with presenilin 1 (*PSEN1*) {hallmark of AD}<br>**Selegiline** and **iproniazid** (target monoamine oxidase (MAO) inhibitors | Local anesthetic | CMap, LINCS (L1000) (LINCS downloaded from Gene Expression Omnibus, GEO), STRING |

**TABLE 9.9** *(Continued)*

| Diseases | Repurposed drugs, identification of transcription factors/targets/MoA | Initial intervention | Approach |
|---|---|---|---|
| **Proposed for anti-AD**[37] | Vinpocetine | Prescribed as a dietary supplement for cognition | GWAS |
| | Omalizumab (recombinant antibody against IgE) | Asthma | GWAS |
| | Naftidrofuryl | Vasodilator-intermittent claudication and peripheral artery disease (PAD) | GWAS |
| | Leflunomide (inhibit activation of T cells) | Rheumatoid arthritis | GWAS |
| | Gemtuzumab ozogamicin (monoclonal antibody, anti-CD33) | AML | GWAS |
| | Celecoxib, Meclofenamic Acid, and Naproxen (NSAIDs) (COX-2 inhibitor) | Fever, pain killer, and anti-inflammatory | GWAS |
| **Parkinson's disease** | Amantadine (anticholinergic)[10,56,57] | Influenza | |
| | Ropinirole (D2 agonist)[10] | Hypertension | |
| | Reduces α-synuclein levels[57,58]: Nilotinib[59], β-agonists, Terazosin, and Lovastatin | Statins reduces cholesterol levels | |
| | Buspirone[57,59] | Anxiety | |
| | Restores mitochondrial function[58,59]: | | |
| | N-acetylcysteine, Ursodeoxycholic acid, and Glutathione | | |
| | Exenatide, Lixisenatide, and Liraglutide (glucagon-like peptide (GLP)-1agonists)[57–59] | Diabetes mellitus Type II | |
| | Decreases inflammation in neuronal cells[57–59]: AZD3241, Sargramostim (granulocyte macrophage colony-stimulating factor (GM-CSF), Azathioprine, Simvastatin, and Lovastatin | Statins reduces cholesterol levels | |

**TABLE 9.9** *(Continued)*

| Diseases | Repurposed drugs, identification of transcription factors/targets/MoA | Initial intervention | Approach |
|---|---|---|---|
| | Ambroxol (increases level of glucocerebrosidase (GCase) enzyme)[57–59] | Respiratory disorders | |
| | Isradipine (calcium channel blocker)[57–59] | Lower blood pressure | |
| | Deferiprone (decreases the excess iron present in patients suffering from Parkinson's disease)[57–59] | | |
| | Inosine[58] (increases urate levels in Parkinson's disease patients) | | |
| | Tetracycline[60] | Antibiotic | Structure–activity relationship (SAR) |
| | Minocycline,[59,60] and creatine.[59] | | |
| | Potency to repurpose as anti-PD[61]: | | Connectivity Map |
| | Tubocurarine chloride, vorinostat, benperidol, and harmaline | | |
| **Schizophrenia** | Drugs that have potential to get repurposed[62]: | | DR: Base Space Correlation Engine |
| | Danazol | Angioedema, endometriosis | Network analysis: Lens for Enrichment and Network Studies of human proteins, GWAS, and DisGeNET |
| | Cinnarizine | | |
| | Antazoline | Sickness while moving & walking, and vertigo | |
| | Cromoglicic acid | | |
| | Acetazolamide | | |
| | Dimenhydrinate | Congestion in nose and conjunctivitis (allergic) | |
| | Tetracycline | | |

**TABLE 9.9** *(Continued)*

| Diseases | Repurposed drugs, identification of transcription factors/targets/MoA | Initial intervention | Approach |
|---|---|---|---|
| | Alfacalcidol | Asthma | |
| | Miconazole | Sickness while climbing mountain, and Glaucoma | |
| | Alendronate | Nausea, moving sickness | |
| | Bepridil | Antibiotic | |
| | | Vitamin D complement | |
| | | Vaginal infection (antifungal) | |
| | | Osteoporosis | |
| | | Chest pain due to reduced blood flow/Angina | |
| | Galantamine | Alzheimer's disease, and to cease smoking | GWAS |
| | Varenicline | To cease smoking | GWAS |
| | Bevacizumab | Anticancer | GWAS |
| | Neratinib | Anticancer | GWAS |
| | Fluticasone | Chronic obstructive pulmonary disorder (COPD), and asthma | GWAS |
| | Zonisamide | Antiepileptic, anti-Parkinson, and bipolar-anxiety problems | GWAS |
| | Memantine | Alzheimer's disease, and other neurological disorders | GWAS |

**TABLE 9.9** *(Continued)*

| Diseases | Repurposed drugs, identification of transcription factors/targets/MoA | Initial intervention | Approach |
|---|---|---|---|
| | Metoclopramide, Trifluoperazine | Antipsychotic, and stomach paralysis caused due to diabetes (Diabetes gastroparesis) | GWAS |
| | Resveratrol | Anti-inflammatory and anticancer | GWAS |
| | Verapamil | Hypertension, chest pain, and arrythmias | GWAS |
| | Dutogliptin, Alogliptin | Diabetes Mellitus Type II | GWAS |
| | Atorvastatin | Dyslipidaemia | GWAS |
| | Xanomeline | Antipsychotic | GWAS |
| | Cinnarizine | Dopamine D2 receptor antagonist, antihistamine, and prevent vomiting & nausea | GWAS |
| | Nifedipine | Hypertension | GWAS |
| | Mecamylamine | Hypertension | GWAS |
| **Bipolar disorder** | Nonsteroidal anti-inflammatory drugs (NSAIDs) – Aspirin (low/high dose), angiotensin agents, statins, and allopurinol[63] | Aspirin: Pain killer and decreases cardiovascular disease (CVD) risk<br>Statins: lowers blood pressure<br>Angiotensin agents- kidney diseases<br>Allopurinol: treatment of gout and decreases uric acid levels | Epidemiology study[54] |

**TABLE 9.9** *(Continued)*

| Diseases | Repurposed drugs, identification of transcription factors/targets/MoA | Initial intervention | Approach |
|---|---|---|---|
| | Tamoxifen[10] | Breast tumor | |
| | Amphotericin B (NSAIDs) | Antifungal[10] | |
| **Depression (D), and attention deficit hyperactivity disorder (ADHD)** | Scopolamine (D) | Prevent vomiting and nausea (antiemetic), and motion sickness | GWAS[37] |
| | Mecamylamine (depression and ADHD)[10] | Hypertension | |
| | Papaverine (D) | | GWAS |
| | Amantadine (ADHD)[10] | Influenza | |
| | Alizapride, mesoridazine (D) | | GWAS |
| | Dexmecamylamine (D)[10] | Hypertension | |
| | Ketoconazole (D) | | GWAS |
| | Zileuton[64] (D) | Initial phase of asthma | Deep and Machine learning approach- Artificial Intelligence (AI) (Google semantic AI universal encoder) |
| | Risperidone, Sulpiride (D) | | GWAS |
| | Levonorgestrel, Diethylstilbesterol (D) | | GWAS |
| | Pregabalin, gabapentin, nitrendipine (D) | | GWAS |
| | Arcaine sulfate, ifenprodil (D) | | GWAS |
| **Huntington's disease** | Selegiline | Alzheimer's disease[65] | |
| | Tetrabenazine | Antipsychotic[65] | |
| | Tiapride | Antipsychotic[65] | |
| | Memantine | Alzheimer's disease[65] | |

**TABLE 9.9** *(Continued)*

| Diseases | Repurposed drugs, identification of transcription factors/targets/MoA | Initial intervention | Approach |
|---|---|---|---|
| | Apomorphine[65] | Parkinson`s disease | |
| | Risperidone | Schizophrenia and bipolar disorder[65] | |
| | Glatiramer acetate | Multiple sclerosis[66] | |
| **Amyotrophic lateral sclerosis (ALS)[65]** | Triumeq | Anti-HIV therapy | |
| | Mastinib (tested in rodents) | Anticancer (dogs) | |
| | Tamoxifen | Breast cancer | |
| | Apomorphine[67] | Parkinson's disease | |
| **Multiple sclerosis[66]** | Amiloride | Hypertension | |
| | Mitoxantrone | Antineoplastic | |
| | Cyclophosphamide | Antineoplastic | |
| | Cladribine | Antimetabolite, anticancer | |
| **Epilepsy[37]** | Equol | Antineoplastic | GWAS |
| | Vinorelbine | Antineoplastic | GWAS |
| | Nornicotine | Insecticide | GWAS |

Note: n represents number of drugs in clinical trial.

## 9.8 CONCLUSION AND FUTURE PERSPECTIVES

Drug repurposing has advanced in the field of cancer, but in neurological disorders, drug repurposing is in infancy and falling behind due to a lack of suitable in vitro and in vivo models, essential for preclinical validation. Unlike cancer, for neurological disorders, proper cell lines are not available and if also present, they cannot mimic the complete neural conditions, like the presence of astrocytes, microglia, other neural cells, and immune cells in the same vicinity. However, the generation and application of 3D-induced pluripotent stem cells (iPSCs) have made a pertinent effort to resolve the issue, but still more work needs to be done. The test compounds that are competent in mice fail in humans, hence, clinicians have to rely mostly on clinical trials for compound validation. Moreover, MODEL-AD consortium (collaboration between Indiana University, Jackson Laboratory, Sage Bionetworks, and University of California Irvine) has been developed to study the late onset of Alzheimer's disease in rodents, IMPRiND (Inhibiting Misfolded Protein Propagation In Neurodegenerative Diseases) consortium, and RIKEN institute are developing macaque and PSEN (Presenilin-1)-mutation-based marmoset model of Alzheimer`s disease, respectively.[70,75,78]

## KEYWORDS

- **drug repurposing**
- **aging**
- **neurodegenerative disorders**
- **computational drug discovery**
- **therapeutic interventions**

## REFERENCES

1. Pina, A. S.; Hussain, A.; Roque, A. C. An Historical Overview of Drug Discovery. *Methods Mol. Biol.* **2009,** *572*, 3–12.
2. Pizza, V.; Agresta, A.; W D'Acunto, C.; Festa, M.; Capasso, A. Neuroinflamm-Aging and Neurodegenerative Diseases: An Overview. *CNS Neurol. Disorders Drug Targets* **2011,** *10*, 621–634.
3. Mayne, K.; White, J. A.; McMurran, C. E.; Rivera, F. J.; dela Fuente A. G. Aging and Neurodegenerative Disease: Is the Adaptive Immune System a Friend or Foe? *Front. Aging Neurosci.* **2020,** *12*, 305.
4. Spittau, B. Aging Microglia-Phenotypes, Functions and Implications for Age-Related Neurodegenerative Diseases. *Front Aging Neurosci.* **2017,** *14*, 194.
5. Kouli, A.; Torsney, K. M.; Kuan, W. L. *Parkinson's Disease: Etiology, Neuropathology, and Pathogenesis*; Exon Publications, 2018; pp 3–26.
6. Pajares, M. I.; Rojo, A.; Manda, G.; Boscá, L.; Cuadrado, A. Inflammation in Parkinson's Disease: Mechanisms and Therapeutic Implications. *Cells* **2020,** *14*, 1687.
7. Gelders, G.; Baekelandt, V.; Van der, P. A. Linking Neuroinflammation and Neurodegeneration in Parkinson's Disease. *J. Immunol. Res.* **2018,** 4784268.
8. Calsolaro, V.; Edison, P. Neuroinflammation in Alzheimer's Disease: Current Evidence and Future Directions. *Alzheimers Dement.* **2016,** *12*, 719–732.

9. Minter, M. R.; Taylor, J. M.; Crack, P. J. The Contribution of Neuroinflammation to Amyloid Toxicity in Alzheimer's Disease. *J. Neurochem.* **2016,** *136*, 457–474.
10. Musella, A.; Gentile, A.; Rizzo, F. R.; De Vito, F.; Fresegna, D.; Bullitta, S.; Vanni, V.; Guadalupi, L.; Stampanoni, B. M.; Buttari, F.; Centonze, D.; Mandolesi, G. Interplay Between Age and Neuroinflammation in Multiple Sclerosis: Effects on Motor and Cognitive Functions. *Front. Aging Neurosci.* **2018,** *10*, 238.
11. Govindarajan, V.; de Rivero Vaccari, J. P.; Keane, R. W. Role of Inflammasomes in Multiple Sclerosis and Their Potential as Therapeutic Targets. *J. Neuroinflammation* **2020,** *17*, 260.
12. Obrador, E.; Salvador, R.; López-Blanch, R.; Jihad-Jebbar, A.; Vallés, S. L.; Estrela, J. M. Oxidative Stress, Neuroinflammation and Mitochondria in the Pathophysiology of Amyotrophic Lateral Sclerosis. *Antioxidants* **2020**, *9*, 901.
13. Mendiola-Precoma, J.; Berumen, L. C.; Padilla, K.; Garcia-Alcocer, G. Therapies for Prevention and Treatment of Alzheimer's Disease. *Biomed. Res. Int.* **2016,** *2016*, 2589276.
14. Mouchaileh, N.; Hughes, A. J. Pharmacological Management of Parkinson's Disease in Older People. *J. Pharm. Pract. Res.* **2020,** *50*, 445–454.
15. Nazari, F.; Shaygannejad, V.; Mohammadi Sichani, M.; Mansourian, M.; Hajhashemi, V. Quality of Life Among Patients With Multiple Sclerosis and Voiding Dysfunction: A Cross-Sectional Study. *BMC Urol.* **2020,** *20*, 62.
16. Ostolaza, A.; Corroza, J.; Ayuso, T. Multiple Sclerosis and Aging: Comorbidity and Treatment Challenges. *Mult. Scler. Relat. Disord.* **2021,** *50*, 102815.
17. De Oliveira, E. A. M.; Lang, K. L. Drug Repositioning: Concept, Classification, Methodology, and Importance in Rare/Orphans and Neglected Diseases. *J. App. Pharmaceut. Sci.* **2018,** *8*, 157–165.
18. Umscheid, C. A.; Margolis, D. J.; Grossman, C. E. Key Concepts of Clinical Trials: A Narrative Review. *Postgrad. Med.* **2011,** *123*, 194–204.
19. Singh, I. P.; Ahmad, F.; Chatterjee, D.; Bajpai, R.; Sengar, N. Natural Products: Drug Discovery and Development. In *Drug Discovery and Development*; Poduri, R., Eds.; 2021, pp 456–517.
20. Zdrazil, B.; Richter, L.; Brown, N.; Guha, R. Moving Targets in Drug Discovery. *Sci. Rep.* **2020,** *10*, 20213.
21. Parisi, D.; Adasme, M. F.; Sveshnikova, A.; Bolz, S. N.; Moreau, Y.; Schroeder, M. Drug Repositioning or Target Repositioning: A Structural Perspective of Drug-Target-Indication Relationship for Available Repurposed Drugs. *Comput. Struct. Biotechnol. J.* **2020,** *18,* 1043–1055.
22. Lee, H. M.; Kim, Y. Drug Repurposing Is a New Opportunity for Developing Drugs Against Neuropsychiatric Disorders. *Schizophr. Res. Treat.* **2016,** *2016*, 6378137.
23. Anighoro, A.; Bajorath, J.; Rastelli, G. Polypharmacology: Challenges and Opportunities in Drug Discovery. *J. Med. Chem.* **2014,** *57*, 7874–7887.
24. Mohapatra, T. K.; Subuddhi, B. B. Repurposing of Aspirin: Opportunities and Challenges. *Res. J. Pharm. Tech.* **2019,** *12*, 2037–2044.
25. Mohammad, J. R.; Sheibani, M.; Nezamoleslami, S.; Shayesteh, S.; Jand, Y.; Dehpour, A. Drug Repositioning: A Review. *J. Iranian Med. Council* **2018,** *1*, 7–10.
26. Li, X.; Rousseau, J. F.; Ding, Y.; Song, M.; Lu, W. Understanding Drug Repurposing From the Perspective of Biomedical Entities and Their Evolution: Bibliographic Research Using Aspirin. *JMIR Med. Inform.* **2020,** *8*, e16739.

27. Gns, H. S.; Gr, S.; Murahari, M.; Krishnamurthy, M. An Update on Drug Repurposing: Re-Written Saga of the Drug's Fate. *Biomed. Pharmacother.* **2019,** *110*, 700–716.
28. Johnson, N. P. Metformin Use in Women With Polycystic Ovary Syndrome. *Ann. Transl. Med.* **2014,** *2*, 56.
29. Sahoo, B. M.; Ravi Kumar, B. V. V.; Sruti, J.; Mahapatra, M. K.; Banik, B. K.; Borah, P. Drug Repurposing Strategy (DRS): Emerging Approach to Identify Potential Therapeutics for Treatment of Novel Coronavirus Infection. *Front. Mol. Biosci.* **2021,** *8*, 628144.
30. Sliwoski, G.; Kothiwale, S.; Meiler, J.; Lowe, E. W. Jr. Computational Methods in Drug Discovery. *Pharmacol. Rev.* **2013,** *66*, 334–395.
31. Pinzi, L.; Rastelli, G. Molecular Docking: Shifting Paradigms in Drug Discovery. *Int. J. Mol. Sci.* **2019,** *20*, 4331.
32. de Oliveira, A. A.; Neves, B. J.; Silva, L. D. C.; Soares, C. M. A.; Andrade, C. H.; Pereira, M. Drug Repurposing for Paracoccidioidomycosis Through a Computational Chemogenomics Framework. *Front. Microbiol.* **2019,** *10*, 1301.
33. Taylor, J. S.; Burnett, R. M. DARWIN: A Program for Docking Flexible Molecules. *Proteins* **2000,** *41*, 173–191.
34. Jamkhande, P. G.; Ghante, M. H.; Ajgunde, B. R. Software Based Approaches for Drug Designing and Development: A Systematic Review on Commonly Used Software and its Applications. *Bull. Faculty Pharm.***2017,** *55*, 203–210.
35. Vincent, K.; Wilfred, G. V. F.; Hünenberger, P. H. A Fast SHAKE Algorithm to Solve Distance Constraint Equations for Small Molecules in Molecular Dynamics Simulations. *J. Comp. Chem.* **2001,** *22*, 501–508.
36. Durrant, J. D.; McCammon, J. A. Molecular Dynamics Simulations and Drug Discovery. *BMC Biol.* **2011,** *9*, 71.
37. Hevener, K. E.; Zhao, W.; Ball, D. M.; Babaoglu, K.; Qi, J.; White, S. W.; Lee, R. E. Validation of Molecular Docking Programs for Virtual Screening Against Dihydropteroate Synthase. *J. Chem. Inf. Model.* **2009,** *49*, 444–460.
38. Carugo, O. How Large B-Factors can be in Protein Crystal Structures. *BMC Bioinform.* **2011,** *19*, 61.
39. Chopra, G.; Samudrala, R. Exploring Polypharmacology in Drug Discovery and Repurposing Using the CANDO Platform. *Curr. Pharm. Des.* **2016,** *22*, 3109–3123.
40. Crupi, R.; Impellizzeri, D.; Cuzzocrea, S. Role of Metabotropic Glutamate Receptors in Neurological Disorders. *Front. Mol. Neurosci.* **2019,** *12*, 20.
41. Minie, M.; Chopra, G.; Sethi, G.; Horst, J.; White, G.; Roy, A.; Hatti, K.; Samudrala, R. CANDO and the Infinite Drug Discovery Frontier. *Drug Discov. Today* **2014,** *19*, 1353–1363.
42. Schuler, J.; Samudrala, R. Fingerprinting CANDO: Increased Accuracy With Structure- and Ligand-Based Shotgun Drug Repurposing. *ACS Omega* **2019,** *4*, 17393–17403.
43. Mangione, W.; Falls, Z.; Chopra, G.; Samudrala, R. cando.py: Open-Source Software for Predictive Bioanalytics of Large-Scale Drug-Protein-Disease Data. *J. Chem. Inf. Model.* **2020,** *60*, 4131–4136.
44. Chopra, G.; Kaushik, S.; Elkin, P. L.; Samudrala, R. Combating Ebola With Repurposed Therapeutics Using the CANDO Platform. *Molecules* **2016,** *21*, 1537.
45. Huang, H.; Zhang, G.; Zhou, Y.; Lin, C.; Chen, S.; Lin, Y.; Mai, S.; Huang, Z. Reverse Screening Methods to Search for the Protein Targets of Chemopreventive Compounds. *Front. Chem.* **2018,** *9*, 138.

46. Kharkar, P. S.; Warrier, S.; Gaud, R. S. Reverse Docking: A Powerful Tool for Drug Repositioning and Drug Rescue. *Future Med. Chem.* **2014,** *6*, 333–342.
47. Bernard, P.; Dufresne-Favetta, C.; Favetta, P.; Do, Q. T.; Himbert, F.; Zubrzycki, S.; Scior, T.; Lugnier, C. Application of Drug Repositioning Strategy to TOFISOPAM. *Curr. Med. Chem.* **2008,** *15*, 3196–3203.
48. Shukla, R.; Henkel, N. D.; Alganem, K.; Hamoud, A. R.; Reigle, J.; Alnafisah, R. S.; Eby, H. M.; Imami, A. S.; Creeden, J. F.; Miruzzi, S. A.; Meller, J.; Mccullum, R. E. Signature-Based Approaches for Informed Drug Repurposing: Targeting CNS Disorders. *Neuropsychopharmacology* **2021,** *46*, 116–130.
49. Lamb, J.; Crawford, E. D.; Peck, D.; Modell, J. W.; Blat, I. C.; Wrobel, M. J.; Lerner, J.; Brunet, J. P.; Subramanian, A.; Ross, K. N.; Reich, M.; Hieronymus, H.; Wei, G.; Armstrong, S. A.; Haggarty, S. J.; Clemons, P. A.; Wei, R.; Carr, S. A.; Lander, E. S.; Golub, T. R. The Connectivity Map: Using Gene-Expression Signatures to Connect Small Molecules, Genes, and Disease. *Science* **2006,** *313*, 1929–1935.
50. Musa, A.; Ghoraie, L. S.; Zhang, S. D.; Glazko, G.; Yli-Harja, O.; Dehmer, M; Haibe-Kains, B.; Emmert-Streib, F. A Review of Connectivity Map and Computational Approaches in Pharmacogenomics. *Brief Bioinform.* **2017,** *18*, 903.
51. Qu, X. A.; Rajpal, D. K. Applications of Connectivity Map in Drug Discovery and Development. *Drug Discov. Today* **2012,** *17*, 1289–1298.
52. Seo, D. Y.; Lee, S. R.; Heo, J. W.; No, M. H.; Rhee, B. D.; Ko, K. S.; Kwak, H. B.; Han, J. Ursolic Acid in Health and Disease. *Korean J. Physiol. Pharmacol.* **2018,** *22*, 235–248.
53. Subramanian, A.; Narayan, R.; Corsello, S. M.; Peck, D. D.; Natoli, T. E.; Lu, X.; Gould, J.; Davis, J. F.; Tubelli, A. A.; Asiedu, J. K.; Lahr, D. L.; Hirschman, J. E.; Liu, Z.; Donahue, M.; Julian, B.; Khan, M.; Wadden, D.; Smith, I. C.; Lam, D.; Liberzon, A.; Toder, C.; Bagul, M.; Orzechowski, M.; Enache, O. M.; Piccioni, F.; Johnson, S. A.; Lyons, N. J.; Berger, A. H.; Shamji, A. F.; Brooks, A. N.; Vrcic, A.; Flynn, C.; Rosains, J.; Takeda, D. Y.; Hu, R.; Davison, D.; Lamb, J.; Ardlie, K.; Hogstrom, L.; Greenside, P.; Gray, N. S.; Clemons, P. A.; Silver, S.; Wu, X.; Zhao, W. N.; Read-Button, W.; Wu, X.; Haggarty, S. J.; Ronco, L. V.; Boehm, J. S.; Schreiber, S. L.; Doench, J. G.; Bittker, J. A.; Root, D. E.; Wong, B.; Golub, T. R. A Next Generation Connectivity Map: L1000 Platform and the First 1,000,000 Profiles. *Cell* **2017,** *171*, 1437–1452.
54. Williams, G.; Gatt, A.; Clarke, E.; Corcoran, J.; Doherty, P.; Chambers, D.; Ballard, C. Drug Repurposing for Alzheimer's Disease Based on Transcriptional Profiling of Human iPSC-Derived Cortical Neurons. *Transl. Psychiatry* **2019,** *9*, 220.
55. Réda, C.; Kaufmann, E.; Delahaye-Duriez, A. Machine Learning Applications in Drug Development. *Comput. Struct. Biotechnol. J.* **2019,** *18*, 241–252.
56. Kuang, Z.; Bao, Y.; Thomson, J.; Caldwell, M.; Peissig, P.; Stewart, R.; Willett, R.; Page, D. A Machine-Learning-Based Drug Repurposing Approach Using Baseline Regularization. *Methods Mol. Biol.* **2019,** *1903*, 255–267.
57. Rodriguez, S.; Hug, C.; Todorov, P.; Moret, N.; Boswell, S. A.; Evans, K.; Zhou, G.; Johnson, N. T.; Hyman, B. T.; Sorger, P. K.; Albers, M. W.; Sokolov, A. Machine Learning Identifies Candidates for Drug Repurposing in Alzheimer's Disease. *Nat. Commun.* **2021,** *12*, 1033.
58. Zhang, D.; Shen, D. Alzheimer's Disease Neuroimaging Initiative. Multi-Modal Multi-Task Learning for Joint Prediction of Multiple Regression and Classification Variables in Alzheimer's Disease. *Neuroimage* **2012,** *9*, 895–907.

59. Moradi, E.; Pepe, A.; Gaser, C.; Huttunen, H.; Tohka, J.; Alzheimer's Disease Neuroimaging Initiative. Machine Learning Framework for Early MRI-Based Alzheimer's Conversion Prediction in MCI Subjects. *Neuroimage* **2015,** *104*, 398–412.
60. Kim, D. H.; Wit, H.; Thurston, M. Artificial Intelligence in the Diagnosis of Parkinson's Disease From Ioflupane-123 Single-Photon Emission Computed Tomography Dopamine Transporter Scans Using Transfer Learning. *Nucl. Med. Commun.* **2018,** *39*, 887–893.
61. Blahuta, J.; Soukup, T.; Čermák, P.; Rozsypal, J.; Večerek, M. In *Ultrasound Medical Image Recognition With Artificial Intelligence for Parkinson's Disease Classification*, 2012 Proceedings of the 35th International Convention MIPRO, Opatija, Croatia, 2012; pp 958–962.
62. Zhao, Y.; Healy, B. C.; Rotstein, D.; Guttmann, C. R.; Bakshi, R.; Weiner, H. L.; Brodley, C. E.; Chitnis, T. Exploration of Machine Learning Techniques in Predicting Multiple Sclerosis Disease Course. *PLoS One* **2017,** *12*, e0174866.
63. Vargas, D. M.; De Bastiani, M. A.; Zimmer, E. R.; Klamt, F. Alzheimer's Disease Master Regulators Analysis: Search for Potential Molecular Targets and Drug Repositioning Candidates. *Alzheimers Res. Ther.* **2018,** *10*, 59.
64. Zhang, M.; Schmitt-Ulms, G.; Sato, C.; Xi, Z.; Zhang, Y.; Zhou, Y.; St George-Hyslop, P.; Rogaeva, E. Drug Repositioning for Alzheimer's Disease Based on Systematic 'omics' Data Mining. *PLoS One* **2016,** *11*, e0168812.
65. Lee, S. Y.; Song, M.-Y.; Kim, D.; Park, C.; Park, D. K.; Kim, D. G.; Yoo, J. S.; Kim, Y. H. A Proteotranscriptomic-Based Computational Drug-Repositioning Method for Alzheimer's Disease. *Front. Pharmacol.* **2018,** *10*, 1653.
66. Corbett, A.; Williams, G.; Ballard, C. Drug Repositioning: An Opportunity to Develop Novel Treatments for Alzheimer's Disease. *Pharmaceuticals* **2013,** *6*, 1304–1321.
67. Elkouzi, A.; Vedam-Mai, V.; Eisinger, R. S.; Okun, M. S. Emerging Therapies in Parkinson Disease - Repurposed Drugs and New Approaches. *Nat. Rev. Neurol.* **2019,** *15*, 204–223.
68. Bortolanza, M.; Nascimento, G. C.; Socias, S. B.; Ploper, D.; Chehín, R. N.; Raisman-Vozari, R.; Del-Bel, E. Tetracycline Repurposing in Neurodegeneration: Focus on Parkinson's Disease. *J. Neural Transm.* **2018,** *125*, 1403–1415.
69. Vargas, D. M.; De Bastiani, M. A.; Parsons, R. B.; Klamt, F. Parkinson's Disease Master Regulators on Substantia Nigra and Frontal Cortex and Their Use for Drug Repositioning. *Mol. Neurobiol.* **2021,** *58*, 1517–1534.
70. Karunakaran, K. B.; Chaparala, S.; Ganapathiraju, M. K. Potentially Repurposable Drugs for Schizophrenia Identified From its Interactome. *Sci. Rep.* **2019,** *9*, 12682.
71. Kessing, L.V.; Rytgaard, H. C.; Gerds, T. A.; Berk, M.; Ekstrøm, C. T.; Andersen, P. K. New Drug Candidates for Bipolar Disorder-A Nation-Wide Population-Based Study. *Bipolar Disord.* **2019,** *21*, 410–418.
72. Kubick, N.; Pajares, M.; Enache, I.; Manda, G.; Mickael, M. E. Repurposing Zileuton as a Depression Drug Using an AI and In Vitro Approach. *Molecules* **2020,** *25*, 2155.
73. Durães, F.; Pinto, M.; Sousa, E. Old Drugs as New Treatments for Neurodegenerative Diseases. *Pharmaceuticals (Basel)* **2018,** *11*, 44.
74. Corey-Bloom, J.; Jia, H.; Aikin, A. M.; Thomas, E. A. Disease Modifying Potential of Glatiramer Acetate in Huntington's Disease. *J. Huntingtons Dis.* **2014,** *3*, 311–316.
75. Auffretm, M.; Drapier, S.; Vérin, M. New Tricks for an Old Dog: A Repurposing Approach of Apomorphine. *Eur. J. Pharmacol.* **2019,** *843*, 66–79.

76. Kwon, O. S.; Kim, W.; Cha, H. J.; Lee, H. In Silico Drug Repositioning: From Large-Scale Transcriptome Data to Therapeutics. *Arch. Pharm. Res.* **2019,** *42*, 879–889.
77. Tanoli, Z.; Seemab, U.; Scherer, A.; Wennerberg, K.; Tang, J.; Vähä-Koskela, M. Exploration of Databases and Methods Supporting Drug Repurposing: A Comprehensive Survey. *Brief Bioinform.* **2021,** *22*, 1656–1678.
78. Paranjpe, M. D.; Taubes, A.; Sirota, M. Insights Into Computational Drug Repurposing for Neurodegenerative Disease. *Trends Pharmacol. Sci.* **2019,** *40*, 565–576.

# CHAPTER 10

# Challenges and Regulatory Issues in Drug Repurposing and Computational Drug Discovery

ANDRÉ M. OLIVEIRA[1] and MITHUN RUDRAPAL[2]

[1]*Department of Environment, Federal Centre of Technological Education, Contagem, MG, Brazil*

[2]*Department of Pharmaceutical Sciences, School of Biotechnology and Pharmaceutical Sciences, Vignan's Foundation for Science, Technology & Research (Deemed to be University), Guntur, Karnataka, India*

## ABSTRACT

The drug repositioning process consists of investigating other uses for drugs developed for a given disease, but which have proven useful in the treatment of other diseases. This approach is advantageous once their physical-chemical, pharmacokinetic and toxicity data are available, which represents a gain in time and a lower cost. This chapter focuses on some important cases of repurposing, emphasizing different strategies for the search for new therapeutic targets for established drugs. The main strategies treated in our discussion as those knowledge-based (target-based drug-repurposing, pathway-based drug repurposing, target mechanism-based drug repurposing, and genome strategy), phenotype-based and computational methods (encompassing machine learning, network models, text mining and semantic inference, established disease–drug pair knowledge and systems biology). In the context of target-based drug-repurposing, a discussion is made about high-throughput screening (HTS) and banks of three-dimensional structures of

Drug Repurposing and Computational Drug Discovery: Strategies and Advances.
Mithun Rudrapal, PhD (Ed)

possible molecular targets for new diseases against which these compounds could be active. Another aspect of the study of drug repositioning that is extensively explored in this chapter is the applicability of machine learning methods, particularly pattern recognition techniques. Statistical validation tools for drug repositioning are also discussed. Area under the receiver operating characteristic (AUROC) metric is one of such popular techniques, whose purpose is telling the model's ability to discriminate between cases (positive examples) and non-cases (negative examples). The chapter ends with case studies involving orphan or rare diseases (ODs) and the pharmacological treatment of SARS-CoV-2.With regard to ODs, the chapter mentions several digital resources for searching information about their nature and genotypic and phenotypic data, what is useful in the development of new therapeutic applications for known drugs. COVID-19 treatments that have been benefited by repurposing strategies involve typically viral main protease (mainly Mpro or 3CLpro) inhibitors that show anti-inflammatory and non-SARS-CoV-2 viral activities (such as darunavir, a former anti-HIV drug).The study of new uses for consecrated drugs has shed light on the various diseases and their seasonal variations, supported by a voluminous amount of pharmacological data that shortens the trail to the discovery of new treatments.

## 10.1 DRUG REPOSITIONING: CONCEPT AND HISTORICAL CONTEXT

The practice of drug repositioning, which consists of discovering new therapeutic applications for drugs already established against certain diseases, has been an object of interest for many decades. However, this practice has only recently acquired the status of systematic research with its own methods and strategies.[1] The development of a new drug is a long and costly process, which does not always yield the desired results. It is estimated that out of 1000 molecules discovered, with potential pharmacological activity, only one will reach the clinical testing stage.[2,3]

The main advantage of drug repositioning in relation to the traditional development process is the availability of physical-chemical, pharmacokinetic, and toxicity data on the studied compounds, which represents a gain in time and a lower cost.[4–6] In addition to substances already marketed for other purposes, the list of candidates for drug repositioning includes:

- Compounds in clinical phase with a relevant mechanism of action for more than one disease.
- Compounds whose development was halted in phase II or III clinical trials for some reason related to their side effects or toxicology, but proved to be adequate in phase I for that specific disease.
- Compounds whose commercialization was interrupted by economic factors.
- Compounds whose patents are close to their expiration date.

Several examples of drugs successfully used in the treatment of different diseases were initially conceived for different purposes. Table 10.1 summarizes some successful cases.

**TABLE 10.1** Some Examples of Successful Drug Repurposing.

| Drug name and former indication | New indication | Ref. |
|---|---|---|
| Infliximab (Crohn's disease and rheumatoid arthritis) | Psoriasis and psoriatic arthritis | [7,8] |
| Beta-blockers (high blood pressure treatment) | Malaria | [9] |
| Cimetidine (peptic ulcer) | Human papilloma virus (HPV) | [10] |
| Glucocorticoids (immunosuppressant agents) | Sepsis | [11] |
| Metronidazole (antibacterial and antiprotozoal agent) | Dermatitis | [12] |
| Sirolimus (fungicide) | Antitumor agent | [13] |
| Thalidomide (antiemetic/insomnia) | HIV | [14] |

*Source:* Adapted from Ref. [15].

Several drug repositioning methods and strategies have been described in the literature.[16] Altogether, the methods fall into three categories: drug-oriented methods, disease-oriented methods, and treatment-oriented methods. According to Figure 10.1 that summarizes the various methods divided into their respective categories, the choice of method depends on the level of knowledge about the particularities of the pharmacological system studied. The different strategies cooperate and reinforce each other, depending on the quantity and quality of information obtained over time.

Computational chemistry and bioinformatics have gained relevance in this context, namely, in drug-oriented methods, with the advance of molecular modeling strategies and biological targets databases. The emerging diseases, with their own challenges, require new therapeutic treatments and drugs, and that represents a rationale for the pursuit of new chemical entities that can be adequately provided by theoretical and computational means (in silico drug repurposing[1]).

The first cases of drug repositioning were the result of "blind method" or serendipity (such as sildenafil), but with the increase in the quantity and complexity of new diseases, it has required systematic approaches. In this scenario, computational repositioning (in silico) acquires great relevance. Some authors choose to classify in silico drug repositioning methods into: knowledge-based methods, signature-based methods, and disease-based methods. This approach was chosen to this chapter.

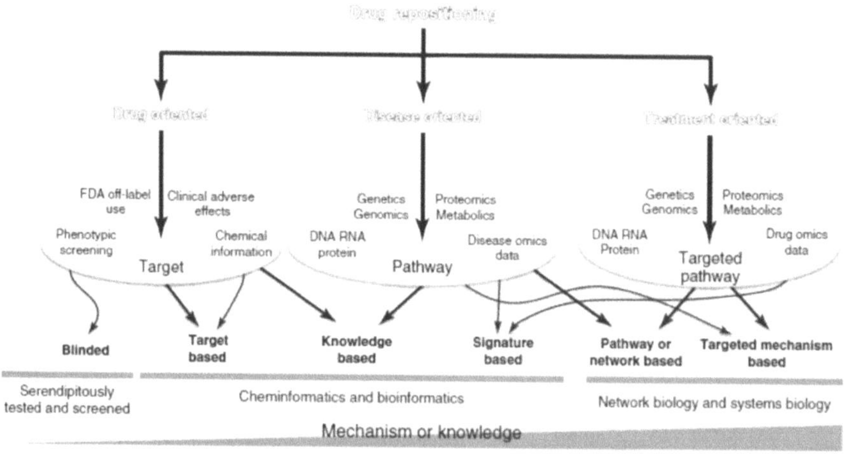


**FIGURE 10.1** Categorization of existing drug-repositioning methods.
*Source*: Reprinted with permission from Ref. [16]. © 2022 Elsevier.

## 10.2 GENERAL STRATEGIES FOR DRUG REPOSITIONING

The more drugs that are developed and patented, the larger the amount of information about their properties, modes of action, biological targets, and adverse effects. This information is stored in databases, many of them available for academic purposes, and it serves as scaffold for the knowledge-based drug-repositioning strategies, such as target-based drug repurposing, pathway-based drug repurposing, target mechanism-based drug repurposing, and genome strategy.

### 10.2.1 KNOWLEDGE-BASED REPURPOSING

- *Target-based drug repurposing:* This strategy is based on the search for new therapeutic applications of drugs known from their affinity with several biological targets associated with other diseases. Several biological target databases are available, such as those exemplified in
- TABLE. If one wishes to know whether a given drug may have a biological activity diverse from that for which it was previously designed, this drug is submitted to a virtual scanning (high-throughput screening, HTS), through which the drug is docked with the set of targets, looking for those with whom there are the best interactions.

**TABLE 10.2** Some Biological Targets Databases.

| Database | Type of targets | Ref. |
|---|---|---|
| Brookhaven protein database | Proteins, enzymes, nuclei acids | [17–19] |
| Nucleic acid database | Nucleic acids | [20] |
| GPCRdb | G protein-coupled receptors (GPCRs) | [21] |
| Therapeutic target database | Target-regulating microRNAs and transcription factors, target-interacting proteins, and patented agents and their targets | [22] |
| TDR targets | Tropical diseases targets | [23] |
| Kyoto Encyclopedia of Genes and Genomes (KEGG) | Genome, metabolic pathways, and biological substances | [24,25] |

*Pathway-based drug repurposing:* This strategy utilizes metabolic pathways, signaling pathways, and protein-interaction networks information as an attempt to stablish connections among diseases and drugs.[26] In this context, the use of bioinformatics resources, such as proteomics, genomics, and metabolomics is essential to establish the bases for the study of the relationship between the drug and the biological environment.

*Target mechanism-based drug repurposing:* This is a strategy that matches the two previous ones in a unique semantic body, establishing connections between the biological targets with the role they play in metabolism and how the drug studied affects those relationships.[27]

*Genome strategy:* When using the genomic information of a disease, available in databases such as NCBI Gene Expression Omnibus (GEO[28,29]), it is possible to map which potential targets can interact with a given drug, leading to the discovery of other diseases related to the same genes.

### *10.2.2 SIGNATURE-BASED REPURPOSING (PHENOTYPE-BASED REPURPOSING)*

This strategy is based on statistical comparison between electronic health records (EHRs), available from various sources, using data mining and pattern recognition methods.[30] Metformin's application as an anticancer agent was discovered through this way.[31]

## 10.3 COMPUTATIONAL CHEMISTRY METHODS APPLICABLE TO DRUG REPOSITIONING

### *10.3.1 MACHINE LEARNING*

Machine learning (ML) can be defined as the process through which an artificial intelligence (AI) is able to make decisions based on information learned during its use, that is, without having been previously programmed for this purpose. This way is especially interesting when dealing with highly complex problems such as the discovery of new drugs (or new therapeutic actions for known drugs[32]). Among the ML methods applicable to drug repositioning, we can mention logistic regression, support vector machine (SVM), neural network (NN), and deep learning (DL).

*Logistic regression*: Logistic regression is a type of mathematical model that makes it possible to obtain the probability of a given result from a discrete input variable, which can admit two values (such as yes or no) or several values.[33] An example of its application to drug repositioning is the PREDICT method, which correlates drug–drug similarities with disease–disease similarities.[34]

*Support vector machine (SVM):* This mathematical approach uses classification and regression analysis within a supervised learning framework. For this purpose, the SVM takes the input data and classifies them into two groups, based on a set of pre-established data used as a test. The algorithm creates a model that assigns points to each of the two categories in order to obtain the best possible separation. An example of this approach is the application of data, such as chemical structure, biological activity, and adverse effects to create an SVM-based classification function that has proven to be very efficient.[35]

*Neural network (NN):* Neural networks are a subset of machine learning and are inspired by the human brain, mimicking the way that biological neurons signal to one another. NN are composed of node layers, containing input and output layers. Each node (or artificial neuron) connects to another and has an associated weight and threshold. If the output of any individual node is above the specified threshold value, that node is activated, sending data to the next layer of the network. Otherwise, no data are passed along to the next layer of the network.[36] An example of the application of neural networks to drug repositioning is the work of Menden and collaborators with antitumor agents,[37] using as input data genomic and chemical structure information of the compounds.

*Deep learning (DL):* The main application of deep learning is in pattern recognition, which is useful in classifying a large volume of data. This makes it possible to identify complex structures in large amounts of raw data, through algorithms that allow the connections between the points to adjust to the information that is assembled alongside the process. A study in 678 drugs across A549, MCF-7, and PC-3 cell lines from the LINCS Program[38] and linked to 12 therapeutic using DL is described by Aliper and co-workers.[39]

### *10.3.2 NETWORK MODELS*

An interesting strategy in drug repositioning is to use algorithms that network nodes, in which each node is occupied by a drug, disease or biological target, and the connections between them are analyzed. From this computational effort, relationships emerge that are not obvious and that may be useful in discovering new applications for existing drugs.[40,41]

### *10.3.3 TEXT MINING AND SEMANTIC INFERENCE*

Taking into account the enormity of data available in databases about the most diverse diseases, the use of intelligent data mining serves very well the purposes of the search for new applications for the drugs currently used. This task presents some challenges that are typical of the search for qualified information in scientific texts, such as:

- The disambiguation of terms used with different meanings depending on the context.
- The treatment required to fragmented data or lacking self-consistent information.

- The unavailability of metadata from old documents, in image format, which requires the use of character recognition tools that are not always efficient.

Semantic inference by its turn utilise a graph-based representation for integrated data, used for finding drug-repositioning opportunities.[42] According to graph-based data, proteins or drugs are represented as vertices and their mutual interactions are edges with specific attributes.

### *10.3.4 ESTABLISHED DISEASE–DRUG PAIR KNOWLEDGE*

The method proposed by Draghici and co-workers[43] employs large-scale gene expression profiles related to human cell lines treated with small molecules, a gene expression profile of a human disease and the known relationship between Food and Drug Administration (FDA)-approved drugs and diseases to make clusters. The search ends assigning a smaller distance among FDA-approved drugs which have already been submitted to government regulatory scrutiny.

### *10.3.5 SYSTEMS BIOLOGY*

Systems biology applied to drug repositioning sets itself the ambitious task of modeling metabolic pathways and regulatory networks from simple molecular systems to entire tissues, organs, and organisms. For this, computational methods segment the biological systems of higher organisms.[44] For this purpose's sake, a study using KEGG Database (TABLE) by means of an impact analysis method[45] led to the proposition of sunitinib, dabrafenib, and nilotinib as repurposing candidates for treatment of idiopathic pulmonary fibrosis.[46]

## 10.4 VALIDATION OF COMPUTATIONAL DRUG-REPOSITIONING MODELS

Whatever the procedure for the development of new drugs (or the discovery of new uses), it is necessary to validate the results, comparing the conclusions obtained with those from other methods. In the computational repositioning of drugs, two types of validation are used: computational and experimental validation.

### 10.4.1 COMPUTATIONAL VALIDATION

Some values may be useful as validity metrics in classification methods, such as:

**Area under the receiver operating characteristic (AUROC):** AUROC is a performance metric for "discrimination": it tells you about the model's ability to discriminate between cases (positive examples) and non-cases (negative examples[47] The plot true-positive rate versus False-positive rate (calculated from confusion matrices, Table 10.3) is shown in Figure 10.2. Table 10.4 summarizes the interpretation of the graph.

**TABLE 10.3** Confusion Matrix, Used to Evaluate the Strengths and Weaknesses of the Model.

| | Actually positive (1) | Actually negative (0) |
|---|---|---|
| **Predicted positive (1)** | True-positives (TP) | False-positives (FP) |
| **Predicted negative (0)** | False-negative (FN) | True-negatives (TN) |

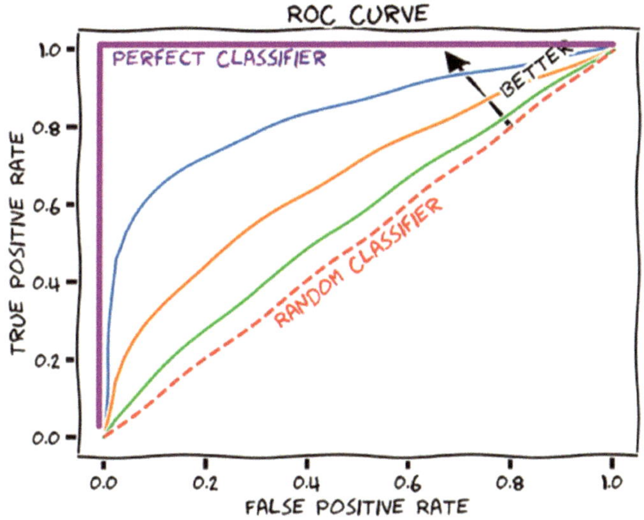


**FIGURE 10.2** AUROC plot.
*Source*: Reprinted from Ref. [70]. Open access.

**TABLE 10.4** Interpretation of AUROC Plot.

| AUROC value | Corresponding area in plot (Figure 10.2) | Meaning |
|---|---|---|
| 0.5 | Area under the red dashed line | A coin flip ( i.e., a useless model) |
| Less than 0.7 | | Sub-optimal performance |
| 0.70–0.80 | | Good performance |
| Greater than 0.8 | | Excellent performance |
| 1.0 | Under the purple line | A perfect classifier |

*Source*: Adapted from Ref. [48].

**Specificity:** The specificity of a test is the proportion of people who test negative among all those who actually do not have that disease, according to confusion matrix (Table 10.2).

$$\text{Specificity} = \frac{\text{TN}}{\text{TP} + \text{FN}}$$

- **Sensitivity:** The sensitivity of a test is the proportion of people who test positive among all those who actually have the disease.

$$\text{Sensitivity} = \frac{\text{TP}}{\text{TP} + \text{FN}}$$

- **Positive predictive value (PPV):** The positive predictive value is the probability that following a positive test result that individual will truly have that specific disease.

$$\text{PPV} = \frac{\text{TP}}{\text{TP} + \text{FP}}$$

**Area under precision-recall curve (AUPRC):** AUPRC is used for imbalanced data in situations where one wishes to avoid finding false-positives. A precision-call (PR) curve is shown in Figure 10.3. Unlike AUROC plot (Figure 10.2), where the baseline is always 0.5, in PR plot the baseline is equal to the fraction of positives, so different classes have different AUPRC baselines.

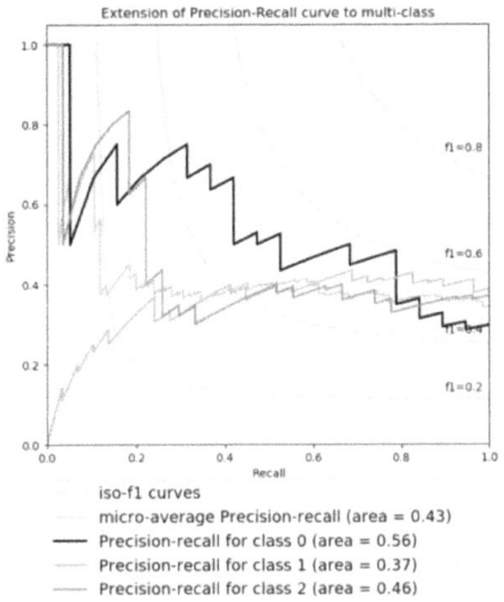


**FIGURE 10.3** PR curves.
*Source*: Adapted, by permission, from Ref. [49].

Comparing targets obtained by computational methods with those available in PubMed, ClinicalTrials[50] or EHRs adds a new layer of validation that can be useful.

### 10.4.2 *EXPERIMENTAL VALIDATION*

Cell-based targeted assays[51] (in vivo and in vitro) and animal experiments are part of this kind of validation. HTS has been described in an adaptation of an adenylate kinase (AK)-based cell death reporter assay to identify members of a FDA-approved drug library with bactericidal activity against *Staphylococcus aureus*small-colony variants.[52]

## 10.5 REPOSITIONING OF DRUGS TO FIGHT RARE OR ORPHAN DISEASES

The Orphan Drug Act (ODA) defines rare or orphan diseases (ODs) as those that affect fewer than 200,000 people in the United States, but that indirectly affect more than 25 million people in that country.[53] More than 6000 ODs are known, although as few as 325 are amenable to treatment (covering only about 5% of the known diseases), and not rarely they lead to death.[54] Most of the known rare diseases are genetic, may appear early in life and about 30% of children with rare diseases die before the age of 5 years. This reality demands many investments in new treatments, and the repositioning of drugs is important in this scenario.

TABLE 10.5 summarizes some useful resources in drug research against ODs. Genetic information databases, considering the nature of most ODs, are especially interesting, such as GARD, in which diseases are classified into the following categories:

- Autoimmune/autoinflammatory diseases
- Bacterial infections
- Behavioral and mental disorders
- Blood diseases
- Chromosome disorders
- Congenital and genetic diseases
- Connective tissue diseases
- Digestive diseases
- Ear, nose, and throat diseases
- Endocrine diseases
- Environmental diseases
- Eye diseases
- Female reproductive diseases
- Fungal infections
- Heart diseases
- Hereditary cancer syndromes
- Immune system diseases
- Kidney and urinary diseases
- Lung diseases
- Male reproductive diseases
- Metabolic disorders
- Mouth diseases
- Musculoskeletal diseases
- Myelodysplastic syndromes
- Nervous system diseases
- Newborn screening
- Nutritional diseases
- Parasitic diseases
- Rare cancers
- RDCRN
- Skin diseases
- Viral infections

**TABLE 10.5** Resources in Orphan and Rare Diseases.

| Resource name | Description and URL |
|---|---|
| NIH Genetic and Rare Diseases Information Center (GARD) | Genetic and Rare Diseases; http://rarediseases.info.nih.gov/GARD/ |
| RDCRN | Rare Diseases Clinical Research Network; https://www.rarediseasesnetwork.org |
| Orphanet | Portal for rare diseases and orphan drugs; http://www.orpha.net |
| EURODIS | European organization for rare diseases; http://www.eurordis.com |
| NORD | National Organization for Rare Disorders; http://www.rarediseases.org |
| OOPD at FDA | Developing Products for Rare Diseases and Conditions; https://www.fda.gov/industry/developing-products-rare-diseases-conditions |
| Orphan drugs at FDA | Rare Disease and Orphan Drug-Designated Approvals; https://www.fda.gov/drugs/nda-and-bla-approvals/rare-disease-and-orphan-drug-designated-approvals |
| List of marketed orphan drugs in Europe | https://www.ema.europa.eu/en/human-regulatory/overview/orphan-designation-overview |
| RAMEDIS | Rare Metabolic Disease Database; https://agbi.techfak.uni-bielefeld.de/ramedis/htdocs/eng/index.php |

*Source*: Adapted from Ref. [54].

We can mention some cases in which drug-repositioning strategies helped in the treatment of ODs. Sardana and co-workers[54] present examples of drug repurposing in divers situations such as: (1) common drug repositioning for orphan diseases, (2) orphan drug repositioning for a common indication, (3) orphan drugs with approval for another orphan disease indication, (4) orphan-designated products with marketing approvals for both common and orphan disease indications, and (5) reviving withdrawn drugs.

### *10.5.1 COMMON DRUG REPOSITIONING FOR ORPHAN DISEASES*

Tretinoin is a metabolite of vitamin A and has been used for the topical treatment of acne vulgaris since its approval by FDA in 1995. An OD named acute promyelocyticleukemia (APL) is caused by an abnormal translocation of chromosome 17 onto chromosome 15 that affects the expression of nuclear retinoic acid receptor alpha (RAR-$\alpha$), resulting in the expression of abnormal messenger RNA (mRNA). Tretinoin exhibited activity against APL to the point of causing remission in many patients.[55]

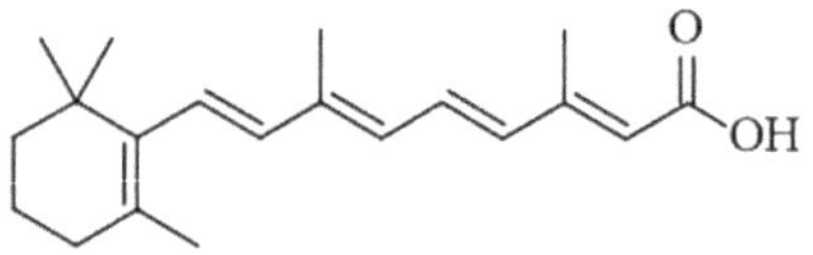


tretinoin

## 10.5.2 ORPHAN DRUG REPOSITIONING FOR A COMMON INDICATION

Albuterol (or salbutamol), a β-adrenergic agonist, is indicated for the relief of bronchospasm, a common condition, and has been redesignated as an orphan drug for prevention of paralysis due to spinal cord injury.[56] This drug stimulates b-adrenergic receptors (predominant in bronchial smooth muscle cells) of intracellular adenyl cyclase, increasing the conversion of ATP to cyclic AMP that by its turn inhibits the release of hypersensitivity mediators from mast cells.

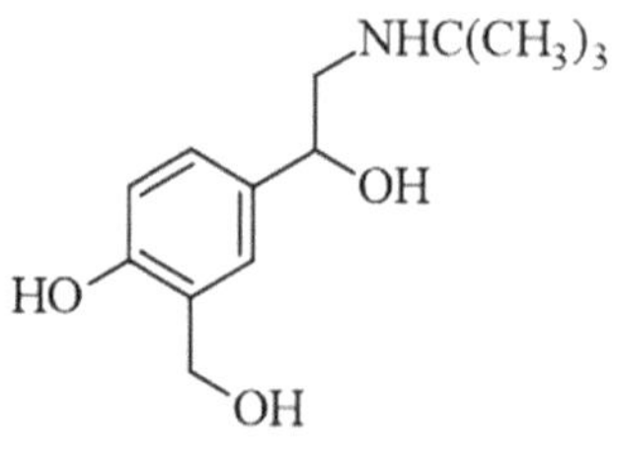


albuterol

## 10.5.3 ORPHAN DRUGS WITH APPROVAL FOR ANOTHER ORPHAN DISEASE INDICATION

Riluzole, used in the treatment of amyotrophic lateral sclerosis, has been designated against Huntington's disease.[57]

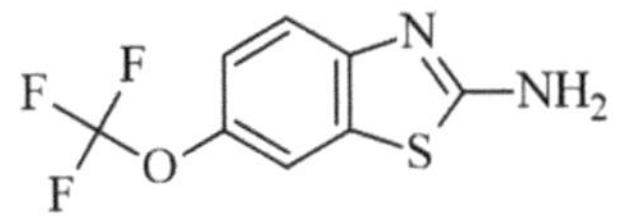

riluzole

### 10.5.4 ORPHAN-DESIGNATED PRODUCTS WITH MARKETING APPROVALS FOR BOTH COMMON AND ORPHAN DISEASE INDICATIONS

Eflornithine was conceived to treat cancer, but was shown to be highly effective in reducing hair growth.[58] This compound inhibits selectively and irreversibly ornithine decarboxylase (ODC), a key enzyme in the biosynthesis of polyamines, catalyzing the conversion of ornithine to putrescine, which plays an important role in cell division and proliferation in the hair follicle.

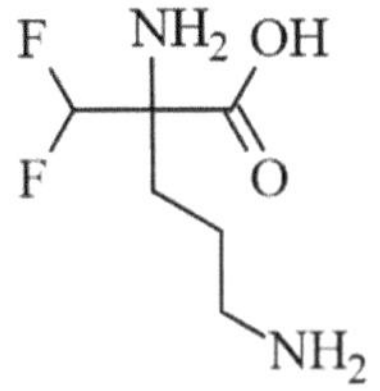

eflornithine

### 10.5.5 REVIVING WITHDRAWN DRUGS

The traumatic history of thalidomide, which caused fetal malformation in the 1960s, for some time relegated a drug to oblivion until the FDA authorized its use in 1998 against erythema nodosumleprosum (ENL), a more severe form of leprosy.[59] This was a fortuitous discovery made in 1964 by a physician Jacob Sheskin at the University Hospital of Marseilles (France). Its immunomodulatory properties have been used against oral and genital ulcers, vasculitis, and rheumatoid arthritis, and many analogues of thalidomide are active against ODs, such as multiple myeloma and myelodysplastic syndromes.[60]

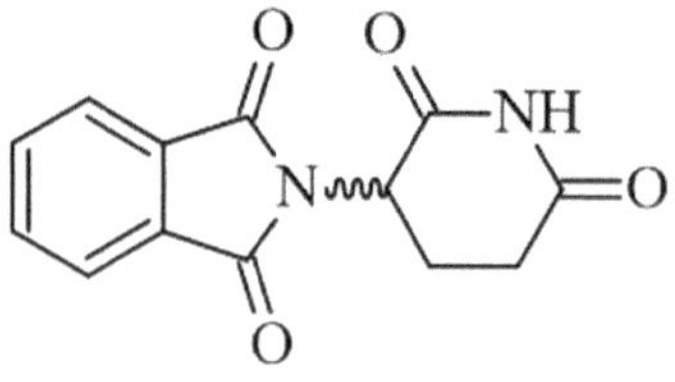


thalidomide

## 10.6 CHALLENGES IN DRUG REPOSITIONING IN THE CONTEXT OF THE SARS-COV-2 PANDEMIC

The COVID-19 pandemic brought immense challenges to the pharmaceutical industry, research centers, universities, and healthcare institutions. Due to its high propagation profile and the lack of knowledge about its mechanisms, new treatments that could cooperate with immunization have been sought. The mortality rate varies from country to country[61] with an overall median value around 2% (Figure 10.4).

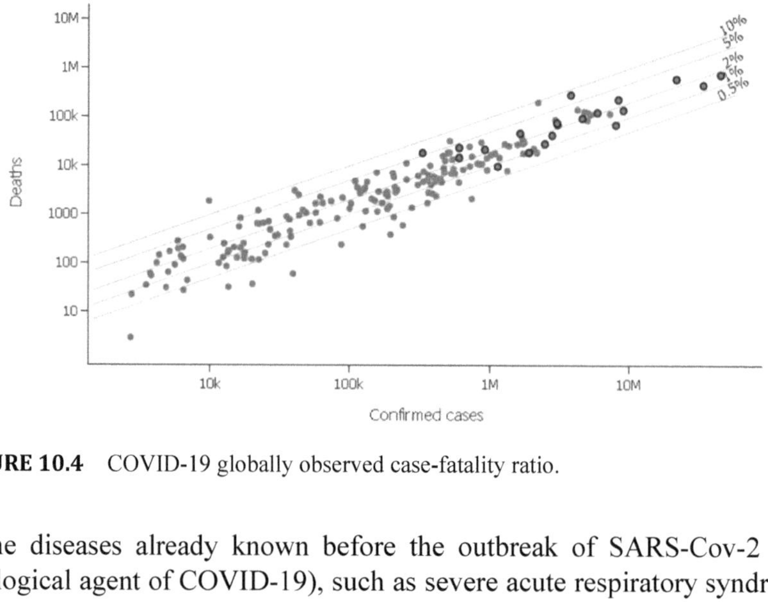


**FIGURE 10.4** COVID-19 globally observed case-fatality ratio.

Some diseases already known before the outbreak of SARS-Cov-2 (the etiological agent of COVID-19), such as severe acute respiratory syndrome (SARS) and Middle Eastern respiratory syndrome (MERS) were the first sources of drug research.[62] A survey of the main drugs being tested (Figure 10.5) shows the predominance of agents linked to unique viral components and processes, such as viral protease and RNA-dependent RNA polymerase.[63]

Among the most important biological targets and mechanisms in the search for therapeutic agents against COVID-19, we have the viral proteases, acetylcholinesterases (ACEs), and polymerases (Figure 10.6). For illustrative purposes, some recent examples of new drug uses that have shown promise in confronting COVID-19 are mentioned.

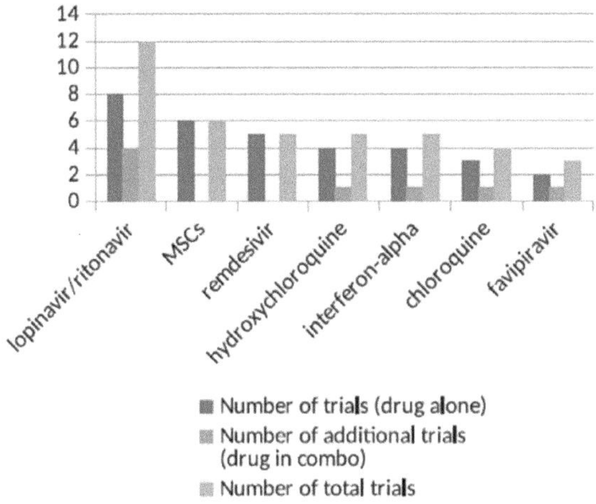


**FIGURE 10.5** Most tested drugs in COVID-19 trials (April 2020).
*Source*: Ref. [63].

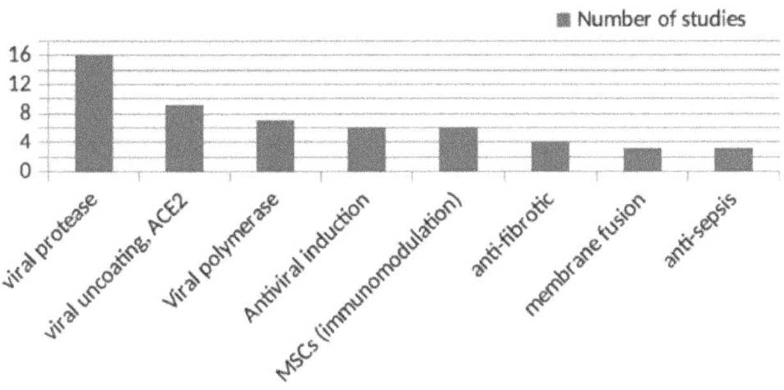


**FIGURE 10.6** Most frequent COVID-19 drug targets and mechanisms.
*Source*: Ref. [63].

Ramanathan and co-workers[64] developed a docking study with Glide algorithm of small molecule inhibitors for protease structure (PDB ID 6LU7).

The binding free energies were obtained by means of the Prime-MM/GBSA algorithm, and a subsequent AutoQSAR algorithm and drug-likeness and toxicity parameters determination were performed, yielding two potential inhibitors, DB02986 and DB08573.

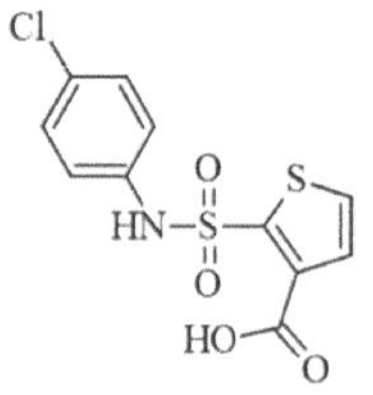

DB08573

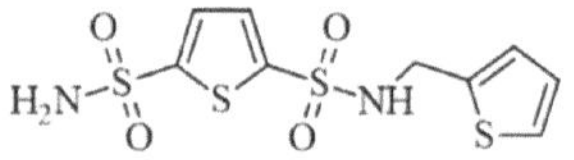

DB02986

Molfetta and co-workers,[65] by means of a molecular dynamics simulation with SAR-CoV-2 main protease ($M^{pro}$ or $3CL^{pro}$) showed an attractive inhibition by darunavir (formerly used to treat HIV patients) and triptorelin (an agonist analog of gonadotropin-releasing hormone).

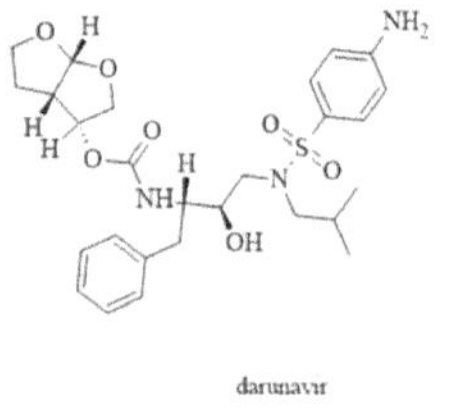

darunavir

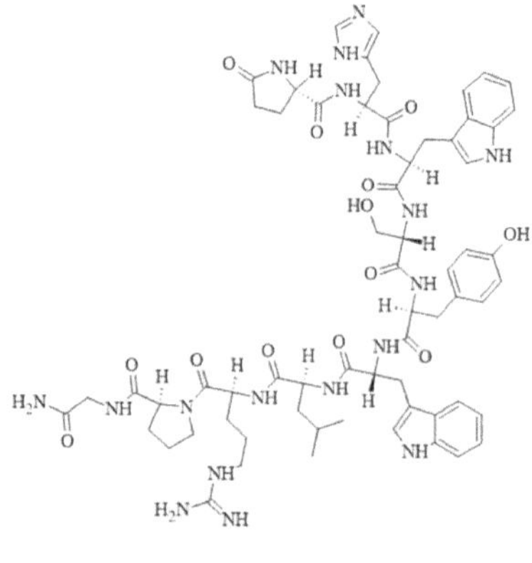

triptorelin

Stojkovic-Filipovic and Bosic[66] described the use of anti-inflammatory chloroquine (CQ) and hydroxychloroquine (HCQ) in the treatment of COVID-19, both are widely used in dermatology. The mechanism of their widespread antiviral activity is associated to spike-glycoprotein/ACE2 interaction inhibition,[67] quinone reductase-2 and biosynthesis of sialic acids inhibition,[68] curbing SARS-CoV-2 ligand recognition and interaction with target cells, among further mechanisms.

## KEYWORDS


- **drug repurposing**
- **high-throughput screening**
- **target-based drug-repurposing**
- **sars-cov-2 drugs**
- **orphan diseases**


## REFERENCES


1. Park, K. A Review of Computational Drug Repurposing. *Transl. Clin. Pharmacol.* **2019,** *27* (2), 59.
2. Plenge, R. M.; Scolnick, E. M.; Altshuler, D. Validating Therapeutic Targets Through Human Genetics. *Nat. Rev. Drug Discov.* **2013,** *12* (8), 581–594.
3. Tamimi, N. A. M.; Ellis, P. Drug Development: From Concept to Marketing! *Nephron Clin. Pract.* **2009,** *113* (3), c125–c131.
4. Chong, C. R.; Sullivan, D. J. New Uses for Old Drugs. *Nature* **2007,** *448* (7154), 645–646.
5. Sleigh, S. H.; Barton, C. L. Repurposing Strategies for Therapeutics. *Pharmaceut. Med.* **2012,** *24* (3), 151–159.
6. Ashburn, T. T.; Thor, K. B. Drug Repositioning: Identifying and Developing New Uses for Existing Drugs. *Nat. Rev. Drug Discov.* **2004,** *3* (8), 673–683.
7. Pipitone, N.; Kingsley, G. H.; Manzo, A.; Scott, D. L.; Pitzalis, C. Current Concepts and New Developments in the Treatment of Psoriatic Arthritis. *Rheumatology* **2003,** *42* (10), 1138–1148.
8. Lebwohl, B.; Sapadin, A. N. Infliximab for the Treatment of Hidradenitis Suppurativa. *J. Am. Acad. Dermatol.* **2003,** *49* (5 Suppl), 275–276.
9. Beta Blockers Effective Against Malaria Parasites - News Center [Online]. https://news.feinberg.northwestern.edu/2003/09/beta_blockers/ (accessed Oct 21, 2021).
10. Micali, G.; Nasca, M. R.; Dall'Oglio, F.; Musumeci, M. L. Cimetidine Therapy for Epidermodysplasia Verruciformis. *J. Am. Acad. Dermatol.* **2003,** *48* (2 Suppl).
11. Bornstein, S. R.; Briegel, J. A New Role for Glucocorticoids in Septic Shock. *Am. J. Respir. Crit. Care Med.* **2012,** *167* (4), 485–486.
12. Koca, R.; Altinyazar, H. C.; Eştürk, E. Is Topical Metronidazole Effective in Seborrheic Dermatitis? A Double-Blind Study. *Int. J. Dermatol.* **2003,** *42* (8), 632–635.
13. Sausville, E. A.; Elsayed, Y.; Monga, M.; Kim, G. Signal Transduction–Directed Cancer Treatments. *Annu. Rev. Pharmacol. Toxicol.* **2003,** *43* (1), 199–231.
14. Ravot, E.; Lisziewicz, J.; Lori, F. New Uses for Old Drugs in HIV Infection. *Drugs* **2012,** *58* (6), 953–963.
15. Verma; U.; Sharma, R.; Gupta, P.; Kapoor, B.; Bano, G.; Sawhney, V. New Uses for Old Drugs: Novel Therapeutic Options. *Indian J. Pharmacol.* **2005,** *37* (5), 279.

16. Jin, G.; Wong, S. T. C. Toward Better Drug Repositioning: Prioritizing and Integrating Existing Methods into Efficient Pipelines. *Drug Discov. Today* **2014,** *19* (5), 637–644.
17. Burley, S. K.; Berman, H. M.; Bhikadiya, C.; Bi, C.; Chen, L.; Di Costanzo, L.;
18. Christie, C.; Duarte, J. M.; Dutta, S.; Feng, Z.; Ghosh, S.; Goodsell, D. S.; Green, R. K.; Guranovic, V.; Guzenko, D.; Hudson, B. P.; Liang, Y.; Lowe, R.; Peisach, E.; Periskova, I.; Randle, C.; Rose, A.; Sekharan, M.; Shao, C.; Tao, Y. P.; Valasatava, Y.; Voigt, M.; Westbrook, J.; Young, J.; Zardecki, C.; Zhuravleva, M.; Kurisu, G.; Nakamura, H.; Kengaku, Y.; Cho, H.; Sato, J.; Kim, J. Y.; Ikegawa, Y.; Nakagawa, A.; Yamashita, R.; Kudou, T.; Bekker, G. J.; Suzuki, H.; Iwata, T.; Yokochi, M.; Kobayashi, N.; Fujiwara, T.; Velankar, S.; Kleywegt, G. J.; Anyango, S.; Armstrong, D. R.; Berrisford, J. M.; Conroy, M. J.; Dana, J. M.; Deshpande, M.; Gane, P.; Gáborová, R.; Gupta, D.; Gutmanas, A.; Koča, J.; Mak, L.; Mir, S.; Mukhopadhyay, A.; Nadzirin, N.; Nair, S.; Patwardhan, A.; Paysan-Lafosse, T.; Pravda, L.; Salih, O.; Sehnal, D.; Varadi, M.; Vǎreková, R.; Markley, J. L.; Hoch, J. C.; Romero, P. R.; Baskaran, K.; Maziuk, D.; Ulrich, E. L.; Wedell, J. R.; Yao, H.; Livny, M.; Ioannidis, Y. E. Protein Data Bank: The Single Global Archive for 3D Macromolecular Structure Data. *Nucleic Acids Res.* **2019,** *47* (D1), D520–D528.
19. Berman, H. M.; Westbrook, J.; Feng, Z.; Gilliland, G.; Bhat, T. N.; Weissig, H.; Shindyalov, I. N.; Bourne, P. E. The Protein Data Bank. *Nucleic Acids Res.* **2000,** *28* (1), 235–242.
20. Burley, S. K.; Bhikadiya, C.; Bi, C.; Bittrich, S.; Chen, L.; Crichlow, G. V; Christie, C. H.; Dalenberg, K.; Di Costanzo, L.; Duarte, J. M.; Dutta, S.; Feng, Z.; Ganesan, S.; Goodsell, D. S.; Ghosh, S.; Green, R. K.; Guranović, V.; Guzenko, D.; Hudson, B. P.; Lawson, C. L.; Liang, Y.; Lowe, R.; Namkoong, H.; Peisach, E.; Persikova, I.; Randle, C.; Rose, A.; Rose, Y.; Sali, A.; Segura, J.; Sekharan, M.; Shao, C.; Tao, Y.-P.; Voigt, M.; Westbrook, J. D.; Young, J. Y.; Zardecki, C.; Zhuravleva, M. RCSB Protein Data Bank: Powerful New Tools for Exploring 3D Structures of Biological Macromolecules for Basic and Applied Research and Education in Fundamental Biology, Biomedicine, Biotechnology, Bioengineering and Energy Sciences. *Nucleic Acids Res.* **2021,** *49* (D1), D437–D451.
21. Berman, H. M.; Olson, W. K.; Beveridge, D. L.; Westbrook, J.; Gelbin, A.; Demeny, T.; Hsieh, S. H.; Srinivasan, A. R.; Schneider, B. The Nucleic Acid Database. A Comprehensive Relational Database of Three-Dimensional Structures of Nucleic Acids. *Biophys. J.* **1992,** *63* (3), 751–759.
22. Munk, C.; Mutt, E.; Isberg, V.; Nikolajsen, L. F.; Bibbe, J. M.; Flock, T.; Hanson, M. A.; Stevens, R. C.; Deupi, X.; Gloriam, D. E. An Online Resource for GPCR Structure Determination and Analysis. *Nat. Methods* **2019,** *16* (2), 151–162.
23. Wang, Y.; Zhang, S.; Li, F.; Zhou, Y.; Zhang, Y.; Wang, Z.; Zhang, R.; Zhu, J.; Ren, Y.; Tan, Y.; Qin, C.; Li, Y.; Li, X.; Chen, Y.; Zhu, F. Therapeutic Target Database 2020: Enriched Resource for Facilitating Research and Early Development of Targeted Therapeutics. *Nucleic Acids Res.* **2020,** *48* (D1), D1031–D1041.
24. Urán Landaburu, L.; Berenstein, A. J.; Videla, S.; Maru, P.; Shanmugam, D.; Chernomoretz, A.; Agüero, F. TDR Targets 6: Driving Drug Discovery for Human Pathogens Through Intensive Chemogenomic Data Integration. *Nucleic Acids Res.* **2020,** *48* (D1), D992–D1005.
25. Kanehisa, M.; Sato, Y.; Kawashima, M.; Furumichi, M.; Tanabe, M. KEGG as a Reference Resource for Gene and Protein Annotation. *Nucleic Acids Res.* **2016,** *44* (D1), D457–D462.

26. Kanehisa, M.; Furumichi, M.; Sato, Y.; Ishiguro-Watanabe, M.; Tanabe, M. KEGG: Integrating Viruses and Cellular Organisms. *Nucleic Acids Res.* **2021,** *49* (D1), D545–D551.
27. Jadamba, E.; Shin, M. A Systematic Framework for Drug Repositioning From Integrated Omics and Drug Phenotype Profiles Using Pathway-Drug Network. *Biomed Res. Int.* **2016,** *2016*, 7147039.
28. Jin, G.; Fu, C.; Zhao, H.; Cui, K.; Chang, J.; Wong, S. T. C. A Novel Method of Transcriptional Response Analysis to Facilitate Drug Repositioning for Cancer Therapy. *Cancer Res.* **2012,** *72* (1), 33–44.
29. Barrett, T.; Wilhite, S. E.; Ledoux, P.; Evangelista, C.; Kim, I. F.; Tomashevsky, M.; Marshall, K. A.; Phillippy, K. H.; Sherman, P. M.; Holko, M.; Yefanov, A.; Lee, H.; Zhang, N.; Robertson, C. L.; Serova, N.; Davis, S.; Soboleva, A. NCBI GEO: Archive for Functional Genomics Data Sets—Update. *Nucleic Acids Res.* **2013,** *41* (D1), D991–D995.
30. Edgar, R.; Domrachev, M.; Lash, A. E. Gene Expression Omnibus: NCBI Gene Expression and Hybridization Array Data Repository. *Nucleic Acids Res.* **2002,** *30* (1), 207–210.
31. Xu, H.; Aldrich, M. C.; Chen, Q.; Liu, H.; Peterson, N. B.; Dai, Q.; Levy, M.; Shah, A.; Han, X.; Ruan, X.; Jiang, M.; Li, Y.; Julien, J. S.; Warner, J.; Friedman, C.; Roden, D. M.; Denny, J. C. Validating Drug Repurposing Signals Using Electronic Health Records: A Case Study of Metformin Associated with Reduced Cancer Mortality. *J. Am. Med. Informatics Assoc.* **2015,** *22* (1), 179–191.
32. Hebbring, S. J. The Challenges, Advantages and Future of Phenome-Wide Association Studies. *Immunology* **2014,** *141* (2), 157–165.
33. Napolitano, F.; Zhao, Y.; Moreira, V. M.; Tagliaferri, R.; Kere, J.; D'Amato, M.; Greco, D. Drug Repositioning: A Machine-Learning Approach through Data Integration. *J. Cheminform.* **2013,** *5* (1), 1–9.
34. Edgar, T. W.; Manz, D. O. Chapter 4 - Exploratory Study. In *Research Methods for Cyber Security*; Edgar, T. W., Manz, D. O., Eds.; Syngress, 2017; pp 95–130.
35. Gottlieb, A.; Stein, G. Y.; Ruppin, E.; Sharan, R. PREDICT: A Method for Inferring Novel Drug Indications With Application to Personalized Medicine. *Mol. Syst. Biol.* **2011,** *7* (1), 496.
36. Wang, Y.; Chen, S.; Deng, N.; Wang, Y. Drug Repositioning by Kernel-Based Integration of Molecular Structure, Molecular Activity, and Phenotype Data. *PLoS One* **2013,** *8* (11), e78518.
37. What are Neural Networks? | IBM [Online]. https://www.ibm.com/cloud/learn/neural-networks (accessed Oct 28, 2021).
38. Menden, M. P.; Iorio, F.; Garnett, M.; McDermott, U.; Benes, C. H.; Ballester, P. J.; Saez-Rodriguez, J. Machine Learning Prediction of Cancer Cell Sensitivity to Drugs Based on Genomic and Chemical Properties. *PLoS One* **2013,** *8* (4), e61318.
39. Stathias, V.; Turner, J.; Koleti, A.; Vidovic, D.; Cooper, D.; Fazel-Najafabadi, M.; Pilarczyk, M.; Terryn, R.; Chung, C.; Umeano, A.; Clarke, D. J. B.; Lachmann, A.; Evangelista, J. E.; Ma'ayan, A.; Medvedovic, M.; Schürer, S. C. LINCS Data Portal 2.0: Next Generation Access Point for Perturbation-Response Signatures. *Nucleic Acids Res.* **2020,** *48* (D1), D431–D439.
40. Aliper, A.; Plis, S.; Artemov, A.; Ulloa, A.; Mamoshina, P.; Zhavoronkov, A. Deep Learning Applications for Predicting Pharmacological Properties of Drugs and Drug Repurposing Using Transcriptomic Data. *Mol. Pharm.* **2016,** *13* (7), 2524–2530.

41. Wu, C.; Gudivada, R. C.; Aronow, B. J.; Jegga, A. G. Computational Drug Repositioning through Heterogeneous Network Clustering. *BMC Syst. Biol.* **2013,** *7* (5), 1–9.
42. Cheng, F.; Liu, C.; Jiang, J.; Lu, W.; Li, W.; Liu, G.; Zhou, W.; Huang, J.; Tang, Y. Prediction of Drug-Target Interactions and Drug Repositioning via Network-Based Inference. *PLoS Comput. Biol.* **2012,** *8* (5), e1002503.
43. Betzler, N.; Van Bevern, R.; Fellows, M. R.; Komusiewicz, C.; Niedermeier, R. Parameterized Algorithmics for Finding Connected Motifs in Biological Networks. *IEEE/ACM Trans. Comput. Biol. Bioinforma.* **2011,** *8* (5), 1296–1308.
44. Sadeghi, S. S.; Keyvanpour, M. R. Computational Drug Repurposing: Classification of the Research Opportunities and Challenges. *Curr. Comput. Aided. Drug Des.* **2019,** *16* (4), 354–364.
45. Butcher, E. C.; Berg, E. L.; Kunkel, E. J. Systems Biology in Drug Discovery. *Nat. Biotechnol.* **2004,** *22* (10), 1253–1259.
46. Draghici, S.; Khatri, P.; Tarca, A. L.; Amin, K.; Done, A.; Voichita, C.; Georgescu, C.; Romero, R. A Systems Biology Approach for Pathway Level Analysis. *Genome Res.* **2007,** *17* (10), 1537–1545.
47. Peyvandipour, A.; Saberian, N.; Shafi, A.; Donato, M.; Draghici, S. Systems Biology: A Novel Computational Approach for Drug Repurposing Using Systems Biology. *Bioinformatics* **2018,** *34* (16), 2817–2825.
48. Guney, E.; Menche, J.; Vidal, M.; Barábasi, A.-L. Network-Based in Silico Drug Efficacy Screening. *Nat. Commun.* **2016,** *7* (1), 1–13.
49. Measuring Performance: AUC (AUROC) – Glass Box [Online]. https://glassboxmedicine.com/2019/02/23/measuring-performance-auc-auroc/ (accessed Oct 30, 2021).
50. Pedregosa, F.; Varoquaux, G.; Gramfort, A.; Michel, V.; Thirion, B.; Grisel, O.; Blondel, M.; Prettenhofer, P.; Weiss, R.; Dubourg, V.; Vanderplas, J.; Passos, A.; Cournapeau, D.; Brucher, M.; Perrot, M.; Duchesnay, É. Scikit-Learn: Machine Learning in Python. *J. Mach. Learn. Res.* **2011,** *12* (85), 2825–2830.
51. Home - ClinicalTrials.gov https://clinicaltrials.gov/ (accessed Oct 30, 2021).
52. An, W. F.; Tolliday, N. Cell-Based Assays for High-Throughput Screening. *Mol. Biotechnol.* **2010,** *45* (2), 180–186.
53. Trombetta, R. P.; Dunman, P. M.; Schwarz, E. M.; Kates, S. L.; Awad, H. A. A High-Throughput Screening Approach To Repurpose FDA-Approved Drugs for Bactericidal Applications Against Staphylococcus Aureus Small-Colony Variants. *mSphere* **2018,** *3* (5), e00422–e00418.
54. Xu, K.; Coté, T. R. Database Identifies FDA-Approved Drugs With Potential to Be Repurposed for Treatment of Orphan Diseases. *Brief. Bioinform.* **2011,** *12* (4), 341–345.
55. Sardana, D.; Zhu, C.; Zhang, M.; Gudivada, R. C.; Yang, L.; Jegga, A. G. Drug Repositioning for Orphan Diseases. *Brief. Bioinform.* **2011,** *12* (4), 346–356.
56. Warrell, R. P.; Frankel, S. R.; Miller, W. H.; Scheinberg, D. A.; Itri, L. M.; Hittelman, W. N.; Vyas, R.; Andreeff, M.; Tafuri, A.; Jakubowski, A.; Gabrilove, J.; Gordon, M. S.; Dmitrovsky, E. Differentiation Therapy of Acute Promyelocytic Leukemia With Tretinoin (All-Trans-Retinoic Acid). *N. Engl. J. Med.* **1991,** *324* (20), 1385–1393.
57. Murphy, R. J. L.; Hartkopp, A.; Gardiner, P. F.; Kjaer, M.; Béliveau, L. Salbutamol Effect in Spinal Cord Injured Individuals Undergoing Functional Electrical Stimulation Training. *Arch. Phys. Med. Rehabil.* **1999,** *80* (10), 1264–1267.
58. Landwehrmeyer, G. B.; Dubois, B.; Yébenes, J. G. de; Kremer, B.; Gaus, W.; Kraus, P. H.; Przuntek, H.; Dib, M.; Doble, A.; Fischer, W.; Ludolph, A. C. Riluzole in

Huntington's Disease: A 3-Year, Randomized Controlled Study. *Ann. Neurol.* **2007,** *62* (3), 262–272.
59. Malhotra, B.; Noveck, R.; Behr, D.; Palmisano, M. Percutaneous Absorption and Pharmacokinetics of Eflornithine HCl 13.9% Cream in Women With Unwanted Facial Hair. *J. Clin. Pharmacol.* **2001,** *41* (9), 972–978.
60. Schulz, M. Dark Remedy: The Impacct of Thalidomide and Its Revival as a Vital Medicine. *BMJ Br. Med. J.* **2001,** *322* (7302), 1608.
61. Zhu, X.; Jiang, S.; Hu, N.; Luo, F.; Dong, H.; Kang, Y.-M.; Jones, K. R.; Zou, Y.; Xiong, L.; Ren, J. Tumour Necrosis Factor-α Inhibition With Lenalidomide Alleviates Tissue Oxidative Injury and Apoptosis in Ob/Ob Obese Mice. *Clin. Exp. Pharmacol. Physiol.* **2014,** *41* (7), 489–501.
62. Mortality Analyses - Johns Hopkins Coronavirus Resource Center [Online]. https://coronavirus.jhu.edu/data/mortality (accessed Oct 31, 2021).
63. Zhu, N.; Zhang, D.; Wang, W.; Li, X.; Yang, B.; Song, J.; Zhao, X.; Huang, B.; Shi, W.; Lu, R.; Niu, P.; Zhan, F.; Ma, X.; Wang, D.; Xu, W.; Wu, G.; Gao, G. F.; Tan, W. A Novel Coronavirus From Patients With Pneumonia in China, 201948. *N. Engl. J. Med.* **2020,** *382* (8), 727–733.
64. Ulm, J. W.; Nelson, S. F. COVID-19 Drug Repurposing: Summary Statistics on Current Clinical Trials and Promising Untested Candidates. *Transbound. Emerg. Dis.* **2021,** *68* (2), 313–317.
65. Muthu Kumar, T.; Rohini, K.; James, N.; Shanthi, V.; Ramanathan, K. Discovery of Potent Covid-19 Main Protease Inhibitors Using Integrated Drug-Repurposing Strategy. *Biotechnol. Appl. Biochem.* **2021,** *68* (4), 712–725.
66. Silva, J. R. A.; Kruger, H. G.; Molfetta, F. A. Drug Repurposing and Computational Modeling for Discovery of Inhibitors of the Main Protease (Mpro) of SARS-CoV-2. *RSC Adv.* **2021,** *11* (38), 23450–23458.
67. Stojkovic-Filipovic, J.; Bosic, M. Treatment of COVID 19—Repurposing Drugs Commonly Used in Dermatology. *Dermatol. Ther.* **2020,** *33* (5), e13829.
68. Vincent, M. J.; Bergeron, E.; Benjannet, S.; Erickson, B. R.; Rollin, P. E.; Ksiazek, T. G.; Seidah, N. G.; Nichol, S. T. Chloroquine Is a Potent Inhibitor of SARS Coronavirus Infection and Spread. *Virol. J.* **2005,** *2* (1), 1–10.
69. Kwiek, J. J.; Haystead, T. A. J.; Rudolph, J. Kinetic Mechanism of Quinone Oxidoreductase 2 and Its Inhibition by the Antimalarial Quinolines. *Biochemistry* **2004,** *43* (15), 4538–4547.
70. Draelos, R., Measuring Performance: AUC (AUROC). Open access. https://glassboxmedicine.com/2019/02/23/measuring-performance-auc-auroc/

# Index

## A

## B



## C